Salt Production, Distribution and Use in the British Iron Age

Janice Kinory

BAR British Series 559
2012

Published in 2016 by
BAR Publishing, Oxford

BAR British Series 559

Salt Production, Distribution and Use in the British Iron Age

ISBN 978 1 4073 0972 9

BAR Publishing is the trading name of British Archaeological Reports (Oxford) Ltd. British Archaeological Reports was first incorporated in 1974 to publish the BAR Series, International and British. In 1992 Hadrian Books Ltd became part of the BAR group. This volume was originally published by Archaeopress in conjunction with British Archaeological Reports (Oxford) Ltd / Hadrian Books Ltd, the Series principal publisher, in 2012. This present volume is published by BAR Publishing, 2016.

Printed in England

BAR
PUBLISHING

BAR titles are available from:

BAR Publishing
122 Banbury Rd, Oxford, OX2 7BP, UK
EMAIL info@barpublishing.com
PHONE +44 (0)1865 310431
FAX +44 (0)1865 316916
www.barpublishing.com

Preface

The initial idea for this book arose after encountering details of briquetage finds within several Iron Age site monographs reviewed while working on assignments for a diploma in archaeology course at Oxford's Department for Continuing Education. I expected that it would be easy to locate a book to quickly learn more about this material which appears in British prehistory. I was shocked to discover that even within the Bodleian Library system no such single volume was to be found. What follows is a synthesis of information about prehistoric British salt production, distribution and use that became my DPhil thesis at Oxford. I hope this will go some distance to filling the gap I found, and will be a useful introduction to the subject to later scholars.

Acknowledgements

Although presented as the work of an individual, that is not the entire truth of how books are written, for without the contributions of many others, it would never be completed. Thus the list of people and organisations to which I owe thanks is quite extensive. Any mistakes in the text, however, are all my own work and not the fault of any of these kind people.

First, I would like to thank my thesis supervisors, Professor Gary Lock and Lisa Brown. Gary shared my interest in the Iron Age and encouraged me to apply to Oxford University despite being a non-traditional student when I visited his office and proposed this thesis topic. His guidance has been invaluable along this journey. Lisa, despite her own enormous personal workload for Oxford Archaeology, has managed to find time to read all of my chapters, attend regular meetings with me, enhance my knowledge of prehistoric ceramics, assist me with personal introductions to key specialists, encouraged me to tackle challenges when I thought I wasn't ready for them and introduced me to the Prehistoric Ceramics Research Group. This list of their assistance is far too short to do their support justice. I also wish to thank my examiners, Professors Chris Gosden and Anthony Harding, for encouraging me to publish my work.

Right behind them on the list of people to be thanked is my husband, Yoni Kinory, who edited my chapters, even those that held no inherent interest to him. Without his help, this text would have been written in a strange mix of British and American English.

I wish to thank Dr Elaine Morris (University of Southampton) for her continuing support of my work. I am grateful to Professor Richard Bradley (University of Reading) for taking the time to meet with me early in my research and set me on a productive path, not to mention disabusing me of some early ideas that were without merit. I also appreciate the interest that Professor Anthony Harding (University of Exeter) has shown in my work, and his willingness to discuss prehistoric salt production in a European context and review chapters for me. Thanks to Dr Paul Sealey (Colchester Museums) for assistance with access to Essex briquetage collections and for reading drafts of several chapters. Dr Zena Kamash (University of Oxford) gave invaluable assistance, critiquing a chapter and providing moral support at difficult moments. My appreciation also goes to Dr Peter Northover (University of Oxford) for advice on Iron Age gilding processes, Professor Stephen Rippon (University of Exeter) for excavation details and Dr Rachel Pope (University of Liverpool) for information on roundhouse interior temperatures.

Compiling my database of briquetage finds required the support of many people, particularly staff from county archaeology departments and SMRs, including Vince Russett and Sarah MacLean (Bath and North Somerset), Teresa Hocking (Berkshire), Gill Dunn and Moya Watson (Cheshire West and Chester), Claire Pinder and Gordon Le Pard (Dorset), A. Keith Elliott (Gloucestershire), Krysia Truscoe (Greater London SMR), Alex Godden (Hampshire), Melissa Seddon (Herefordshire), Victoria Brown (Hull Archaeology Partnership), John Giblin (Isle of Wight), Andrew Mayfield (Kent), Ken Davies (Lancashire), Alexandra Thornton (Lincolnshire), Sarah-Jane Farr (Liverpool Museum, Merseyside SMR), Katherine Daws (Northamptonshire), Sarah Howard (Norfolk), Nick Boldrini and Louisa Matthews (North Yorkshire), Esther Cameron and Susan Lisk (Oxfordshire), Rebecca Casa-Hatton (Peterborough), Dr Richard Brunning (Somerset), Jude Plouviez (Suffolk), Giles Carey and Ben Wallace (Warwickshire) and, finally, Emma Hancox, Derek Hurst, David Kendrick and Simon Woodiwiss (Worcestershire). I am also appreciative of those busy individuals working in the commercial archaeology sector who took the time to answer my emails or meet with me, including Jim Stevenson (Archaeology South-East), Andrew Young (Avon Archaeological Unit), Grahame Appleby, Alison Dickens, Chris Evans, Jacqui Hutton and Samantha Smith (Cambridgeshire Archaeological Unit), Mary Alexander (Cotswold Archaeology), Hugo Lamdin-Whymark (flintwork.co.uk), Tim Allen, Paul Booth, Stuart Foreman and Cynthia Poole (Oxford Archaeology Unit), Stephen Sherlock (Stephen Sherlock Services) and Margaret Bunyard (Wessex Archaeology). The information about the unpublished Poulton site provided by Kevin Cootes has been most enlightening. The resources at the Lincolnshire County Library in Lincoln were also used, and my thanks to Kevin Best and Gillian Wallhead there, for tracking down an out-of-print local journal for me which had important information about a site for the database. Anne-Marie McCann of the University of Southampton Hartley Library was also most helpful.

I would also like to thank those who gave me the opportunity to publicly try out my ideas by listening to and questioning my presentations of work at the Oxford University Archaeology Graduate Student Symposium, the Prehistoric Ceramics Research Group 2009 AGM, Kellogg College Archaeology/English Local History Symposium, Oxford University Archaeology Society and the Later European Prehistory Group at Cambridge University. You have helped me see connections in my work to the work of others that I had missed, and helped me see that my work had wider implications than I had thought.

My final note of thanks goes to the staff at the Institute of Archaeology, University of Oxford. Thanks for providing workspace and lots of coffee.

Janice Kinory

Copyrighted Illustrations Used with Permission

I wish to thank the many individuals and organisations who have so kindly granted permission for the use of their photographs, illustrations, datasets or plan and section drawings in the text:

-Tim Allen and Oxford Archaeology for the use of Figures 68, 69 and 70.
-The Archaeological Institute of America for the use of the photograph in Figure 119.
-The Colchester Archaeological Group for the use of Figures 2, 3, 5, 8 and 9.
-The Colonial Williamsburg Foundation for the use of the photo in Figure 114.
-The Council for British Archaeology and Nicholas Thomas for the use of Figures 46 through 61 inclusive.
-The Council for British Archaeology, the Worcestershire County Council, Anne Crone and Simon Woodiwiss for the use of Figures 14, 15 and 16.
-Sir Barry Cunliffe for the use of Figures 107, 108, 109 and 110.
-Sir Barry Cunliffe and Cynthia Poole for the use of Figures 77 to 106 inclusive and Figure 111.
-The Dorset County Museum for the use of Figures 11, 12 and 13.
-Dover Publications for the use of Figure 17.
-Essex County Council for the use of Figues 72, 73, 74 and 75.
-Andrew and Annelise Fielding for the use of the photograph in Figure 7.
-The Food and Agriculture Organisation of the United Nations for the use of Figures 112, 113 and 115.
-George Lambrick, Tim Allen and Oxford Archaeology for the use of Figure 67.
-Tom Lane, Elaine Morris and Heritage Lincolnshire for the use of Figure 118.
-Elaine Morris for the use of Figures 20 and 21.
-The Ordinance Survey for the use of the © Crown Copyright datasets used in composition of the maps made by the author.
-The Oxfordshire Architectural and Historical Society for the use of Figures 64, 65 and 66 reproduced from MYRES, J. N. L. (1937) A Prehistoric and Roman Site on Mount Farm, Dorchester. *Oxoniensia,* II, 12-40.
-The Prehistoric Society for the use of Figures 25, 26, 27, 28 and 29.
-Thames and Hudson Ltd. for the use of Figure 30.

Abstract

The study of salt during British prehistory has experienced an awakening during the past 40 years led in large measure by the work of Dr Elaine Morris of Southampton University. The earliest salt production in Britain using an industrial ceramic known as briquetage is now firmly dated to the Middle Bronze Age and its use extends to the early Roman period. While prehistoric salt production in Lincolnshire was addressed in 2001 by Lane and Morris, other regional production areas have not had similar modern syntheses. This book will explore the evidence for the production of salt in the coastal regions of Essex, along the south coast and at the Droitwich salt springs in the prehistoric period. The evidence for the distribution of salt from Essex, the south coast and the salt springs of Droitwich and Cheshire will be reviewed based upon discussion of briquetage finds. Morris' Specific Salt Container Index (SSCI) will be utilised to support this analysis. The models for and implications of salt distribution networks will be considered. A more speculative discussion of non-archaeologically visible distribution will also be presented. Four case studies comparing Iron Age sites in salt producing and salt using regions are included to establish the relative presence of salt evidence in the archaeological record and its value, if any, as a social status discriminator. Finally, information is presented on how salt may have been used in the Iron Age. While food preservation is the usage most commonly cited by archaeologists for this mineral, it would have had been used in numerous other ways, including usage in animal diets, medicine and metallurgy. The social and ritual uses of salt will also be discussed. A gazetteer and bibliography of 519 Bronze and Iron Age briquetage find sites is included as a supplemental CD.

Please note that the CD referred to throughout the text has now been replaced with a download available at www.barpublishing.com/additional-downloads.html

Contents

Please note that the CD referred to throughout the text has now been replaced with a download available at www.barpublishing.com/additional-downloads.html

Table of Figures

Table of Tables

Chapter 1
Introduction

> *"There is a need to raise the overall profile of briquetage as an important material type within the archaeological profession in order that excavators and museum curators be aware of its presence and significance."* (E. Morris in Lane & Morris, 2001: 473)

Until the nineteenth century, when refrigeration was developed, salt had been the key component in most food preservation methods. The history of this fundamental role for salt is demonstrated even now through the way it is woven into the idioms, economy, rituals and customs of British life; although it rarely is at the forefront of our consciousness unless it is in short supply for winter roads or our physician has told us to reduce our intake of the mineral. In sharp contrast, ensuring an adequate supply of salt would have been a major concern in prehistory. Given the clear link between the availability of food and human survival, it would seem that the archaeology of salt ought to be more central in archaeological thought. Archaeologists attempt to use all of the available evidence to construct an understanding of past societies, hoping to reach a semblance of a worldview akin to the perspective of the individuals living in those periods. Yet in prehistoric archaeology in Britain there is a relative dearth of salt evidence to investigate and consider, no Neolithic salt production as found in Poland (Hopkinson, 1975) or Germany (Foster, 1990) nor Early Iron Age salt mines such as those known from Hallstatt, Austria (Wells, 1980). The archaeology of prehistoric salt in Britain has been hard won, coming together slowly, based on relatively sparse evidence.

> *"There comes a time in research when the accumulation of data that had seemed too sparse to allow any generalisations suddenly reaches a point at which a coherent picture begins to emerge* (Salway, 2000: ix)*."*

For more than 160 years antiquarians and archaeologists have been adding to the store of knowledge about the production, distribution and use of salt during the British Iron Age. This gradual process has been based largely upon the recognition of a specialised industrial ceramic, generally called briquetage, and its association with salt production and, in some regions, salt distribution along with the discovery of a limited number of prehistoric salt production sites. As a result of the hygroscopic nature of salt this secondary evidence is all that is available; pure salt in its crystalline form does not survive in the archaeological record in Britain. The majority of early published information about briquetage finds and prehistoric salt production appeared in regionally focused journals, and even in the modern era the most comprehensive treatments of Iron Age salt production have been regional in their approach, such as the publications on Essex and Lincolnshire (Fawn et al., 1990, Lane and Morris, 2001). Although the picture is still far from complete, it is now possible to produce a wider view of the production and distribution of salt during this period using information gathered on more than 500 sites where briquetage has been discovered. Further, the available briquetage evidence can be combined with other types of archaeological information and insights from other disciplines, such as cultural anthropology and economics, to develop an extended commentary upon the possible roles of salt in society in the period from roughly 800 BC to AD 43. In addition to the text which follows, a Microsoft Access database which includes details about each British Bronze or Iron Age briquetage find site used in this research and a bibliography of references for each of these sites is also included on a CD as part of Appendix 8.

The initial section of the work in Chapter 2 will review the history of the antiquarian and archaeological study of prehistoric salt production and distribution in Britain, establishing a context for the synthesis which follows. By the mid nineteenth century it was recognised that finds of a crude ceramic being made in Britain had similarities with material recovered in the Channel Islands (Franks, 1850). Archaeologists were aware of continental parallels for the material as well at least as early as the second decade of the twentieth century (Bulleid, 1914) though the purpose of the material was still being debated at that date. The issue remained the subject of argument for several more years (Smith, 1918), until the Society of Antiquaries, who had supported early century twentieth work on the Essex salt production sites, or Red Hills, agreed on a single theory that the material was associated with salt production. In the modern era, the existence of European parallels for briquetage do not necessitate a retreat to a cultural diffusionist position such as that which prevailed until the mid-twentieth century in British archaeology (e.g. Smith, 1948) for an explanation of the origins of briquetage here, however, as its production is accomplished using the same underlying technology as the manufacture of other types of pottery, a technology that dates as far back as the British Neolithic period (e.g. Woodward, 2002).

Four basic questions drove the research that will be presented here. The first, and perhaps the most obvious question, asks what is presently known about the production of salt in prehistoric Britain. Although the possibilities for salt production methods leaving little archaeological evidence will be considered, the answer will focus primarily on the evidence from the British Iron Age and relate to the use of briquetage, and will also address the auxiliary questions about when, where and how this production technology first appeared in Britain. These related questions serve as the basis for Chapter 3, where the range of possible methods of salt production is discussed, the technical challenges of using ceramics in salt production reviewed and information on the regional nature of salt production in prehistoric Britain is presented through case studies. These case studies cover Essex, the south coast between Dorset and East Sussex, and the Droitwich salt springs in detail. The extensive Iron Age salt industry in the Fenlands is not the subject of a case study here, though such might be anticipated, as it was the subject of a comprehensive monograph published in 2001 (Lane and Morris). As an actual prehistoric salt production source in Cheshire has not yet been located, it is impossible to provide a review of the production methodology in use there, despite the significant role the output from the source or sources has in discussions of Iron Age salt distribution. Preparation for writing this chapter included, in addition to the literature survey, a review of the large briquetage collections from the Old Bowling Green and Friar Street sites in Droitwich, Worcestershire, and briquetage from multiple Red Hill sites in Essex held at their respective county museum stores to insure familiarity with the unique fabrics and forms of this material.

The second question is an inquiry into the present level of information available regarding the distribution of salt in Britain during the Iron Age. As noted above, salt does not survive in the archaeological record, leaving only briquetage as the primary evidence for salt distribution at sites which are remote from salt production areas. Since the late 1970s, largely due to the ongoing work by Dr Elaine Morris (e.g., 1979, 1985), the pivotal role of briquetage as distribution evidence for salt originating at the Droitwich and Cheshire salt springs has been more widely recognised. The information presented in Chapter 4 will include new briquetage find distribution maps that reflect the review of books, journals, "grey literature" written by commercial archaeology firms for clients and not always widely disseminated in the public domain, county historical environment records (HERS) and English Heritage records. The chapter will also review the multiple models for Iron Age exchange networks, drawing on models based in economics and anthropology, which have been proposed and possible methods for the transport of salt across the landscape. It will also consider possibilities that are associated with the absence of such evidence in the archaeological record.

The two chapters described above focus primarily on sites which have yielded briquetage, yet these are numerically a tiny subset of all known Iron Age sites in Britain. Chapter 5 takes a wider perspective on the Iron Age and compares and contrasts sites where briquetage has been found to the more numerous excavated Iron Age sites where it has not been recorded through the presentation of four case studies. The study areas, 20 by 20km in size, are located in Essex, Worcestershire and Oxfordshire and on the Hampshire-Wiltshire border. These more detailed reviews of archaeological data reinforce the regional distinctiveness of prehistoric salt evidence which is also evidenced in the study of salt production methods and salt distribution patterns. Such regionalism is a key theme in the British Iron Age that repeats itself in multiple aspects of material culture, showing up in pottery, structural evidence, metal working and burial practices (Cunliffe, 2005), to give but a few examples.

The final two questions to be addressed by this research relate to the possible uses for salt in the Iron Age and the impact that salt had on the daily life in that period. While Chapters 2 and 3, as syntheses, are by their nature highly derivative, Chapter 6, where these questions are considered, draws not only on the standard range of archaeological material but also integrates into the analysis relevant information from other fields, including Classical texts, information on food preservation methodology, cookbooks, domestic animal care manuals, ethnographic literature and a range of specialist monographs which are not normally found in archaeological bibliographies. The inclusion of this non-traditional material allows for the proposal of relatively detailed theories about the Iron Age diet in Britain as mediated by salt availability.

Although archaeozoologists have confirmed the occurrence of autumn culls of domestic animals during the Iron Age (e.g. Hamilton, 2000), far less attention has been given to the actual handling of the meat and hides that would have resulted from these slaughter periods. For example, while it is frequently stated in archaeological literature that the primary use of salt was for food preservation (e.g. Cunliffe, 1991: 466), little information about the actual technology of food preservation has been included in archaeological literature. While this omission would have caused little problem in the Britain of 1930, where many people, even in urban areas, were still directly involved in home food preservation or dined on home preserved food on a regular basis, and where thirty pound blocks of salt, sufficient to preserve a typical pig, were still widely sold (Kirkham, 2001: 410), few contemporary archaeologists share this personal knowledge and firsthand experience. This unfortunate disjunction between the knowledge of actual food preservation process requirements and the interpretation of site evidence can result in suggestions which, while sounding plausible, are physically impossible within the parameters of salt preservation technology. Similarly, much has been written about the range of dairy products in prehistoric diets but without much apparent awareness of the technical aspects of their creation and attention to the limits of their storage life. These oversights can have significant negative impacts on the accuracy of interpretations of archaeological evidence.

A second aspect of Chapter 6 will be consideration of the range of uses for salt beyond the food preservation stereotype. For example, its role in animal husbandry will be reviewed, as will evidence for its use in Iron Age social and ritual life. A range of industrial processes potentially using salt will also be reviewed. While there is only limited archaeological evidence in Britain for these other uses of salt, theoretical support for these potential applications will be provided. In some cases, ethnographic analogies and Classical texts have been used to assist in the creation of reasonable models for use in the discussion.

It is hoped that in addressing the four central questions posed here relating to salt in the British Iron Age the true importance of briquetage evidence within the archaeological record will have been demonstrated. Further, it is hoped that the risks to archaeological interpretation accuracy related to the failure to consider such evidence will have been made apparent, thus allowing for future improvements.

Chapter 2
The study of prehistoric salt production and distribution in England and Wales

2.1 Introduction

Until the nineteenth century, salt was the main basis for most food preservation. Despite salt's central role in sustaining life, the study of prehistoric salt production and distribution was generally neglected by most British antiquarians. In this they differed significantly from their continental peers, trailing by more than a century publications on the subject in France. The relative tardiness of British study of this key commodity may have resulted from the absence of highly visible sites clearly linked to early salt production like those of the Bronze and Iron Age salt mines of Hallstatt and Hallein (Nenquin, 1961, Jodłowski, 1977, Coles and Harding, 1979). Unlike this region, British salt beds occur at great depth and were not worked until the 17th century (Fielding and Fielding, 2006), so there were no prehistoric salt mines to investigate. In comparison, the evidence for prehistoric salt production at British coastal sites is unspectacular. Even large sites such as the Red Hills of Essex had little to recommend them, beyond their unusual soil colour and poor quality fragmentary ceramics, to the attentions of antiquaries digging barrows or searching for Roman villas on a weekend. These inferior ceramics, however, would ultimately prove to be the strongest evidence available for prehistoric salt production and distribution in Britain.

The earliest known paper dealing with ancient salt production was published in France by de la Sauvagère in 1740. It mentions that the briquetage of Marsal had been studied for some time and refers to earlier papers, now lost, on this subject (Gouletquer, 1974). The term 'briquetage' used by de la Sauvagère is now used broadly to describe the range of ceramics associated with early salt production and distribution in Britain and in continental Europe. The word itself seems to derive from traditional use in the Lorraine region, where it was applied to the deposits of burnt clay and ashes found in the Seille valley (Gouletquer, 1974). In Britain, the term "briquetage" coexists with the phrase "Very Coarse Pottery" or "VCP", first published by Gelling and Stanford (1965 77) to describe "... quantities of crude brick-red pottery ..." recovered from secure Iron Age contexts at hillforts in Wales and the Marches. In a subsequent paper, Stanford (1974) noted that the term VCP was not their invention, but was actually coined by Wing Commander T. W. Ellcock, a volunteer working with the pottery from the Croft Ambrey hillfort. In modern texts the material is sometimes referred to as Stony VCP, a designation that characterises the fabric as well as the forms, and the term is applied only to Iron Age salt distribution ceramics from the brine springs in Cheshire.

The nineteenth century antiquarian study of briquetage on the continent continued to build upon the earlier work. Publication of briquetage found at Halle on Saale dates from 1825 (Riehm, 1961). In comparison, the earliest known British publication of an artefact now associated with prehistoric salt production dates to 1850 (Franks, 1850). A second article that same year, noting similarities between earlier finds in Lincolnshire and Kent and more recent ones in the Channel Islands, however, suggests that antiquarian interest in British sites and finds associated with ancient salt production may have begun decades earlier, though this does not seem to have resulted in any surviving publications (Lukis, 1850). From the antiquarians through to contemporary British archaeology, the pattern has been one of varying levels of academic awareness, shifting between periods of intense research and relative disinterest, driven forward by a small number of key individuals with a deep interest in the subject of prehistoric salt production and its associated briquetage.

It is now accepted, however, that briquetage associated with salt production and distribution has been found at British sites dating from the Middle Bronze Age to the early Roman Period, the period *c.*1400 BC to the second century AD. The majority of finds, however, date to the British Iron Age, the period roughly 800 BC to AD 43. The process which lead to the accumulation of this knowledge is discussed below.

2.2 Early publications and theories

The study of prehistoric salt production and distribution in Britain has developed with a regional focus. Individual antiquarians and archaeologists have tended to concentrate on a single area of the country, a practice that continues to the present day. For many of the early investigations, the starting point was a briquetage find. The ceramics associated with early British salt production are functional rather than artistic. They lack decoration, usually are crudely formed and finished and occur most commonly as undiagnostic broken fragments at sites. Before the advent of radiocarbon methodology, sherds, particularly at salt production sites, were frequently found with little other material that was closely datable. This constellation of characteristics led to much wild speculation about the age and function of briquetage well into the twentieth century.

As noted above, published British reports on these ceramics in scholarly journals began in 1850, when several discussions appeared within a few months of each other. The earliest mentions of what would now be known as briquetage refer to the finds as 'hand bricks', a designation that suggests both their method of production and their imagined use. Finds from Guernsey (Franks, 1850) and Lincolnshire (Nicholson, 1850, Lukis, 1850) were published by the Royal Archaeological Institute of Great Britain and Ireland (RAI). Lukis comments that as

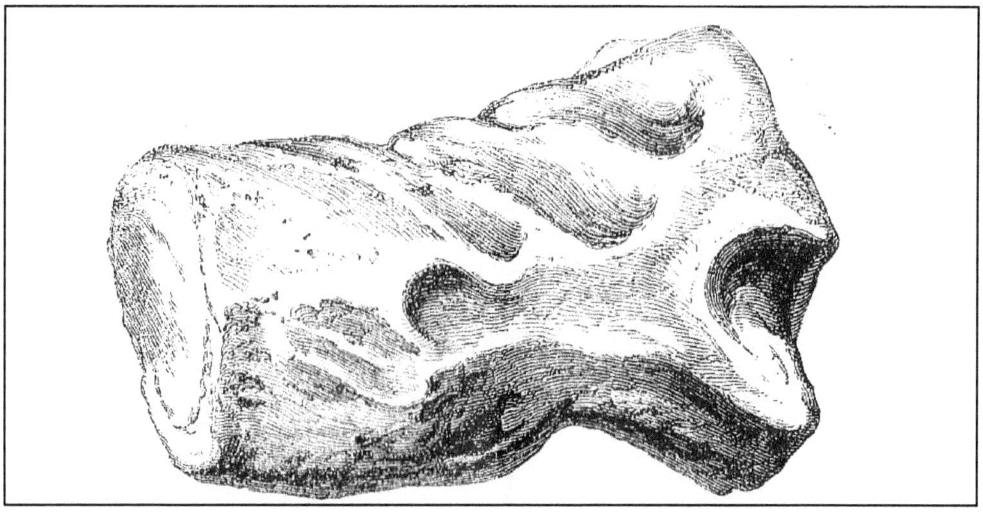

Figure 1. Briquetage misidentified as medieval material.
Original caption: "'Hand brick' or clump of terra cotta, found with pottery on the coast of Lincolnshire. " (Reynardson, 1850).

early as 1839 or 1840, finds of hand brick material had been made in the area of Ingoldmells, Lincolnshire. He may have been the first person who reported publicly on briquetage in Britain, noting in the same article that he had discussed the finds at the RAI meeting in Canterbury in 1845 (Lukis, 1850: 176).

The catalogue of antiquities exhibited at the 1848 AGM of the RAI included, among the display of medieval pottery and ceramics, a 'hand brick' from Wainfleet, Lincolnshire, shown in Figure 1. Franks (1850), whilst not venturing a guess at their age, speculates on their function, suggesting that the finds may have been kiln furniture. Lukis (1850: 176-77) wrote that originally he had regarded the material as medieval, but had come to believe it to be either of Bronze or Roman age and evidence of human migration or marauding invaders. He promoted a suggestion he had received that the purpose of the items may have been for use as fishing net weights, due to their presence at a coastal site.

The next published prehistoric site was located in Essex. Henry Stopes, a geologist, published a paper on an 1878 excavation of a 'Red Hill' at Peldon (1879). He described the Red Hills as a common feature in Essex and Suffolk along rivers and estuaries, of up to 30 acres (12.14ha) in size and 6 feet (1.83m) in depth, and composed of red clay, broken pottery, ash and wood charcoal (Stopes, 1879: 369). He also noted the crude finish of the ceramics he had recovered, "... ruder and rougher than the early British urns of the oldest type", the large quantity of grass temper in the clay and the fragmentary state of all his finds (ibid.: 371). Although he reported that local people thought the mounds were the remains of Roman brickworks, he offered no personal opinion on their purpose (ibid.: 370-71). An article the following year commenting upon the Stopes paper highlighted the physical similarities of the Essex Red Hills to similar mounds in Teeside known to be associated with medieval saltworking (Atkinson, 1880). In this article Atkinson asserted that the Red Hills "may be, and most likely are, all that is left of ancient saltworks once carried on ... along no small extent of the Essex coast ..." (ibid.: 198).

Seven years after Atkinson's remarks were published, Stopes despaired publicly at the continued lack of interest in exploration of the Red Hills since his earlier paper (1887). He acknowledged Atkinson's idea as to the origin of the mounds while citing a theory put forward by a fellow member of the Royal Geographical Society that the mounds were ancient camp sites. On balance, he was unwilling to accept Atkinson's idea as the mounds were generally located along freshwater rivers "... where a whole summer's evaporation would not yield much salt from the water," (ibid.: 101). This comment demonstrates that his mental model of the salt production process was based on solar evaporation methods used in sunny, warm climates, as well as a failure to consider possible sea level changes over time and their possible effect on tidal river salinity, a surprising lapse for a geologist.

Two years later the result of fieldwork on a Red Hill that may be the subject of the illustration in Figure 2, was reported in the *Essex Naturalist* (Cole, 1889), again failing to develop a single agreed theory as to the original purpose of the site. Following the initial presentation of the paper at a meeting, the minutes report that at least five competing theories were presented to explain the purpose of the site: "...that they are the remains of pottery, salt or glass works, refuges for cattle during the overflow of the waters, the terminals of Roman roads..." (Cole, 1889: 164), with all seeming equally plausible and the matter left open to further investigation. A footnote reports that there was an ongoing effort at the time to identify and catalogue Red Hills by a club member, Dr Henry Laver, (Cole, 1889: 163), possibly the first attempt to compile a British gazetteer of prehistoric salt production sites.

Figure 2. Excavation of a Red Hill at Burnham on Crouch in 1888.
From Fawn et al., 1990: 1.

2.3 The twentieth century

The new century did not immediately bring a conclusive answer to the riddle of the Red Hills. The topic reappears in 1904, reporting on a 1903 field excursion. The article published in the *Essex Naturalist* demonstrates that Red Hill research had continued during the intervening period with "hundreds" of these sites documented as of that date (Cole, 1904). Little progress, however, had been made on defining their original function:

> *One explanation was that they were old salt works; but that was discountenanced by their position inland; whilst another view, supported by the fact that they were always found above a high water mark, was that they were the sites of refuges made to protect cattle against floods. Mr Cole, however, favoured the hypothesis that they were the sites of some very early pottery kilns, for in them were found three or four kinds of pottery, some of which appeared to be Romano-British* (Cole, 1904: 244).

Dr. Laver, who had been researching Red Hills for at least 14 years by this date, was present and pointed out the occurrence of "Celtic" pottery at a Red Hill site in Mersea. He argued that an even earlier date for the Red Hills was possible. He also recommended an application to the British Association for funds to continue the research to a final conclusion (ibid.: 245). Essex was not unique in having early twentieth century archaeological interest in prehistoric salt production. The first detailed description of a Lincolnshire salt production site was also published in 1904 (Kirkham, 1981: 5).

Two years later, a paper on the history of saltmaking in Essex referred briefly to Atkinson's suggestion from 1880 that the mounds may be pre-Roman, without accepting the accuracy of this claim (Christy, 1906). Later that same year Cole (1906), having seemingly changed his position since 1904, went even further, suggesting that the ancient pottery was probably used for brine boiling. This echoes a similar theory published in a French paper in 1904 (Nenquin, 1961), although it is unknown whether Cole had read this paper or had developed his theory in isolation upon examination of the local evidence.

Perhaps the 1903 suggestion by Dr Laver to seek outside funds to further the investigation of the Red Hills had been addressed, as 1906 saw the beginning of a two-year multi-disciplinary excavation programme by the Society of Antiquaries designed to resolve the mystery of the Red Hills. The results of this effort were published in 1908 (Reader and Wilmer). A typology of the various ceramics recovered was developed, with the objects designated supports, firebars, wedges, t-pieces and luting, (Reader and Wilmer, 1908: 166). A *terminus ante quem* of the first century AD was proposed for their production (ibid.: 180). Similarities between the Essex finds and material known from Kent and Peterborough were noted (ibid.: 178-79). Due to the absence of saggars, the sites were discounted as potteries, but the report fails to reach a definitive conclusion on their function:

> *Salt-making perhaps is the simplest suggestion to meet some of the conditions, but this in no way explains the vast quantities of burnt earth nor the shape and nature of the mounds. The objects of briquetage also seem unsuited to this industry, there being a great absence of pieces which might have formed circular pans or vessels for holding liquids* (ibid: 180-81).

The above quotation contains a telling comment on the absence of circular pans or vessels. As subsequent research has established, the vast majority of briquetage brine evaporation pans used in Essex were usually rectangular or subrectangular in shape (e.g. Rodwell, 1979, Fawn et al., 1990). What had been termed luting and saggar sherds were, in fact, the missing pan sherds (Reader and Wilmer, 1908: Fig. 15, 16 & 17). This was not the first time that archaeologists had come to wrong conclusions trying to force evidence to meet unrecognised preconceived expectations, and certainly not the last. In retrospect, it is possible to state that the bias in expectation towards circular pans seems to have extended the Red Hill purpose debate by more than a decade.

With the conundrum of the Red Hills still unresolved, a second Society of Antiquaries expedition in 1908-09 excavated additional sites. Two reports, with finds as inconclusive as those of the first excavations, failed to clearly define the purpose of the mounds (Reader, 1910a, Reader, 1910b). Also appearing that year, a response to the chemical analysis of the red soil from a site which had been included in an appendix to the earlier report (Jenkins, 1908) was presented by Flinders Petrie. Petrie (1910) suggested that the high level of potash and soda noted on some small samples of fused material from the mounds was evidence that the Red Hills were the remains of Celtic kelp burning, as discussed by Pliny. Jenkins dismissed this idea, commenting that the ratio of potash to soda was not indicative of that process, while also highlighting the potential shortage of suitable seaweed at the Red Hill sites (1910). Such efforts to make British sites conform to Classical commentary were a practice that dogged British archaeology well into the twentieth century (e.g. Wheeler and Wheeler, 1936).

Shortly before the outbreak of the First World War, Reginald A. Smith was mentioned in a Society of Antiquaries meeting report as considering the Red Hills to be salt production sites, similar to those recently found in France (Wilmer, 1914). Four years later Smith (1918) presented a paper to the Antiquaries on his theory of the Red Hills' origin. He was convinced that the sites were evidence of "... a vast salt-making industry in the pre-Roman period ..." (ibid.: 37). His argument, drawing on knowledge of chemistry as well as archaeology in Britain and continental Europe, and supported undoubtedly by his status as Keeper of British and Medieval Antiquities at the British Museum, seems to have been persuasive. The Red Hill debate was over for most archaeologists. Seven years later the paper was cited to help explain briquetage finds in Kent (Burchell, 1925). A subsequent discussion of Iron Age saltmaking in Kent, based on the Essex model formulated by Smith, was included in a 1930 synthesis of the archaeology of the county (Jessup, 1930). This archaeological consensus did not preclude old ideas reappearing however, particularly among amateurs, and as late as 1961 it was still possible to find alternate theories for the Red Hills appearing in print (Edwards, 1961).

While Smith had correctly interpreted the former function of the Red Hills, his thoughts on the nature of the production process itself were incorrect (1918: 39-45), clearly being driven by Pliny's comments on salt production in Gaul and Germany (*NH*, XXXI, vii, 39, 82) and experiments then underway in Europe to replicate this process. This Classical commentary on the supposed barbarian salt production process seems to have led to further variations on the myth, expanded now to include Britain and legitimised by its presence in an HMSO publication:

> *It is stated that in Pre-Roman times the natives of Cheshire evaporated briny spring-water by pouring it over faggots of charcoal, and that the Romans introduced the improved method of evaporation in open pans, a method supposed to have been first applied at a small brine-spring near Sheath Street, Northwich* (Sherlock, 1921: 2).

In the period up to the end of World War Two, finds of briquetage, sometimes referred to as 'hand bricks', were recognised at other, primarily coastal, sites in Norfolk (Hawkes, 1932, Rainbird-Clarke, 1939), Lincolnshire (Swinnerton, 1932, Warren, 1932, Phillips, 1934), Hampshire (Fox, 1935) and Suffolk (Rainbird-Clarke, 1939). Swinnerton (1932) reported similarities between the finds from Lincolnshire and Essex, while Hawkes (1932) thought the finds from Lincolnshire, Norfolk and Essex were comparable. By 1939, Rainbird-Clarke (op. cit.: 21 & 23) was able to expand his comparison to include Lincolnshire, Essex, Norfolk and Suffolk finds. Working outside East Anglia, Fox found similarities between Hampshire finds and those from Essex (1935).

The list in the prior paragraph strongly suggests awareness among East Anglian archaeologists, along with Mr Fox in Hampshire, of the finds of briquetage and their significance. However, not all coastal regions seemed to share this knowledge. Dorset, for example, seems to have been bypassed by this research stream. A review of the annual journal of the Dorset Antiquarian Field Club, later the Dorset Archaeological Society, from the late nineteenth century onwards, found no description of possible briquetage before 1930 (Smith, 1930), and the material went unrecognised as such for at least the next six years (Davies, 1936). This society, though having 300-400 antiquarian or archaeologist members, seems to have lacked an individual with an interest in prehistoric salt production to drive this research forward, suggesting it as a subject for local field excursions or papers while significant early progress was being made elsewhere.

In the pre-World War II and early war years, material found at inland sites was more likely to be misidentified, such as Verulamium's 'Belgic bricks' (Wheeler and Wheeler, 1936), or described as inferior local Iron Age pottery (e.g. Leeds, 1931, Myres, 1937, Bradford, 1942) than properly classified as briquetage. Wildly inaccurate dating of misidentified briquetage persisted into the second half of the century, presumably due to the lack of absolute dating techniques. For example, during the 1934 excavation of Pant-y-Saer, Anglesey, briquetage finds were mis-characterised as Dark Age pottery (Varley, 1950, Gelling and Stanford, 1965).

As was most British archaeology, the study of prehistoric salt production and distribution was disrupted by the

Second World War. The pre-war excavation findings were not forgotten, however. Hawkes and Hull's volume on Camulodunum (1947) features a discussion of the Red Hill explorations and Smith's explanation of these sites. One of the first post-war articles on briquetage, published in 1948, covered work conducted between 1931 and 1948 in the Isle of Purbeck area in Dorset (Calkin, 1948). Calkin was certain that he had found evidence of a major salt working industry in the area, like those known from elsewhere in Britain and on the continent (ibid.: 58). Smith's 1918 work is cited in a subsequent article endorsing Calkin's findings, suggesting its seminal contribution to the field of prehistoric salt production studies in Britain (Frend, 1949). North eastern Essex was surveyed in the mid-1950s and 19 previously unrecorded Red Hills were identified (Farrands, 1959).

What might be called the next pivotal year in European salt studies, 1961, saw the publication of two significant papers on prehistoric salt. A doctoral thesis providing a synthesis of the study of ancient European salt production from the Neolithic onwards appeared (Nenquin, 1961), followed closely by a paper documenting 10 years of experiments in Germany attempting to replicate the production of salt using ceramic briquetage (Riehm, 1961). Nenquin's thesis provided scholars with a bibliography of earlier studies from Britain, France, Belgium, Austria and Germany, and a gazetteer of European salt production sites to review. Riehm, however, was a medical doctor with an antiquarian-like interest in archaeology. His experimental archaeology led to a detailed explanation of a prehistoric salt production process utilising briquetage. A 1955 find of numerous broken but replicable briquetage containers at Halle led ultimately to a successful reproduction of the saltmaking process using these components (ibid.: 183). Starting with accurate vessel forms, Riehm was able to reverse-engineer the brine salt production process. These two publications were the foundation for much of the briquetage-related archaeological research of the next 15 years in Britain, with Riehm being frequently cited in the interpretation of the prehistoric British process (e.g. Farrar, 1962b, Gelling and Stanford, 1965).

At roughly the same time, Farrar published a synthesis of Dorset Iron Age and Romano-British salt working, pointing out numerous differences between Dorset briquetage and that from Essex, Lincolnshire and Norfolk, highlighting the absence of vegetable temper in the fabric and the cut edge of many vessels (1962b). The article was clearly written after a thorough review of previous literature on the subject of briquetage and prehistoric salt production. As with Fox before the war, his broader view of the subject was somewhat unusual in Britain during this period.

Continuing the pre-war pattern, the misidentification of briquetage being found on inland sites as poor quality pottery or fired clay continued into the early 1960s (e. g. Richardson and Young, 1951, Fell, 1961). At this time the recognition of briquetage, even where it had been previously identified, was not guaranteed. For example, an Iron Age salt production hearth and accompanying ceramics were found at Wyke Regis in Dorset (Bailey, 1962). Bailey described the presence of a poorly made container body sherds and bases in the assemblage (ibid.: 134). In a pattern soon to be repeated elsewhere, the ceramic assemblage from the site was reviewed by Farrar and the crude pottery was found to be briquetage (1962b). With a continuing interest in this material, Farrar may have been the first in Britain to recognise briquetage at an inland Dorset site, Gallows Gore, as possible evidence of the salt trade (1962a).

The same problem of find misidentification also continued in Wales and the Marches into the 1960s (e.g. Gelling and Stanford, 1965). Sherds found at the Croft Ambrey, Herefordshire, and Dinorben, Denbighshire, hillforts, whilst properly dated to the Iron Age, were described as coarse red pottery (Gardner and Savory, 1964). In their synthesis of VCP finds the following year, Gelling and Stanford (1965) discount the ideas that VCP was used for cooking pots, crucibles or the salt trade, preferring to interpret it as forming vase-like containers that were inverted to cover small ovens. This unique interpretation was repeated in subsequent articles on Croft Ambrey (Stanford, 1967), Credenhill Camp (Stanford, 1970) and a synthesis of Welsh border hillforts (Stanford, 1972). The theory was not universally accepted, however, and the alternate explanation of VCP as coarse Iron Age pottery persisted (Savory, 1971).

The incidental briquetage finds from these hillforts were in contrast to the contemporaneous efforts focusing on early salt production sites on the other side of the country. The Colchester Archaeological Group, inspired by amateur archaeologist Kay de Brisay, embarked on a renewed series of explorations of the Essex Red Hills (de Brisay, 1972, Richardson, 1972). Almost all Essex briquetage was found at Red Hills or near potential production sites, with the scale of these sites suggesting production greatly exceeding local consumption levels (e.g. Drury and Rodwell, 1973, Rodwell, 1974, de Brisay, 1978, Rodwell, 1979).

In 1974, the Colchester Archaeological Group hosted a conference on salt production archaeology. The published volume of conference papers, covering ancient sites and contemporary production methods from around the world, is still a key reference for those interested in salt production archaeology (de Brisay and Evans, 1975). British speakers at the event reported on work being done in Dorset (Farrar, 1975), the Hampshire and Sussex border (Bradley, 1975), Kent (Miles, 1975), and Lincolnshire (Baker, 1975, Kirkham, 1975) among other locations. Bradley's contribution included a discussion of the seasonality of saltmaking within the wider Iron Age agricultural calendar, a theme he expanded in a subsequent book (1978). Droitwich, not then recognised as a major Iron Age production site, was discussed with respect to its medieval saltmaking (Berry, 1975).

The long-term Fenland landscape survey project beginning in the early 1980s located many prehistoric salterns in the region (Hall, 1987, Hayes and Lane, 1992, Lane and Morris, 2001). Further details of Iron Age salt production emerged in Lincolnshire. Ingoldmells, site of some of the earliest recorded briquetage finds, was the first site that yielded nearly complete briquetage troughs, the containers likely to have been used on hearths for brine evaporation, from within a peat layer thought to be near their production site (Ambrose and White, 1981, Kirkham, 2001, Crosby, 2001).

In parallel to the work discussed above, the study of ceramics and their distribution took a new direction with the application of thin-section methodology as an analytical tool by Peacock in the late 1960s (1968). Application of this technique allowed the production source and method to be determined more accurately, defining with varying degrees of precision the site where the clay had been obtained. His work set the stage for that done by E.L. Morris; in her master's dissertation and doctoral thesis, she identified Stony VCP as salt distribution briquetage originating in Cheshire and identified numerous other VCP finds from the Severn–Avon drainage basin as briquetage coming from Droitwich (Morris, 1979, Morris, 1983). In her research, Morris incorporated the evidence available from the large quantity of Iron Age briquetage excavated in Droitwich between 1973 and 1979, although these sites were not published in full until 1992 (Woodiwiss). Since the publication of Smith's 1918 paper, the existence of a coastal Iron Age salt production industry, evidenced by briquetage, had become widely accepted. Morris' work (e.g., 1985) led to widespread recognition that Iron Age briquetage from inland salt springs might be found in many ceramic assemblages recovered from sites far from salt production areas.

The first evidence that this ceramic-based production technique had also been used in the Bronze Age came from Fengate, Peterborough and Billingborough in the early 1970s (Pryor, 1980, Jones, 1977). At Mucking, Essex, a 1977 briquetage find further expanded the geographic area known to be involved in Bronze Age salt production (Jones, 1977, Jones, 1978, Jones, 1980). Subsequently, finds of similar or earlier dates demonstrating salt production and distribution using briquetage have been reported in Berkshire (Bradley et al., 1980), Cambridgeshire (Gurney, 1980, Hall, 1987), Cheshire (Fairburn et al., 2003), Devon (Quinnell, 2006) East Yorkshire (Faulkner, 2008), Essex (Barford, 1984-85, Fawn et al., 1990), Hampshire (Wainwright and Davies, 1995), Kent (e.g. Masefield et al., 2003, Griffin, 2006), Lincolnshire (Palmer-Brown, 1993, Lane and Morris, 2001), Somerset (Foster, 1990) and Surrey (Proctor, 2002). Features shared by these sites include the lack of evidence of their existence in aerial photographs (Hall and Coles, 1994), an absence of related artefacts in ploughsoil near the site and an inability to identify such sites through fieldwalking (Lane and Morris, 2001). Of the production areas in this group, only Devon has thus far failed to produce evidence of continuity of briquetage use for salt production between the Bronze and Iron Ages.

2.4 Conclusion

It is now possible to look back and see that Morris' work marked a new direction in briquetage studies during the late twentieth century. Since the publication of her doctoral thesis in 1983, the evidence for prehistoric salt production and distribution has continued to accumulate at an accelerating rate. When briquetage is present, Iron Age excavation reports published since the late 1980s generally include specialist discussion of the finds, and possible briquetage is reported even when found in small quantities (e.g. Fulford, 1987, Elsdon, 1994, Jackson et al., 2000), with the evidence taken as proof of the existence of salt trade networks. A full reappraisal of the Essex Red Hills was undertaken and an expanded gazetteer of sites issued (Fawn et al., 1990). As noted above, the major landscape survey project undertaken in East Anglia during the 1980s greatly increased the number of known Iron Age salt production sites in the area (e.g. Hayes and Lane, 1992, Lane and Morris, 2001). Reappraisal of old ceramic assemblages has increased the number of known find spots in some areas such as Essex (e.g. Sealey, 1995), and has led to the recognition of an Iron Age production and distribution network in the Tyne-Tees region (Willis, 1995). Greater briquetage awareness among archaeologists means that the number of known inland briquetage find spots continues to increase (e.g. Hardy and Barclay, 2007), as does the number of Bronze Age finds being reported (e.g. Faulkner, 2008, Patten, 2009).

More than a century after the Society of Antiquaries' first expedition to the Red Hills of Essex, different questions relating to salt in the British Iron Age have been proposed and remain under discussion. An understanding of salt production and distribution in those portions of northern and western Britain that were largely aceramic in prehistory is still to be achieved. Wider discourse and debate continues as to the details of salt production, distribution and use within the economy and social structure of the Iron Age (e.g. Serjeantson, 2007, Morris, 2007b). Undoubtedly, much yet remains to be said on these related topics.

Chapter 3
Prehistoric salt production in Britain

3.1 Introduction

The archaeology of prehistoric British salt production is problematic in that the subject under study – salt – does not survive in the archaeological record, making this the archaeology of the invisible. Archaeologists have only indirect evidence to study, namely the artefacts and structures that seem to relate to salt manufacture, putting them at a disadvantage compared to those, for example, who study prehistoric metalworking and have available to them finished objects and also the technological refuse to review. Despite this evidentiary shortcoming, a composite picture of the technology of salt production in later prehistoric Britain, obtained through the study of multiple sites, has been developed by archaeologists.

In this work, the British Late Bronze Age is considered to be the period *c*.1150 - 800 BC (Haselgrove and Pope, 2007, Needham, 2007), although these dates should not be interpreted as finite cut-off points that define clearly distinct types of cultural behaviours. The Early Iron Age is the period from *c*.800 to 400 BC. *Contra* Moore (2007), the more traditional separate terms Middle Iron Age and Late Iron Age will be used here, to align with the previously defined time periods associated with various production sites and to assist in delimiting more closely the dating of changes in technology. Here the Middle Iron Age is defined as *c*.400 BC to 100 BC, and the Late Iron Age as the period from *c*.100 BC to AD 43, the Roman Conquest; although once more the dates are approximate boundaries rather than absolute ones. For the sites in southern and eastern England that will be discussed below, the constraints on the use of the tripartite Iron Age system as pointed out by Harding (2004), particularly problematic when applied in northern Britain, are far less of an issue.

As in the prior chapters, the term "briquetage" will be used to describe all of the crude ceramics associated with salt production or distribution from any British site, while "Stony Very Coarse Pottery" or "VCP" is a term applicable only to briquetage associated with the Cheshire salt springs.

3.2 Prehistoric salt production processes

Humans crave salt, and even have specialised taste buds on their tongues for this flavour (UM, 2004). In northern Europe, humans need as little as 1 gram per day of salt in their diet to maintain health (Carter, 1975), a minimum easily met without supplementary salting of food when meat or animal blood are the main components of the diet. This physiological requirement suggests that humans were driven to locate salt in their environment millennia before the earliest recognised salt production evidence was created. The possibility of early salt production leaving no archaeological evidence has long been recognised by archaeologists (e.g. Bradley, 1975). Based on modern ethnographic analogies, Gouletquer (1974) suggested the following methods for meeting the need for dietary salt: "...earth-eating, salt-water drinking, the eating of plants, blood and urine drinking, necrophagy and, of course, salt eating." While most of the activities on this list would leave little - if any - archaeological evidence, distinguishing specific salt-obtaining behaviour from general food and drink intake is virtually impossible even where evidence exists. This means that the earliest recognised archaeological evidence for the production of salt in Britain dates from a period far later than the earliest probable point in time at which people would have been engaged in behaviours designed to add salt to their diet.

Sodium chloride, more commonly referred to as salt, is abundantly available in the natural environment. Typical seawater, covering 70% of the planet, is approximately 3.5% salt. Salt also exists on land within saliferous geological beds. Water flowing through these beds dissolves the salt, and may then rise to the surface as salt springs. Although seawater, salt beds and salt springs are all found in or around Britain, only two of these forms are germane to discussions of prehistoric salt production. Whereas salt mining in continental Europe may have begun as early as the Late Bronze Age-Early Iron Age period (Jodłowski, 1977) the depth of salt-bearing strata in Britain means that mining here began as late as 1670 (Sherlock, 1921, Fielding and Fielding, 2006, and *contra* Mulville, 2008).

A wide range of methods for producing salt are known. Probably the earliest system, given that it does not rely on sophisticated technology, is solar evaporation. Pliny (*NH*, XXXI,vii, 39, 73) describes the process as does Agricola (1556). This ancient system is still in use (e.g. Nenquin, 1961, Andrews, 1980); and figure 3 illustrates a modern industrial application of the process. A prehistoric production system based on the harvesting of naturally occurring salt crystals along shorelines or those created in ephemeral shallow oceanside depressions, is likely to be archaeologically invisible. The obvious process flaw, however, with this method is that simple evaporation mixes all dissolved salts present in seawater, including undesirable bitter compounds of magnesium or calcium, into the finished product. This problem can be corrected by prolonged storage of the finished crystals, allowing natural hygroscopic action of the bitter salts to separate them from the far more desirable sodium chloride (Multhauf, 1978). While effective, this process is slow, with storage of up to a year required to obtain maximum purity. Palmer-Brown (1993) found rare evidence suggesting that solar evaporation may have been the first stage in salt production at Bronze Age Tetney, Lincolnshire. Based on this example, it can

Figure 3. Modern solar salt production at Lanzarote.
From Fawn et al., 1990: Plate 10.

reasonably be speculated that evaporative technology had a wider role in the prehistoric British salt industry than has been demonstrated to date, but its full extent and duration are impenetrable. It is clear, however, that the British climate would be less amenable to this system than sunnier, warmer and drier regions.

A second salt production system which might have been used in British prehistory, and which would also leave little archaeological evidence, is the production of salt from the ashes of halophytic plants. Compared to solar evaporation, this is a complex, multi-stage process. The plants are burned and the ashes gathered, then washed with water. The runoff is collected and then concentrated, ultimately yielding a mixture of sodium chloride and potassium chloride, with the end product often referred to as bitter salt. Detailed descriptions of this process, still used in the savannah and forest areas of western Africa and New Guinea during the twentieth century, were provided by Gouletquer (1975) and Godelier (1977). The method is mentioned also by both Jodłowski (1977) and Hopkinson (1979). The process was used in Europe as well: salt-impregnated peat was burnt and the ashes were treated in this way in medieval Holland (Fawn et al., 1990). These widespread reports suggest the possibility that similar activity might have occurred in prehistoric Britain, but such has not yet been documented. Gouletquer (1975: 51) notes that the residue from this production method includes large heaps of washed ashes. Large ash spreads without obvious purpose are sometimes encountered on prehistoric archaeological sites in Britain. Perhaps some of these were related to plant-sourced salt production, and a more nuanced evaluation of the ashes in likely locations might prove this to be so. For example, Cleal and Chowne (2001) note an Iron Age clay context at Billingborough, Lincolnshire, the presence of which they attribute to the burning of salt-impregnated plant material without considering the possibility of salt production from the resulting ashes.

As noted above, the British climate, like that of the rest of northern Europe, is unsuited to the mass production of salt by solar evaporation. This problem is overcome by applying heat to the brine, speeding up the natural evaporative process that results in salt crystallisation. The first step in the process seems to have been allowing particulates to settle out of the seawater, as illustrated by structures interpreted as settling tanks found in Essex (Fawn et al., 1990) and Lincolnshire (Lane and Morris, 2001). The production process itself relied on the differential precipitation rates of the naturally occurring dissolved salts of calcium, sodium and magnesium in seawater, of which sodium chloride makes up approximately 72% (Multhauf, 1978). Calcium carbonate and calcium sulphate would precipitate most readily (Encyclopaedia Britannica cited in Bridbury, 1955), sometimes forming an insoluble white crust found on briquetage, followed by sodium chloride crystal formation when the solution reached a 27-28% concentration (Fawn et al., 1990: 20). Crystal size was determined by the speed and temperature at which the solution was evaporated (Sherlock, 1921); large crystals result from slow evaporation. The crystals could be removed from the solution as they formed and the remaining warm liquid in the evaporation pan topped up (Rodwell, 1979) probably using buckets, with fresh brine and the process continued until the concentration of bitter salts in solution became sufficient for their precipitation, yielding a contaminated end product and halting the saltmaking process.

Two methods using indirect heating of the brine, accomplished either by dropping heated stones into organic or ceramic containers holding brine or by pouring brine directly on to heated stones (Gouletquer, 1974), leave limited archaeological evidence and none which would link the evidence unambiguously to salt production. Definitive archaeological evidence for the use of either of these methods is absent in Britain, but their possible use is suggested by the discovery of fire-cracked stones along with briquetage in salt production furnaces at Wyke Regis, Dorset (Bailey, 1962), and Bays Meadow, Droitwich, Worcs. (Barfield, 2006). While the presence of heated stones, thought to be used for cooking in domestic contexts, has been widely noted, the limited use of heated stones with briquetage may be evidence that they impeded the saltmaking process: the presence of heating stones in the brine would interfere with the removal of salt crystals as they formed in the concentrating vessel. The various possibilities for salt production mentioned above mean that Morris' comment (2001b) "Without briquetage, there would have been no salt...," should perhaps have been "Without briquetage, there would have been no archaeological evidence of salt...".

A Neolithic site at Lower Halstow, Kent, now destroyed by erosion, was claimed as a very early sea salt production site in 1957, though the evidence for this suggestion is little more than geographic proximity of hearths to the ocean (Brothwell and Brothwell, 1969, KCC, 2009). The earliest briquetage evidence for British salt production dates to *c.*1400 BC, in the Middle Bronze Age, from a site at Brean Down, Somerset (Foster, 1990) and Pode Hole Quarry, Cambridgeshire (Morris, 2009) may be of a similar date. Northey, Cambridgeshire, and Billingborough, Lincolnshire, are also thought to date to the Middle Bronze Age (Morris, 2007b). Whether this was an independent invention or a transfer of technology from Germany (Foster 1990: 172; Matthias 1976 cited in Barford, 1995) or France (Barford, 1988), where briquetage of similar date and somewhat similar form has been found, is still a matter for speculation.

The idea of Bronze Age salt production using briquetage has been controversial since the first finds at Fengate in the early 1970s (Pryor, 1980), whose existence ran counter to the prevailing theory that salt production utilising briquetage was primarily an Iron Age phenomenon. The discussion is rendered more complex by the changing interpretation during the twentieth century of the starting date for the British Iron Age. As new evidence accumulated, the starting date of this period shifted back from *c.*550 BC (Hawkes, 1959) to the currently suggested date of *c.*800 BC (e.g. Haselgrove and Pope, 2007). Thus, a briquetage find at Almondbury hillfort was reported originally as possibly dating to the Late Bronze Age (Varley, 1976), but is now considered to date to the Early Iron Age. Despite this revision in the prehistoric chronology, the number of Bronze Age sites has continued to increase. A 1998 map of British Bronze Age salt production sites had only 7 definite and one possible production location (Harding, 1998), a sharp contrast to the 13 production sites and 18 non-production finds shown in Figure 4 and Table 1, compiled from publications, grey literature and Historic Environment Record (HER) reports, all of which are currently interpreted as dating to *c.*800 BC or earlier. The number seems likely to continue to expand, with Sharples suggesting recently that Langstone Harbour, Hampshire, may have been the site of Late Bronze Age salt production though acknowledging that the present evidence for this is inconclusive (2010). The importance of rivers for distribution of salt inland during that period becomes ever clearer as newly located sites are added to the map.

While the dating of some of these sites has been contested (e.g. Darvill, 1987, Needham, 2007), the use of briquetage in the Bronze Age should not be considered improbable. Even without recourse to the problematic discussion of technological diffusion, British pottery use dates back to the Neolithic, with the Grimston-Lyles Hill series dating to the fourth and third millennia BC (Gibson and Woods, 1997), which allows ample time for technological innovators to discover independently the appropriate ceramic forms and fabrics for this application. As the number of securely dated find sites continues to increase, there is now little reason to doubt that briquetage was being used during the Bronze Age at a small number of locations in Britain to reduce brine, dry salt crystals and transport salt. While these sites have the presence of briquetage in common, the individual briquetage forms and fabrics found are all highly localised (*contra* Henderson, 1991).

The salt production process using briquetage containers and supports pioneered in the Bronze Age had much to commend it. The successful adaptation of ceramic technology to salt production required evolutionary fine tuning of existing production methods for cooking pots and metal moulds over time, rather than a major technological breakthrough. The technical characteristics of vessels required for this process were similar to those needed in ordinary ceramic cooking vessels, such as resistance to fracture through thermal shock (Arnold, 1985, Morris, 2001b).

Direct, continuous heating also allowed for greater temperature control of the brine. The relatively low temperatures needed for the salt crystallisation (*c.*100°C) and drying (*c.*60-70°C) processes (Riehm, 1961, Kleinmann 1975: 45 cited in Foster, 1990) were helpful also in that they permitted the utilisation of a wide range of fuels: "... wood and/or charcoal, cereal processing waste, probably peat and other herbaceous plants such as rushes and reeds ... apparently, used in various combinations," (Gale, 2001). Additionally, oil shale, also known as Kimmeridge coal or blackstone, served as a fuel at salt production sites near the Isle of Purbeck, Dorset (Calkin, 1948, Frend, 1949, Bailey, 1962).

Evidence that salt crystals were removed from the brine concentrating pans as they formed is provided by the attritional wear patterns (Schiffer, 1987) seen sometimes on the inner surface of putative brine evaporation pan sherds (Morris, 2001a).

A question that had been posed with regard to ancient salt production is its place within the annual agricultural and pastoral calendar in Britain. For many, this issue of

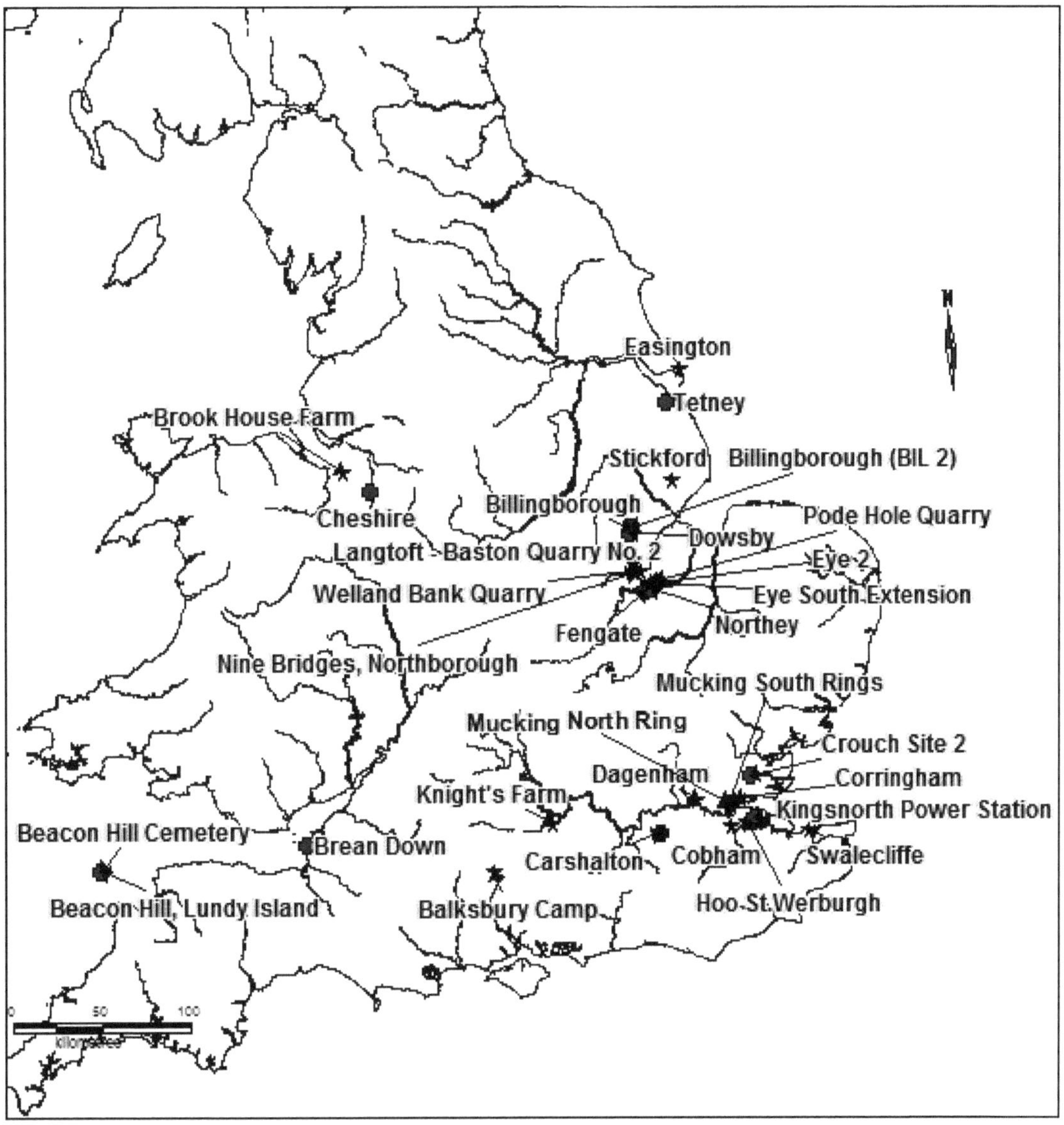

Figure 4. Bronze Age briquetage find sites.

Key: The map shows production sites as dots, briquetage finds as stars. The "Cheshire" dot represents an unknown Bronze Age production site in that county, which is the source of the Brook House Farm find. Site details are shown below in Table 1, with further information available in Appendix 8. Underlying map © Crown Copyright Ordinance Survey.

seasonality was addressed fully by Richard Bradley (1975), who used climatic information, tide patterns, historical and ethnographic analogies, the agricultural cycle and grain impressions on briquetage to deduce that salt production would most likely have been a task best performed between May and August. Additional support for the seasonality theory was provided later by Foster (1990), whose report on the briquetage from Brean Down highlighted the research on variable salinity in the Bristol Channel (Bassingdale, 1943) that demonstrated that the June to August period is the time of greatest salinity in the Bristol Channel. Gurney (1986: 141-142) said that the optimal salt production period in the Fens falls during the months of March through June, due to variation in rainfall, temperature, hours of sunshine and wind frequency, which he thought would place the labour demand of saltmaking in a relative lull in the agricultural cycle "between ploughing and sowing spring crops and harvesting the winter crop." If one accepts a degree of environmental determinism as a factor in prehistoric salt production and believes that saltmaking took place only in the climatically optimal months, this would suggest that different social arrangements existed in Britain during the Iron Age allowing the saltmakers to cope with the labour requirements of salt production and agriculture regardless of the time of year saltmaking occurred. In general, though, in the time since Bradley's paper was published, the seasonality argument has stood up well to the accumlated evidence from numerous British Iron Age salt production sites (e.g. Willis, 2007).

Iron Age 'full-time' production, however, is not the same as twenty-first century industrial production; an expectation of continuous salt production for 12 months of the year is probably erroneous. For comparison, medieval Droitwich which used a similar saltmaking

Site Name	County	Site Type	OS Grid Ref	Date
Balksbury Camp	Hampshire	hillfort	SU 350 445	LBA, EIA
Beacon Hill, Lundy Island	Devon	hut circles	SS 1322 4421	LBA
Beacon Hill Cemetery, Lundy Island	Devon	hut circle	SS 1325 4425	LBA
Billingborough	Lincolnshire	saltern	TF 127 332	LBA, EIA
Billingborough (BIL 2)	Lincolnshire	saltern & enclosure	TF 1260 3335	LBA, EIA
Brean Down	Somerset	saltern	ST 29575 58725	MBA
Brook House Farm, Bruen Stapleford	Cheshire	settlement	SJ 4975 6385	LBA, M-LIA
Cheshire unknown production site	Cheshire	saltern	***SJ 652 523***	LBA, LIA
Cobham	Kent	settlement	TQ 6918 6950	LBA
Corringham	Essex	find site	TQ 7105 8332	LBA
Crouch Site 2	Essex	saltern	TQ 8020 9646	LBA
Dowsby	Lincolnshire	saltern	***TF 115 295***	LBA, EIA
Easington	East Yorkshire	find site	***TA 396 199***	BA
Eye South Extension	Cambridgeshire	saltern & settlement	TF 236 019	LBA
Eye 2, Peterborough	Cambridgeshire	saltern	TF 217 015	LBA
Fengate, Peterborough	Cambridgeshire	saltern	TL 2135 9900	BA
Hoo St Werburgh Vicarage	Kent	pipeline trench	TQ 7810 7168	LBA
Kingsnorth Power Station, Hoo	Kent	saltern	TQ 817 734	LBA-EIA, RB
Knight's Farm, Burghfield	Berkshire	settlement	SU 678 700	BA
Langtoft – Baston Quarry No 2	Lincolnshire	settlement	TF 131 130	LBA
Mucking North Ring	Essex	settlement	***TQ 6755 8112***	LBA, EIA
Mucking South Rings	Essex	settlement	TQ 675 804	LBA
Nine Brideges, Northborough	Cambridgeshire	settlement	TF 1506 0757	LBA
Northey	Cambridgeshire	saltern	TL 2365 9881	E or MBA
Pode Hole Quarry	Cambridgeshire	find site	TF 265 036	M-LBA
School Road, Dagenham	London	find site	TQ 49569 84015	LBA
Stickford	Lincolnshire	find site	***TF 355 595***	BA?
Swalecliffe	Kent	saltern	TR 134 674	LBA-EIA
Tetney (near Cleethorpes)	Lincolnshire	saltern	***TA 315 015***	LBA
Welland Bank Quarry, Deeping St James	Lincolnshire	find site	***TF 181 085***	LBA
Westcroft Road, Carshalton	Surrey	ritual pit	TQ 283 647	LBA

Table 1. Details of Bronze Age sites shown in Figure 4.

NB: Ordinance Survey (OS) Grid References shown in boldface italic are estimated site locations.

process to the prehistoric method, albeit one utilising metal vessels for brine condensation, had a saltmaking season running from June until December due to winter freshwater contamination of the brine wells (Morris, 2001b), a situation equally likely to have arisen during the winter months of the Iron Age. Stormy, windy coastal areas are likely to have been equally insalubrious for saltmaking during Iron Age winters as they are now. This suggests that 'full-time' is not synonymous with 'full year' in terms of actual salt production even at its prehistoric maximum output, although it is possible that the non-production months might have been devoted by the salt makers to ancillary activities. A list of necessary adjunct activities that might have filled the other months was provided by Lane and Morris (2001) and includes briquetage production, creation of pits, ditches and hearths, and possibly even leaving the area to exchange salt for other goods. Other activities such as clay digging, charcoal production or other fuel gathering activities, not to mention agricultural work, are equally probable for the non-production months. The shift from part-time, seasonal production of salt to full-time production, therefore, meant that more labour was devoted to both production and production support activities throughout the year. It is not unreasonable to speculate that this shift may well have necessitated a range of changes in other social behaviours to support the intensification of these activities. For example, a change in the use of child labour or an increase in slavery may have accompanied the shift, though such activities would not necessarily be visible in the archaeological record.

3.3 Regional styles of briquetage

The eastern Bronze Age coastal salt production areas using briquetage, unlike those in the southwest, persisted and expanded during the Iron Age; the Fenlands, Essex and Kent all became prominent salt production areas. Although briquetage finds suggest its continuity, the extent of the Teesside Iron Age salt industry awaits further archaeological investigation. In Hampshire, the Bronze Age briquetage find from Balksbury Camp (Rees, 1995), although not identified as to a particular production source, implies the early existence of a south coast salt production tradition that expanded during the Iron Age. With the exception of the Tees Lowlands, the other sites in Table 1 have been the subject of long term archaeological study, and much is known of their respective production processes.

The use of briquetage at inland salt springs in Cheshire and Droitwich seems to have begun later than in the coastal areas. A single site with Bronze Age Cheshire Stony VCP (Fairburn et al., 2003) implies that briquetage may have been in use at Cheshire salt springs prior to the widely accepted Early Iron Age date for its incorporation in the production process, although more finds will be needed to secure the new chronology (Morris, 2003a). Finds of Droitwich briquetage in Wales, the Marches and the Cotswolds suggest that the introduction of briquetage at this salt springs source occurred later, during the sixth-fifth century BC (Morris, 1985).

Each of these areas developed unique forms and fabrics of briquetage tailored to the regional production and distribution process. A synthesis of these variations known from Essex and the south coast - contrasting regions of coastal production - and the Droitwich salt springs is presented below. Information on the major Lincolnshire Iron Age salt production region was excluded as the recent volume on Fenland salt production (Lane and Morris, 2001) contains far more detail than could be presented here. The briquetage of the less well known industries in other East Anglian counties such as Cambridgeshire, Suffolk and Norfolk, is comparable to that found in Lincolnshire documented in the Lane and Morris volume. The exclusion of a review of Kentish saltmaking was also intentional. Kentish Bronze and Iron Age briquetage has similarities with that of Essex (e.g. Burchell, 1925, Jessup, 1930, Miles, 1975), but Kentish salt production was most extensive during the Roman period (Griffin et al., 2005), a matter beyond the parameters of this discussion. The Tees Lowlands industry is best known from inland finds rather than from the exploration of Iron Age production sites. Similarly, the well-known Stony VCP associated with the Cheshire salt springs was excluded, as all the known finds are related to salt distribution, rather than production.

Recovered briquetage from any region may range in colour from cream to shades of red, orange, grey, black or purple, although shades of orange, pink and red are the most common. The core colour can be the same as the surface or a shade of grey, depending on whether it is fully, partially or un- oxidised, reflecting the variability associated with clamp firing (Rice, 1987). The inconsistency in surface colour is due to the combination of the original clay's chemistry, temper selection and firing, the variability in salt production processes in different localities and the number of reuse cycles, if any, of each piece of briquetage. Occasional spots of accidental green salt glaze (Fawn et al., 1990: 7) or patches of insoluble white scale (Morris, 1991) may occur and affect the surface colour.

3.4 Essex

The Essex salt production process was already well developed by the Middle Iron Age, with settling tanks used to hold brine before heating and structured hearths used to produce salt (Morris, 2001b, Germany, 2004). However, the vast majority of Essex salt production evidence seems to date to the Late Iron Age or Early Romano-British period, as suggested by pottery finds recovered from the more than 300 'Red Hills' known in the county (Fawn et al., 1990, Jefferies and Barford, 1990). The descriptive name 'Red Hill' was already in use by 1879 (Stopes, 1879), although the term is likely to be far older.

Even though all Red Hills were located in tidal areas when in use, sea level changes have left some in modern dry areas and shore erosion has destroyed other sites that were documented in the nineteenth or early twentieth centuries (Eddy, 1982, Wilkinson and Murphy, 1986). The Red Hills in dry fields have suffered extensive damage due to agriculture, and are now more likely to appear as red patches in ploughed fields then as distinct topographic features. The best preserved Red Hills are those in marshy areas. The majority of Red Hills are now roughly 1/3 hectare in size, although they vary from *c.* 200 square metres to about a hectare (Fawn et al., 1990: 6). The scale of the damage to these sites from agriculture can be approximated by comparing that description to one from the late nineteenth century. An early report of a Red Hill excavation gave its size as 30 acres, approximately 12 hectares, with a deposit depth of 5 feet or *c.*1.5 metres (Stopes, 1879: 369-70). Chemical analysis carried out during the first Society of Antiquaries investigation of Red Hills during 1906-07 determined that the red earth of the hill, the briquetage and clay from a bed beneath the excavation were all essentially one material (Jenkins, 1908). Figure 6, drawing on publications, grey literature and HER reports, shows Red Hills whose locations could be determined based on the available information. Site details are listed in Appendix 1 and Appendix 8, the supplemental database.

A second product of the Society of Antiquaries investigation was a typology of Essex briquetage: "... firebars, wedges, T-pieces, pedestals, etc. while for the more plentiful flat and curved pieces of indefinite form the term 'luting' has been given" (Reader and Wilmer, 1908: 166). The t-piece form was subsequently recognised as the top of the pedestal rather than a separate class of briquetage objects (Fawn et al., 1990: 12) and most 'luting' was recognised as portions of vessels. A complex system for classifying Essex briquetage sherds encompassing finds after that early date was proposed by de Brisay (1978), although later archaeologists have simplified her system (Rodwell, 1979) to include only vessels, pedestals, firebars, wedges, packing pieces and hearth components. As the Red Hill briquetage includes sherds from hearth structures, Fawn et al. (1990: 10) suggest that the term 'briquetage' in Essex represents fabric descriptions, rather than the usual form-related definition of vessels and supports. As the term 'briquetage' had already been very broadly defined for some time when this comment was made and there has been no subsequent trend in British archaeology to restrict its meaning, there is no reason to reject this suggestion retrospectively. The briquetage sherd sizes in Essex are highly variable, ranging from partial vessels, supports or firebars large enough to suggest the size and shape of the complete object, down to particles of fired clay found in the Red Hill soil and only visible under magnification (Fawn et al., 1990).

Rodwell's typology (1979: 136-43, 153 & 155) also separated Essex briquetage into two broad form classes having non-overlapping distributions: Type A represents roughly the sites from north east Essex found with Late Iron Age or Belgic pottery, and Type B those on or near Canvey Island found with early Romano-British pottery. This suggests a temporal distinction as well as a geographic one. However, changing water levels have submerged the Iron Age shoreline of Canvey Island, so this apparent geographic distinction has been questioned (Eddy, 1982). Almost 25% of the catalogued Red Hills are undated and few Red Hills have been excavated to modern standards, suggesting that a programme of planned excavations is needed to clarify this point. Without this further work, it will be difficult to resolve the precise nature of the relationship between prehistoric and early Romano-British salt production in Essex.

Type A briquetage, when found with pottery, is usually accompanied by Late Iron Age or Belgic pottery (Rodwell, 1979: 155). As described in Fawn et al. (1990: 11), the Type A brine boiling vessels were generally shallow, rectangular or subrectangular, with either vertical or outwardly sloping sides, often constructed with joined slabs of clay or from coils, with the base added later. In some cases, a flat slab was formed into a 'U' shape with the end and sides added afterwards. This type of sherd represents up to 90% of most assemblages. A typical end sherd is shown in Figure 5. Although vessel sherds may be up to 4cm thick, the average thickness is 1.9cm, these modern measurements being surprisingly consistent with those reported for briquetage in 1908 (Reader and Wilmer). Plain rims outnumber decorated examples considerably, and knife trimming is frequently evident. Rodwell (1979: 144) noted internal staining 1-4cm below the rim, evidence of brine being replenished during saltmaking. A much rarer Type A form, described in Fawn et al. (1990: 11), consists of circular vessels 30-40cm in diameter and of uncertain height.

A second diagnostic form of Essex briquetage is the pedestal, a typical example of which is shown in Figure 8. According to Fawn et al. (1990: 12), the pedestal bases are usually circular, although square examples have been found. Examples up to 7.5cm in diameter have been discovered. Many pedestal surfaces exhibit pitting, heat spalling or vitrification. The t-section generally has a flat surface. Pedestal fragments suggest a typical overall height between 20 and 30cm, although shorter examples are known. Whereas the pedestal exteriors are known in gray, black or deep purple due to overfiring, the core colour may be red, suggesting that the pedestals were reused until they fractured. The actual function of pedestals is unclear. They may have supported the vessels during heating (de Brisay, 1972) or acted as supports for the flat topped, longer firebars (Rodwell, 1979).

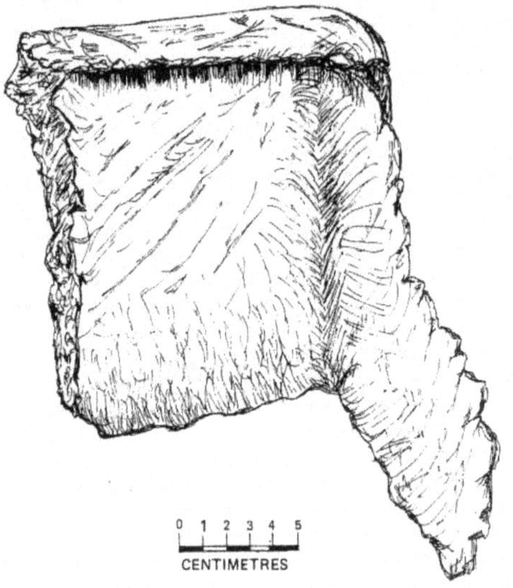

Figure 5. Essex briquetage trough end and side sherd.
From de Brisay, 1975: 9.

Firebars, as illustrated in Figure 8, are known from fragmentary examples and are estimated to have been between 20 and 45cm in length. Firebars resemble low isosceles triangles, although often with the upper vertex truncated to form a flat surface (Fawn et al., 1990: 13). They appear rectangular in cross-section, varying in thickness from 1.4 to 4cm, consistent with the briquetage vessels and suggesting that clay preparation for firebars was the same as that for the other forms of briquetage. Examination by the author of Essex firebars in the collection of the Colchester Museum gave the impression that they had been produced in batches from flattened slabs of clay and cut to shape with a thin knife blade, supporting a suggestion to this effect made by Fawn et al. (1990). They were far more consistent in shape than the obviously individually handmade pedestal fragments examined at the same time. While firebars are obviously some kind of support, the precise details of their use remain somewhat obscure.

Pedestals and firebars are defining characteristics of Type A assemblages, but seem to have become technologically obsolete and are not normally found in Type B assemblages (Rodwell, 1979; Fawn et al., 1990: 12-13). Both the circular vessel and the wedge form of briquetage are rarely found in Type A assemblages but are common in Type B. This pattern may suggest that the technological evolution away from pedestals and firebars

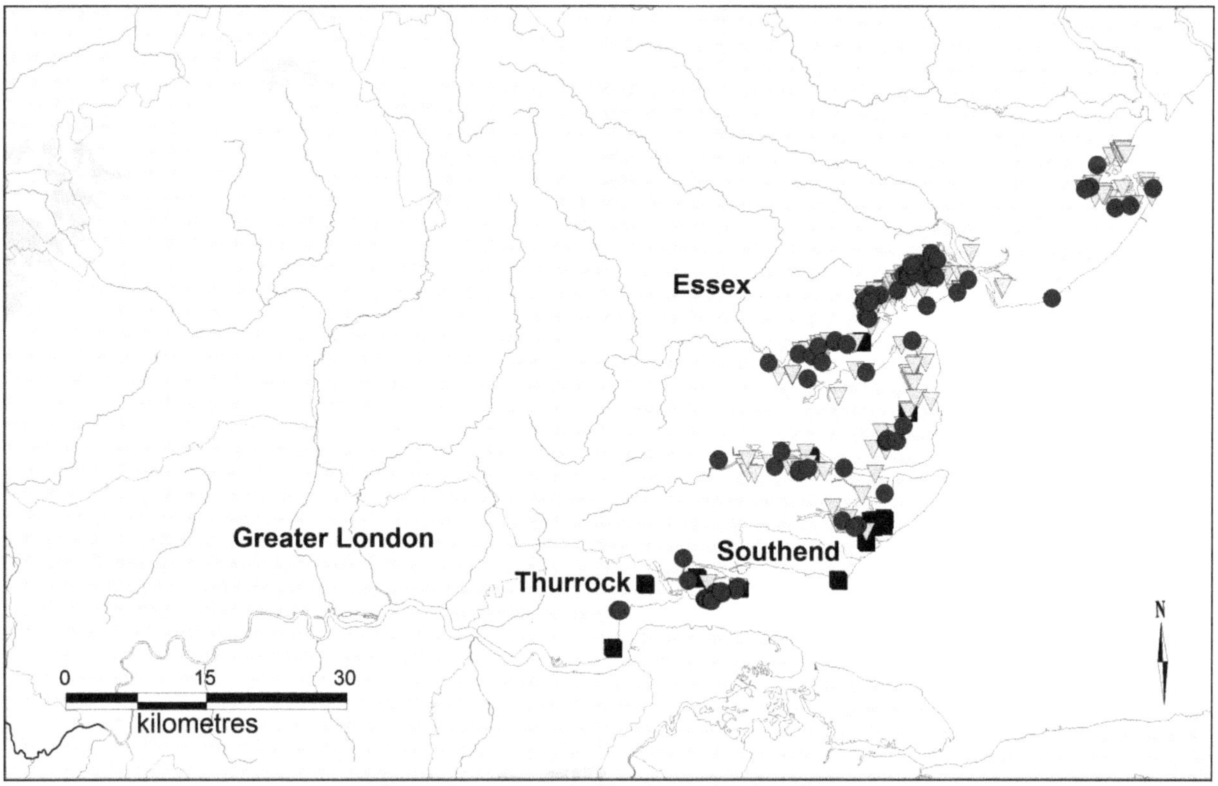

Figure 6. Essex Red Hill salt production sites.

Key: Bronze or Iron Age Red Hills defined by dots; sites thought to be purely Romano-British are shown as squares; undated Red Hills shown as triangles. Site names are listed in Appendix 1. Underlying map © Crown Copyright Ordinance Survey.

had begun during the Late Iron Age in Essex, although evidence for this is obscured by the overall difficulty of dating Red Hills and the paucity of sites clearly spanning both periods.

As discussed above, and as is typical of all British briquetage, the fabric used in Essex is formed of local clay. The clay had not been pretreated as it would have been for pottery production by weathering, puddling or kneading (Fawn et al., 1990: 10), although it had been mixed with a tempering material that improved its thermal shock resistance (Morris, 1983, Arnold, 1985). Usually the temper is organic material, often chopped vegetable matter, but samples with inorganic temper such as granular quartz are known also (Fawn et al., 1990: 10). When chopped vegetable matter was used as organic temper, a feature of briquetage in most British regions, it would burn out during firing, leaving voids in the fabric. The resultant fabric was lightweight and had a technical advantage:

> *Initially the vessels would leak slightly owing to the porosity of the fabric. However, when the temperature of the outside surfaces rose above the boiling point as it must have done for sufficient heat to be conducted to the water inside, solid salt would form in the pores, blocking them and reducing the leakage* (Fawn et al., 1990: 19).

As was noted earlier, the brine solution in Essex was not composed solely of dissolved sodium chloride, but contained small amounts of other, more soluble, minerals such as calcium sulphate and magnesium chloride, as well (Sherlock, 1921). The porous vessels also facilitated the migration of these bitter salts, as shown in Figure 7, to the surface of the container, thus producing a purer finished product (Hopkinson, 1979).

Figure 7. Salt dried in porous (left) and non-porous (right) vessels.

From Fielding and Fielding, 2006: 9

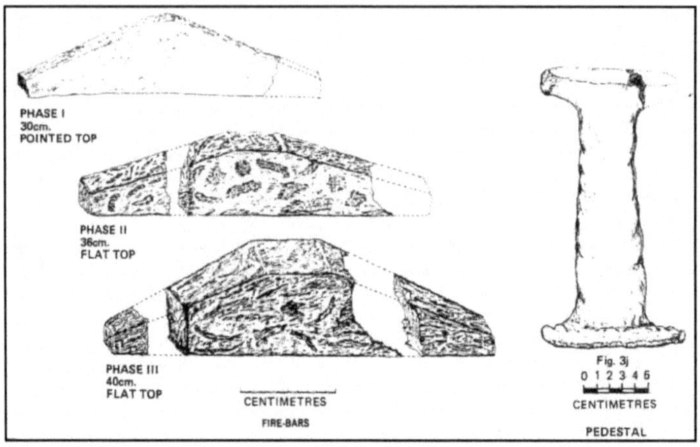

Figure 8. A typical Essex pedestal and stylistic evolution of firebars.
From de Brisay, 1975: 8.

The end of briquetage use in Essex appears to have been a gradual change. It seems to have continued in use during the mid- to late first and possibly even early second centuries AD well into the Roman period, although apparently in the Type B forms mentioned briefly above. Miles (1975: 26) notes similarities between the Type B briquetage and that in use from known Romano-British salt production sites in Kent, suggestive of a possible 'industry standard' being imposed and replacing local forms. As noted earlier, however, Romano-British salt production is a topic beyond the scope of this text.

Knowledge of briquetage forms, while useful, is insufficient to provide a complete understanding of the salt production process used in Essex. A much clearer picture of the process resulted from Red Hill excavations at Goldhanger X and Leigh Beck (Fawn et al., 1990: 8), Osea Road (de Brisay, 1972), Lauriston (de Brisay, 1975) and Peldon (de Brisay, 1978). While Goldhanger X provided the first examples of brine settling tanks (Fawn et al., 1990: 8), each of these sites had multiple sets of either 3 or 4 deep clay-lined tanks dug into the ground, some capable of holding up to 1000 litres of liquid (ibid.: 19). In some instances, the tanks seem to have been fed via a network of channels or ditches (de Brisay, 1978), although alternatively they could have been filled by means of buckets. Transfer of the clarified brine to empty, shallow boiling vessels, probably already sitting on supports above a hearth prior to initial heating to prevent cracking, was probably accomplished with buckets, (Fawn et al., 1990: 19). At Osea Road, de Brisay (1972: 26 & 33) found the impressions of pedestal bases in clay lumps and within a rammed clay area she interpreted as a working floor. Such an arrangement clearly would have acted to support the pedestals. Salt hearth evidence followed shortly after finds of the tanks. The first confirmed archaeological evidence of an Essex salt production hearth, shown in Figure 9, came from Peldon in 1973 (Fawn et al., 1990: 8), while other possible hearths were discovered shortly thereafter from Red Hills at Tollesbury and Leigh Beck (ibid.: 9).

The fate of the freshly precipitated crystals is something of a mystery in Essex. Fawn et al. assert that

> *...if the crystals are not removed from the evaporating vessel but allowed to accumulate until the vessel is full, and the salt is then dried at a moderate temperature, say 50 or 60 degrees C., the result is a single mass of salt cake in the shape of the vessel* (1990: 17).

Figure 9. Iron Age salt production hearth and brine settling tank at Red Hill 117, Peldon, Essex.
From Fawn et al., 1990.: Plate 4.

While technically true that a large salt cake would result, for the reasons noted earlier, this process would yield salt with high levels of contaminants, resulting in both a less palatable taste and a product unsuitable for food preservation (e.g. Bridbury, 1955, Fielding and Fielding, 2006). By removing salt crystals from the solution as they formed, contamination by the less soluble bitter salts would be avoided. The wet crystals may have been placed in a second shallow pan, identical to the brine boiling pan, and then heated to 50-70°C to dry them completely (Riehm, 1961 Fawn et al., 1990: 17). This proposed method would be consistent with the archaeological evidence known from Essex. None of the briquetage finds from the Red Hills suggests production of small, standardised salt cakes, as there are no small vessels for forming such cakes in either the Type A or B assemblages. Alternatively, the salt may have been placed on a flat slab in a dry area to form a cake (Fawn et al., 1990: 17), although this presents obvious difficulties given the British climate. A further method yet to be demonstrated archaeologically is placing of the wet salt into an organic container, such as a basket, to finish the drying process. With Essex salt production often occurring in marshy areas where reeds and sedges would have provided readily available raw materials for basketry, this is a viable possibility albeit an unproven one.

3.5 The south coast

Unlike Essex, the south coast is a geographic region characterised by small-scale salt production during the Iron Age. The term "south coast" as used here, as shown in Figure 10, means the coastal area roughly from Poole Harbour in Dorset, to the Ouse Valley in East Sussex, including the Isle of Wight. The sites are listed in Appendix 2 and further details about these locations appear in the Appendix 8 supplemental database. The inclusion of Bishopstone, East Sussex, in this group may be questioned by some; however, the use of shelly-tempered briquetage for Iron Age salt production at this site has led to the author's decision to expand the south coast group beyond the limits initially defined by Bradley (1975). This decision was also taken with the intention of highlighting the archaeological potential for the discovery of additional Iron Age salt production sites in this region.

This area has traditionally been considered by archaeologists (e.g. Cunliffe, 1984c) as being populated by diverse groups – one hesitates to use the word "tribes" with its sociological and Classical studies baggage – although with respect to salt production, it seems to be a coherent area of technological convergence, however one not lacking in diversity of briquetage forms. It might be argued that some commonality in the production method might be expected, as the physical parameters of the process such as the chemistry of seawater and its boiling point are the same throughout the region and this alone could lead to convergent technological solutions. However, there is more similarity in this region, and distinction from others such as Kent and Essex, than can be explained exclusively by environmental determinism. Like Essex, this area has all the natural resources for briquetage production, including a supply of littoral clay near the shore and a range of possible fuels for saltmaking hearths, but no Red Hills occur in this region.

A second factor in the commonality of this region is our knowledge, or lack thereof, of prehistoric salt production. A seeming bias among archaeologists working in this area towards the study of the Neolithic, Bronze Age and Romano-British monuments apparently blinded them to the possibilities of prehistoric saltmaking. No discussion of possible saltmaking appears in early papers on sites that have since become known for prehistoric saltmaking, as exemplified by the papers on the Hampshire mudflats (Shore, 1892), and the Dorset sites of Hamworthy, described by Smith (1930), and Kimmeridge, published by Davies (1936). The summer field trips of the Dorset Natural History and Antiquarian (later Archaeological) Field Club visited many sites where salt making could have been discussed, at least as a possibility. The report of the group's visit to Poole and Poole Harbour (Fletcher, 1926) is typical, managing to omit discussion of saltmaking during any period. This blind spot did not go completely unnoticed, as in this comment after the publication of the Hook, Hampshire finds: "I am not sure that of late years the interest of our inland country has not led our Club rather to neglect our Coast (Williams-Freeman, 1933: 215)."

It was only from the 1930s onward that the existence of this industry seems to have been recognised. Reports from this period and shortly after the Second World War liken the south coast sites to the Essex Red Hills and the Lincolnshire red mounds in composition, if not in scale, and they note that they share with these sites an absence of occupation evidence (e.g. Williams-Freeman, 1933, Fox, 1935, Calkin, 1948). Despite this evidence, some continued to ignore the potential significance of prehistoric saltmaking in this region. A 1948 history of the Poole area (Smith) thanks Calkin for his many contributions to local Bronze Age archaeology, but enumerates Poole's Iron Age industries as pottery production, "... iron-smelting, the manufacture of beads and armlets from Kimmeridge shale, and spinning and weaving ..." (ibid.: 54), neglecting any mention of Calkin's finds that relate to saltmaking.

Hampshire briquetage had some similarities to Essex finds while at the same time exhibited some unique forms. A wide range of handmade briquetage from Hook, Warsash, identified collectively as struts by Fox (1935), but including objects called firebars or pedestals in other regions, was interpreted as having been used to support pans holding seawater for salt evaporation. Only one sherd in this assemblage, however, appeared to him as clearly coming from a pan, so his interpretation was based heavily on its similarity to the Essex and Lincolnshire finds. After World War II, Calkin (1948: 56) reported the recovery of fragments of large, rectangular trays 1 inch (2.5cm) thick with other briquetage from the shores of Hobarrow Bay, shown in Figure 11. The Hobarrow tray walls exceed 1.5cm in thickness (Hawkes, 1987). Hawkes (ibid.) also states that the vessels were either oval or subrectangular in shape and appear to have had straight walls, although he does not estimate their height. The cut edge is a common feature of south coast briquetage, occasionally found in Essex as well, as discussed above (Fawn et al. , 1990: 11), but virtually unknown in Iron Age Lincolnshire (Cleal and Bacon, 2001), both areas having been used as comparators by Calkin (1948: 58).

More intriguingly, Calkin also described finds of crudely made cylindrical jar sherds, shown in Figure 12, found in the remains of an eroding red mound near Kimmeridge:

> *... some 3 to 4 inches [7.5-10cm] in diameter, and a 1/4 inch to 1/2 inch [0.6-1.2cm] thick. Their height is not known, but it often exceeded*

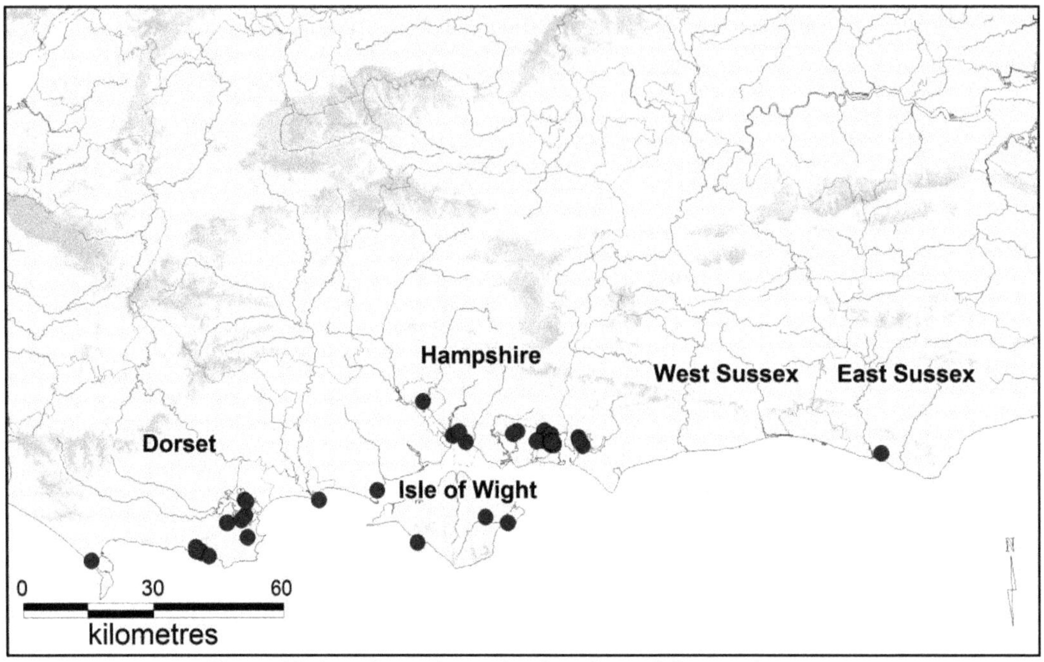

Figure 10. Iron Age salt production sites of the south coast.
Site details listed in Appendix 2. Underlying map © Crown Copyright Ordinance Survey.

> *4 inches [10cm]. Before firing they must have been inverted, and were then cut in half with a knife across the base and down the sides. Presumably the cut continued to the end, but no rim piece is known which shows the severed edge. The clay was not always cut right through, as the jagged inner edges sometime indicate* (Calkin, 1948: 56).

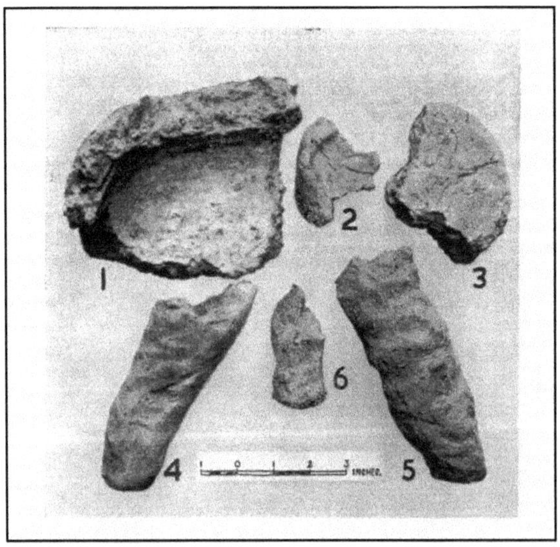

Figure 11. Trays and supports from Hobarrow, Isle of Purbeck, Dorset.
From Calkin, 1948: 57.

Higher in the same mound, a sherd appeared to come from a bowl with a 1 inch (2.5cm) hole in its base that had been made before firing (as shown in Figure 13), along with other sherds that Calkin likened to sugar basins. He suggested that the bowls might have been used for the same purpose as the cylinders, but is vague as to what that might be (ibid.: 57). Subsequent finds of further rectangular pans and bowls led to speculation that an additional stage in the production process had existed, namely packing wet salt crystals into the bowls for drying and for the formation of salt cakes (Bailey, 1962). Farrar (1962a) seems to have been the first to suggest that it was the small containers, poorly shaped for salt crystal formation, which might prove to be evidence of an inland salt trade from the south coast.

The two stages of the process need not be carried out at a single site. At both Hobarrow and Kimmeridge, Dorset, finds suggest that the wet salt may have been taken to nearby cliff tops for repacking into smaller containers (e.g. Calkin, 1948, Cunliffe, 2005). The use of briquetage for a salt crystal drying and/or packing stage was unlike anything known from Essex or Lincolnshire, where the crystals seem to have been placed into archaeologically invisible containers. The firebars, cut cylinders, pans and bowl are now seen as typical briquetage forms along the south coast (Poole, 1987). The two forms of Iron Age briquetage known from the Isle of Wight, "... a flat-based trough thought to have a rounded end ..." and "... a wide-mouthed jar with a thin irregular rounded rim of diameter approximately 20cm ...." (Tomalin, 1989: 95) are said to resemble those found on the Dorset coast. Tomalin compares the Isle of Wight forms to Poole's Danebury form 2 and Hengistbury Head form 5 respectively (Poole, 1984b, Poole, 1987), an interesting potential extension of trade links with the Danebury hillfort. The south coast briquetage producers applied a range of solutions to the problem of selecting tempering material, undoubtedly influenced by the nature of local clays and temper availability. Arnold's research (1985) suggests that tempering materials would usually come from within 6-9km of the pottery production site, allowing for great variability in choice along the more than 200km of the south coast considered here as a grouping.

As noted above, shell was the main tempering material found at Bishopstone, although chalk and vegetable matter were also used (Bell, 1977). Although Fox reported that the fabric of his finds from Hampshire exhibited evidence of grass tempering (1935), this was not reported originally in briquetage from Dorset, where small angular pebbles or sand fulfilled this function (Farrar, 1962b). Later Hampshire finds include briquetage with either organic or flint tempering (e.g. Hughes, 1973), and the use of both organic and inorganic tempers was also subsequently identified in the Iron Age briquetage from Hengistbury Head, Dorset (Poole, 1987). When organic temper was used, it could provide up to 40% of the original fabric in some sherds, creating a highly porous and lightweight ceramic (ibid.: 178). As in Essex, the organic tempered fabrics become more common over time during the Middle Iron Age, possibly as their desirable functional qualities became more widely known to briquetage makers and salt producers alike, who may or may not have been the same individuals.

Figure 12. Kimmeridge, Isle of Purbeck, Dorset cylinder sherds.
Note diagnostic cut edges. From Calkin, 1948: 57

Given the far smaller number of sites in this region than are found in Essex, there have been proportionally fewer Iron Age hearth finds. These coastal sites generally pale in comparison with the largest known in Essex, suggesting a pattern of seasonal hearth use with limited periods of reuse. An exception to this pattern was found at Gaulter Gap, Dorset, where the ash and briquetage layer was described as being 10 feet (3.05m) thick (Frend, 1949). Bailey (1962) described a second Dorset hearth or oven in a site near the river Fleet. The base of the hearth was filled with ash from shale chippings, accompanied by large quantities of broken briquetage and Iron Age B pottery sherds, which would be identified as Middle Iron Age pottery in more modern chronologies. The entire site only occupied an area of approximately 30 feet by 60 feet, or roughly 9.1 by 18.3 m, and even this size was interpreted as having been enlarged by wave action, as it was under water at high tide. The absence of Dorset Red Hill equivalents was remarked upon by Farrar (1962b: 140), a comment equally applicable anywhere in the south coast region.

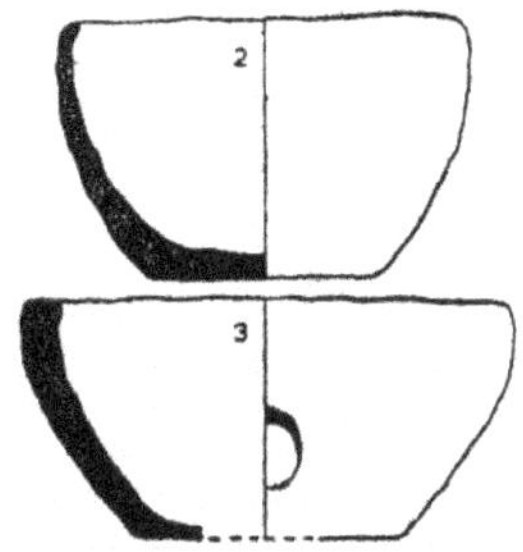

Figure 13. Kimmeridge "sugar basin" briquetage sherds.
From Calkin, 1948: 57.

More recent excavation of a group of sites on the south side of Poole Harbour also failed to reveal massive salt production sites, locating instead multiple small sites of Middle or Late Iron Age and Early Romano-British date, with shale-fuelled hearths that could be used either for saltmaking or for pottery production (Cox and Hearne, 1991). A report in the Hampshire Archaeology and Historic Buildings Register (AHBR) (HCC, 2006b) discusses the 1971 find of multiple hearths at Langstone Harbour, while another record (HCC, 2006a) reports the details of a 1930 find at Fareham, reinvestigated in 1985, of two rectangular platforms, interpreted as possible working floors associated with salt production. The Southampton HER (2006) has a 1971 find report of a possible working area related to salt production, albeit noting that this interpretation has been questioned. The burnt daub present at the site may once have been part of a hearth. This slim evidence is some of the best available for early hearth descriptions in this region.

Hearths with briquetage are known also from the early Romano-British period in this region, and their numbers suggest a possible intensification of production at this time. On the Isle of Wight, for example, the two hearths of certain Late Iron Age date are complemented by six of the early Romano-British period (Trott and Tomalin, 2003). Other production sites of similar date are known from elsewhere on the coast (e.g. Cox and Hearne, 1991, Allen and Gardiner, 2000). There is sufficient evidence available from these sites to demonstrate that briquetage use was part of the salt production process on the south coast for at least a portion of the first century AD, as it was in Essex.

3.6 Inland brine spring production

Up to this point, the discussion has focused exclusively on coastal production, but inland brine springs were also exploited for salt production during the Iron Age, if not before. The springs result from water flowing through subsurface salt beds, so that on reaching the surface the water has a dissolved salt concentration of 24-26%, very close to its saturation point of 27% (Smith, 1918). Water from brine springs would, therefore, require far less evaporation to yield salt crystals than seawater, which is only 3% to 3.5% saline. Although both liquids contain sodium chloride, the overall chemistry of the brine from the inland springs differs considerably from that of seawater. The latter has greater concentrations of magnesium and potassium salts than does salt spring brine, which characteristically contains more sulphates and calcium salts (Hopkinson, 1979). As the calcium salts are more soluble than sodium salts, they would tend to migrate more readily to the surface of porous containers, leaving purer salt crystals behind (Hopkinson, 1979). Although any natural salt spring, large or small, would have been a potential salt production site during the Iron Age, salt production with briquetage evidence is known only from the major springs at Droitwich, Worcestershire, and in Cheshire.

Salt production using briquetage at Droitwich dates to at least the sixth-fifth century BC (Woodiwiss, 1992), the transition point between the Early and Middle Iron Age. A radiocarbon date of 1045 - 838 cal BC obtained from the lowest excavated level of the Old Bowling Green site in Droitwich did imply some form of human activity at the salt springs during the Late Bronze-Early Iron Age period, although no explicit salt production evidence was recovered from these Phase 1 contexts (Woodiwiss, 1992), and Droitwich briquetage of Bronze Age date has not been encountered anywhere. As has been pointed out above, however, absence of briquetage does not preclude the possibility that some form of saltmaking was being undertaken.

Unlike the prehistoric coastal production locations which were the subject of early archaeological investigations, the Iron Age salt production industries of Droitwich and Cheshire were first revealed indirectly though petrologic analysis of briquetage sherds found in the Welsh Marches and in the Avon and Severn River valleys (Sawle, 1978, Morris, 1979). The redevelopment of central Droitwich during the 1970s led to the recognition of three Iron Age production sites in the town centre (Hopkinson, 1979, Sawle, 1978, 1984, Woodiwiss, 1992, Barfield, 2006), with a fourth site at the Upwich spring located in 1983-84 (Hurst, 1997). All the sites were exploited during both the Iron Age and early Romano-British periods. The Cheshire Iron Age production site is now thought likely to be Middlewich, but this is as yet unconfirmed through excavation (Morris, 2007a). In the absence of archaeological evidence from the Cheshire Iron Age production source, Droitwich evidence is our only current model for salt spring production in Britain. When the Iron Age Cheshire production site is finally located, its production method may differ considerably from what is presently known of Droitwich.

The first two sites located during the redevelopment of central Droitwich, the Old Bowling Green, which covered approximately 1,340 square metres, and Friar Street, a 400 square metre site (Woodiwiss, 1992) have provided a great deal of the information about Iron Age and early Romano-British salt production there. Dendrochronological dating of preserved wood from the sites shows that salt production was taking place concurrently at both places (Woodiwiss, 1992). Unlike sites in Essex or the south coast, at the Old Bowling Green site the briquetage fragments were not mixed with large quantities of ash or burned soil, but often made up complete fills and layers (Rees, 1986). Morris (1985) has described Droitwich briquetage as being typified by coarse fabrics and crude forms, having an orange-red or pinkish to pinkish-orange colour, although other colours do occur. Morris (ibid.) defined two types of briquetage fabrics. Following a numbering system originated by Morris, the briquetage is now identified widely as Type 1, sandy, and Type 2, organic briquetage. Sandy briquetage, with quartz sand temper, was in use throughout the Iron Age, with the organic tempered forms appearing in the Middle Iron Age and predominating from that point until the end of the period (Allen, 1990). Morris (1983) noted that the organic temper would have provided superior shock resistance, and combining organic temper with marly clay, as found in the geologic layer above the Droitwich salt beds (Sherlock, 1921), would also improve its mechanical shock resistance. The change in temper seen in Droitwich briquetage, therefore, should be viewed as deliberate technological improvement rather than simply practical recycling of available organic waste.

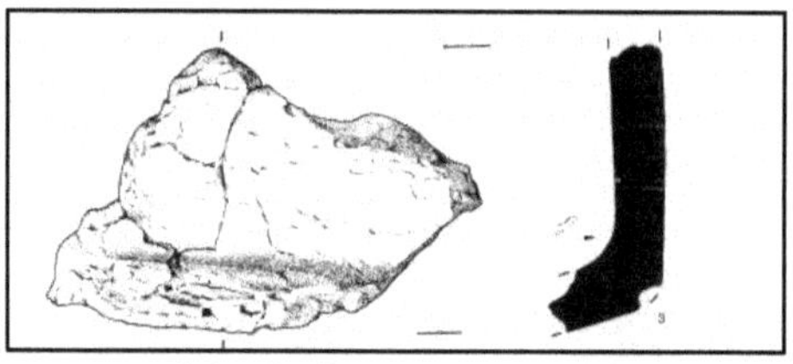

Figure 14. Possible shallow pan sherd from Droitwich. From Rees, 1992: 53. Reproduced by permission of Simon Woodiwiss and the Council for British Archaeology.

Although the Old Bowling Green site yielded more than 21,000 sherds of briquetage (Rees, 1986: 47) weighing nearly 1,000kg (Sawle, 1984), with a further 18kg of briquetage found at Friar Street (Hurst, 1992a), only one example of a briquetage sherd possibly from a shallow pan suitable for brine condensation, shown in Figure 14, has been identified (Woodiwiss, 1992). In this matter, the Droitwich assemblages are completely distinct from those at coastal sites, where shallow pan fragments and various types of supports and kiln furniture predominate (Sawle, 1984, Woodiwiss, 1992). The apparent absence of flat boiling pans is also anomalous in the context of their use at Droitwich from the late Roman period until the nineteenth century. The possible use of either organic containers or lead pans for salt boiling at Iron Age Droitwich was discounted by Woodiwiss (ibid.: 17) due to lack of evidence for either in the waterlogged Friar Street and Old Bowling Green sites. An alternative interpretation of the existing assemblage, although perhaps a less likely one, would be that all of the "coastal-type" shallow pan briquetage forms are simply buried under an unexcavated area of central Droitwich, and that the Old Bowling Green and Friar Street sites were areas used for salt drying and packing in the past, rather than brine boiling. The fragmentary nature of sherds from the Old Bowling Green has caused Rees (1986) to speculate that many of the containers were broken on site to remove the dried salt cakes, which may support this theory.

With the exception of the sherd discussed above, all Droitwich briquetage appears to come from only two forms of crudely finished coil built containers (Morris, 1985), taller than they are wide (Rees, 1986), highly suitable for drying salt crystals, although possibly also used for brine boiling (Hughes and Hunt, 1992). Most container base sherds from the Old Bowling Green (Rees, 1986: 49), or Friar Street (e.g. Hurst, 1992a) have an indented lower surfaces as shown in Figures 15 and 16-1. This form occurs in both sandy and organic tempered fabrics. A second style, with a flat base and made of sandy fabric 1, shown in Figures 16-2 through 16-6, is known from the Friar Street site (Hurst, 1992a) and is rarely found outside of Droitwich (Hurst and Pearson, 1996). Rees (1986: 49) reported that most indented bases had an external diameter of 8-12cm with an internal diameter generally of 6-10cm, with Hurst (1992a: 132-33) and Barfield (2006: 151) reporting similar dimensions regardless of base configuration. No complete vessel profiles have been found in Droitwich, but two at the Old Bowling Green site had partial sides extending up to 40cm and a third was complete up to 30cm (Rees, op. cit.: 49).

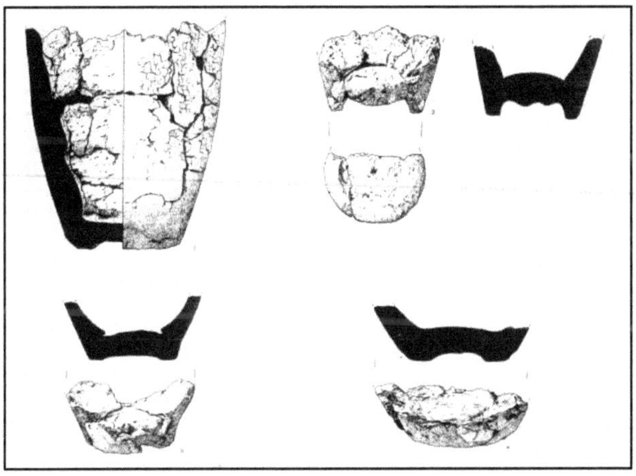

Figure 15. Old Bowling Green drying vessels.

From Rees, 1992: 51. Reproduced by permission of Simon Woodiwiss and the Council for British Archaeology.

Further information about the production process in use at Droitwich came from the discovery, in Middle and Late Iron Age levels at these production sites, of multiple hearths and pits that are likely to have been brine settling tanks. The hearth remains were pits with charcoal fill, but no remaining superstructures (Woodiwiss, 1992). The steep-sided tanks or pits cut into the natural surface, lined with either clay or wood, much of which had been preserved by waterlogged conditions, were not integrated into any obvious network of feeder channels (ibid.: 8-13, Hughes and Hunt, 1992: 116-118, Barfield, 2006: 86-93). While a few clay lined pits thought to act as settling tanks for seawater are known from Lincolnshire (Lane and Morris, 2001) and others, as noted above, have been found in Essex Red Hills holding up to 1000 litres, these are dwarfed by the two largest tanks at the Old Bowling Green which were estimated to hold 5,300 and 14,000 litres respectively (Woodiwiss, 1992: 13). In the absence of evidence for feeder channels or ditches connecting the settling tanks to the springs, filing the tanks must have required considerable effort. Moving the brine by hand from the spring to the settling tanks, possibly using wooden buckets or the briquetage containers, with a similar method being used to remove the brine once it had settled, would have been both labour intensive and time consuming.

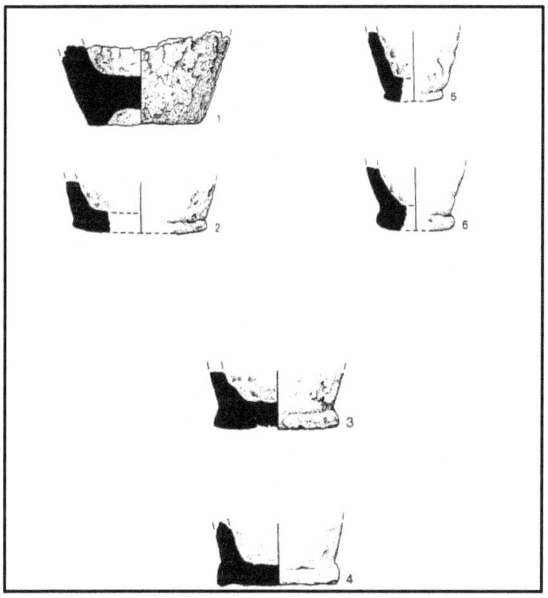

Figure 16. Friar Street, Droitwich, indented and flat vessel bases.

From Hurst, 1992: 134. Reproduced by permission of Simon Woodiwiss and the Council for British Archaeology.

Rees (1986) also reports the comments of excavators at the Vines Park site, who remarked on the ease with which salt crystals formed through solar or wind evaporation on site. This would suggest that salt production through the containment of brine in shallow soil scoops might have been possible there, although no direct evidence for this was discovered at this site. The flat, shallow, interlinked trenches found at the Bays Meadow site in Droitwich (Barfield, 2006) do suggest this possible interpretation in the light of Rees' comments. At the present time, however, it would appear that most of the Iron Age production at Droitwich may have involved filling the tall briquetage vessels with the settled brine, then heating the vessels on hearths to crystallise the salt without the use of boiling pans (Rees, 1986: 50-1). The finds at Friar Street, where a large number of bases were recovered within the ruins of a hearth, support the proposed link between the vessels and controlled heating (Hughes and Hunt, 1992).

3.7 The Iron Age salt industry

Despite all the variation described above, there is a surprising degree of uniformity in the evolution of the salt production industry at all these sites. Bronze Age production using briquetage seems to be rare and appears to occur on what Peacock has described as a 'household production' scale (1981), with the exception of Crouch 2 in Essex where production seems to have been more intensive (Morris, 2001b: 397). Individual Bronze Age production sites rarely appear to have an Iron Age production phase, although Billingborough (Hall and Coles, 1994), Billingborough 2 (Chowne et al., 2001) and Dowsby, Lincs. (Lane, 1994, Lane and Morris, 2001), as well as the site at Kingsnorth Power Station in Hoo, Kent (Griffin, 2006) may have continued into the Early Iron Age. The Mucking North Ring site in Essex (Jones, 1977) and Swalecliffe, Kent (Masefield et al., 2003) have more ambiguous production evidence than the sites just named, but do also cross the LBA-EIA boundary. The number of known Bronze Age salt production sites is, however, still quite small and such generalisations may prove to be unwarranted. It could be that changing sea levels during the early Iron Age necessitated a relocation of coastal salterns, which is why so few sites seem to span this transition (e.g. Palmer-Brown, 1993, Gardiner, 2000). Strong evidence for Bronze Age to Iron Age continuity may also be demonstrated when the Cheshire production source is located.

Production during the Early Iron Age is more widespread than during the Bronze Age, possibly to support a limited domestic British salt trade best known from inland finds of Droitwich, Cheshire and south coast briquetage. This is one of many social changes that seem to occur at or near the Bronze Age-Iron Age transition (Haselgrove and Pope, 2007). During the course of the Middle Iron Age, all the production regions show evidence of an increase in production intensity, probably reaching Peacock's 'household industry' level by the end of the period (Peacock, 1981), the implication being production in excess of domestic consumption requirements. The reasons for the expansion of production need not be linked exclusively to increasing demand for salt consumption (Barrett, 1989). Refinement of briquetage forms and the introduction of vegetable tempered fabrics are characteristics of this period, although the clays in Cheshire and Dorset made vegetable temper unnecessary for thermal shock resistance (Morris, 2001b). However, as technology can be inextricably interlinked with cosmology (Miller, 1987), the changes in salt production methodology and intensity may be more than just variations in economic activity. Our inability to conduct detailed archaeological research into prehistoric cosmology should not negate its consideration. Salt production may have been transitioning during the Middle Iron Age from being the province of magicians and sorcerers to being the work of ordinary people.

By the Late Iron Age, production at many sites, particularly the Red Hills of Essex, expands to an industrial scale, similar in many ways to Peacock's 'workshop industry' phase (1981). The expansion of salt production has parallels in other sectors of commodity production at this time. Mining and pottery production also become more intensive, probably operating on a full time basis at some locations (Haselgrove, 1999). Other social and economic changes in Britain that occurred at approximately the same time include the rise of *oppida* (Haselgrove and Moore, 2007) and the increased circulation of coinage (de Jersey, 2001), which suggest a major change in exchange patterns during this period. The use of coins could signal a change in the perception of value associated with commodities such as salt. As discussed by Simmel, "Value ... is never an inherent property of objects, but is a judgement made about them ..." (Simmel and Frisby, 1990). Salt, with its increasing availability, may have completed its transition from being an exotic, special substance to a more mundane one. Collectively, these coincident changes suggest fundamental social changes emerging in the century or so prior to the Roman conquest of Britain. There might, of course, have been a more prosaic reason for the apparent increase in salt production at this time. Offering such a view, Maltby (2006) suggests that the Late Iron Age increase in salt production was linked directly to an increase in demand for salt for food preservation.

Beyond the long term intensification of production, a second similarity observed at Iron Age salt production sites across Britain is their typical segregation from habitation, resulting in the general absence of what is termed 'occupation debris' finds from most of these sites. This may be purely functional, reflecting directly the difficulty of living for extended periods on exposed coastlines or next to salt springs without a ready supply of fresh water for domestic purposes. Alternatively, it might reflect the existence of social controls, possibly a taboo, keeping people from living permanently at the salt sites. Even in the recent past there are ethnographic examples, such as among the Baruya of New Guinea (Godelier, 1977), where saltmaking was the province of magicians, and their work area was to be avoided by others. An absence of defensive features at any of the salt production sites, as pointed out by Morris (1983), highlights a curious disregard for a valuable resource, unless the observed social norms made it unnecessary. There are, for example, no Iron Age fortifications protecting the salt springs at Droitwich, yet the first Roman marching camp there is dated to shortly after the conquest (Hurst et al., 1988).

The end of the era of salt briquetage use seems to have come gradually, but inexorably, after the Roman conquest. Control of salt production, as with any other mineral resource, would have fallen to the provincial Roman government after the conquest (Hurst, 2006). Briquetage use in general seems to have ended during the second century AD, although it was still being deposited into contexts at Droitwich during the third century AD, where it has been interpreted as residual material

(Woodiwiss, 1992). At Droitwich, the use of briquetage for salt transport ends earlier than its use in the production process (Morris, 1983). This gradual change does not imply that the Romans did not rapidly exert influence over the salt works. As mentioned earlier, in the late first century AD the most intense salt production seems to move away from northeast Essex, near Camulodunum, to Canvey Island; new forms of briquetage are introduced, although even these sites cease to use briquetage early in the second century AD (Fawn et al., 1990). Droitwich, a hub on the Roman road network (Woodiwiss, 1992, McAvoy, 2006), seems to have become an important site for meat preservation at the same time as briquetage use for salt production was being phased out during the late first and second centuries AD (Sawle, 1978).

This chapter has focused primarily on aspects of technology associated with salt production from the Bronze Age to the early Roman period in Britain. It would be incomplete, however, without touching upon the ongoing academic discussion of gender with respect to briquetage and salt production. Arnold (1985), in his wide ranging study of pottery production, concluded that the typical potter's gender was a socially controlled decision, affected by the ability to merge production with household routine, the potential social and economic benefit for women of making pottery, and the availability of men's labour when not engaged in other activities that remove them from the household environment. Peacock (1981) highlighted other factors, and linked female pottery production to the use of limited technology in its production and low output for domestic purposes. Application of these same criteria to prehistoric salt production with briquetage would suggest that at least during the period before saltmaking became a full time activity, the production of briquetage and salt was an activity performed by women, a position supported recently by Morris (2007b). Peacock's view is that the next major change in production intensification, to what he termed a 'workshop industry', also marks a change in the main salt producers' gender (1981). Such a view is amply supported by the processualist analysis of gender and work conducted by Murdock and Provost (1973). In the post-processualist school, however, the view is different. Weedman (2006: 268), for example, would see Peacock's ideas as being rooted in Western views of the role of men as industrial workers outside the home and the ascription of "... the most labour-intensive, dull and unskilled occupations to women ...", capable of making only minor contributions to the household economy.

While accepting the association of women with small-scale salt production as has been proposed, the author, *contra* Peacock, sees no rationale for women abandoning salt production after being its main producers for centuries, and would argue that full time production of salt may have become a family activity, employing men and women as well as children of both sexes, whenever their labour was available, working on the wide range of tasks required for salt production. Godelier (1977) noted gender as a critical factor in task assignment among the Baruya during salt production, thus providing a model for a work system of this type. In a historical western context, a late medieval illustration from Agricola (1556), shown here as Figure 17, includes a youth and a wife among the list of equipment necessary for salt production, demonstrating the model of family-based production in an industrial setting as proposed here.

After more than 160 years of research, a good deal of information has been assembled about the production of salt using briquetage in Britain, although not all questions have been answered. Locating the unknown Cheshire Iron Age salt production site would be a major accomplishment. Resolving the questions about the actual production process at Droitwich, with its seeming absence of shallow brine concentrating pans, is another mystery for the future. An exploration of coastal East Sussex, seeking further examples of Iron Age salt production, would be useful. Ample room remains for further archaeological research into possible prehistoric salt production where briquetage use has not been demonstrated. Areas such as the south coast of Cornwall, which became a Roman salt production area, may yet reveal a prehistoric salt production industry. Information is also lacking on early salt production in north-western England and Scotland, areas that were largely aceramic in the Iron Age. Unfortunately, few of these issues are likely to be a priority for developer-funded archaeology. It appears that prehistoric salt production research in Britain is headed into another quiescent period.

BOOK XII. 553

A—Wooden dipper. B—Cask. C—Tub. D—Master. E—Youth. F—Wife. G—Wooden spade. H—Boards. I—Baskets. K—Hoe. L—Rake. M—Straw. N—Bowl. O—Bucket containing the blood. P—Tankard which contains beer.

Figure 17. Medieval German salt production.
From Agricola, 1556: 553.

Chapter 4
Salt distribution: inland briquetage finds and their absence in the landscape

4.1 Introduction

While the regional case studies in Chapter 3 illustrated that salt production was a large-scale activity in Britain, particularly during the Middle and Late Iron Ages, this chapter will explore a range of issues associated with the prehistoric movement of salt away from production locations. Briquetage evidence accumulated since the 1960s has established that the salt was moving away from the production areas to other sites. The basis for much of this archaeological analysis derives from ideas initially put forward in the seminal research done by Elaine Morris that was published in her PhD thesis (1983). Her work will be a starting point for the present analysis, considering her initial conclusions about salt distribution from Droitwich and Cheshire at a landscape level in the light of new data. While this chapter will primarily consider issues at a landscape level, Chapter 5 will focus on detailed analysis of some of the individual sites within the distribution networks of the Droitwich, Essex and south coast production centres. These site-specific case studies of salt distribution will allow for subtle evidence, which can be gathered only when finds are considered at a context level, to come to the fore.

To be consistent with terminology used in earlier chapters, "briquetage" can indicate material from any salt production source while "Stony VCP", "VCP" or "Cheshire VCP" are interchangeable terms for briquetage associated with the as-yet unlocated Iron Age salt spring or springs in that county.

4.2 Limitations of the data

The archaeological evidence available to demonstrate movement of salt within Bronze or Iron Age Britain is restricted due to the solubility of salt and the limited number of ways in which it can be detected in the archaeological record. This is similar to the problem of archaeological study of imported wine in Late Iron Age Britain, whose presence can be detected indirectly through finds of amphorae sherds and bronze and ceramic vessels associated with wine drinking (Carver, 2001). Unfortunately there are no equivalent "salt using vessels" to track in the archaeological record. A comparison with the study of prehistoric textiles, also considered an aspect of material culture with limited evidence in prehistory, may usefully illustrate the depth of the problem. Table 2 illustrates the constricted range of salt evidence sources, comparing the potential archaeological prehistoric data sources available for textile studies (Wild, 2003) and salt studies. The hygroscopic nature of salt assures that there can be no preservation of salt in any damp or wet contexts, and Britain lacks the necessary desert climate to provide the low moisture levels required for preservation of salt crystals through desiccation. The long term survival of salt at burned sites is precluded by the greater likelihood of contact with moisture. Salt is not known to have been produced from plant ashes in Britain, so there are no indicative plant remains for study by archaeobotanists.

Archaeologists now routinely use briquetage as evidence of salt distribution networks, but as was mentioned in the previous chapter, not all of the major salt production areas, including the Fens, Essex and much of the south coast, routinely used briquetage salt distribution containers. It is quite possible to transport salt in block form without any packaging at all, as is done in the Sahara Desert where large blocks of unprotected salt are carried on camels to Timbuktu from salt mines in the desert (Rainer, 2010a, Rainer, 2010b). Britain's climate, however, would suggest that this method was probably not widely used. Even where briquetage containers were in use for transport, as at Droitwich and Cheshire, it is likely that other containers were also involved in the distribution process (*contra* Lambrick and Robinson, 2009). A belief in the use of organic containers for salt distribution, such as woven cloth sacks, baskets, leather buckets, bark or wooden containers, or possibly even Malvernian pots that apparently shared a distribution network with Droitwich salt (Morris, 1983), is an essential, if indemonstrable, part of a realistic review of Iron Age salt distribution (Lane, 2001). If one believes in the existence of these containers, as does this author, the discussion below which limits itself to ceramic evidence, must now be recognised as demonstrating the minority of salt distribution in the British Iron Age.

Having raised the issue of archaeologically invisible salt exchange, it is worth pointing out that many of the find sites, especially domestic sites at a distance from Droitwich or Cheshire, actually yield very small amounts of briquetage. For example, Wellington Quarry, Herefordshire, a site thought to be active from the late Middle Iron Age into the Romano-British era, a period in excess of 200 years, had two sherds, with a total weight of 20gm, of Droitwich briquetage, and 7 sherds of Cheshire Very Coarse Pottery (VCP) with a total weight of 36gm (Jackson et al., 2000). Hill (1995) citing an undated work by Pierpoint, estimated that as little as 1-10% of ceramic material may survive in Iron Age assemblages, and for a poorly fired ceramic like briquetage, survival would be expected to be towards the lower end of this range, strongly suggesting that the absence of briquetage at sites does not necessarily mean it was never present. To quote an archaeological maxim, the absence of evidence is not evidence of absence. As archaeologists, we must make a choice: either we can accept this absence of briquetage as evidence that these sites had little or no salt; or, if we believe the oft-repeated comment that salt was essential for food preservation in the Iron Age (e.g. Cunliffe, 2005), then we have to regard the total absence of salt at domestic sites that were in use for any length of time as highly unlikely, even in the absence of strong archaeological evidence.

Type of Evidence	Textiles	Salt
Direct evidence		
Survivals at dry sites	Possible	Only outside UK
Charred material from burned sites	Possible	Unlikely
Survivals at waterlogged sites	Possible	Not possible
Fibres adhering to corrosion on metal objects	Possible	Not possible
Gold thread	Possible	Not applicable
Woven fabric/salt crystal impressions on ceramics	Possible	Unlikely
Production artefacts	Possible	Possible
Distribution artefacts	Not applicable	Possible
Indirect evidence		
Fibre-producing plant pollen	Possible	Not applicable
Cut bones from animal culls	Not applicable	Possible
Comments in Classical texts	Possible	Possible

Table 2. Prehistoric textile and salt archaeological evidence sources.

As might be anticipated from the preceding paragraph, the briquetage evidence record is variable among Iron Age sites even in areas where it is frequently found, a point noted by Morris (1983, 1985) and still relevant. It is not difficult to replicate the pattern she illustrated; for example, the author reviewed the more than 8,000 sherds of Iron Age and Romano-British pottery gathered at the settlement discovered during the construction of the Ashville Trading Estate in Abingdon, Oxfordshire, (Parrington, 1978) and confirmed it to be an example of a briquetage-free site, although briquetage was found at the nearby Abingdon Town Centre *oppidum* (Allen, 2000), a Late Iron Age to Early Romano British site, and the late Middle Iron Age Vineyard site (Lambrick and Robinson, 2009). All three sites have large ceramic assemblages considered typical for the region, yet briquetage is not a consistent presence within them. As with any archaeological evidence, it may be the case that the effort to compile briquetage distribution patterns from published reports suffers from the possibility of material being misidentified in ceramic assemblages from areas where it has not been previously noted, as was the case, for example, in north eastern England prior to the work of Willis in the 1990s (1994, 1995).

A further analytical constraint is that imposed by the quality of the archaeological data presented for individual sites. For example, some Droitwich distribution site reports mention the presence of briquetage and provide a sherd count but not the total weight of the material (e.g. Brown, 1991a), with others providing a weight but no sherd count (e.g. Barfield, 2006), thus preventing any further analysis based on mean sherd weight. It is also possible to find published examples with neither exact sherd count nor weight. Perhaps to save space, one report provided briquetage weights and piece counts as a percentage of assemblage totals, with the potential for rounding errors when attempting to calculate numerical values for these characteristics (Napthan et al., 1997). More frustrating to the author were those references that mention the find without any quantification whatsoever. This is most common in very early reports (e.g. Varley, 1950), a situation sometimes subsequently resolved when E. Morris used these sites in her PhD research, but it also occurred in more recent non-specialist reports (e.g. Selkirk, 1981), where only a passing reference to a briquetage find was included, and in qualitative HER records.

The non-standardised data presentation problem is not restricted solely to Droitwich finds. While Stony VCP may sometimes comprise the bulk of the ceramic assemblage in largely aceramic areas such as Wales and central or northwest England, the same issues of key data absence and variability noted above were found in some of the VCP find reports. Problems can occur even where the find is given an extended discussion within an article. For example, no weight is reported for the single sherd of briquetage found at Normanton le Heath, Leicestershire, (Elsdon, 1994) which is both illustrated and accompanied by a two paragraph statement of its archaeological significance with respect to salt distribution. In the Enderby, Leicestershire, report, the amount of briquetage found is said to consist of less than 1% of a ceramic assemblage weighing close to 35kg (Elsdon, 1992), therefore describing the find as some quantity weighing up to 350gm and consisting of an unspecified number of sherds.

Since Morris published her thesis, there has been a visible change in much subsequent archaeological literature reflecting a de-emphasis of quantitative analysis. This may be due, in part, to changes in the theoretical background of archaeological practice in that time, but also reflects the time- and funding-constrained reality of commercial archaeology after the implementation of Planning Policy Guidance 16 (PPG 16) in the United Kingdom in 1990, only recently replaced by Planning Policy Statement 5 (PPS 5) in March 2010. For example, the absence of a standardised statistical reporting format for sites, as discussed above, clearly make inter-site comparisons more difficult. The author has found that replication of the statistical methodology utilised by E. Morris, particularly the calculation of the Specific Salt

Container Index or SSCI, discussed at length below, has been problematic despite the greater number of find sites now known. Consider, for example, the finds from Madeley Heath, to the northeast of Droitwich, where the ceramic assemblage "...was composed almost entirely of briquetage" from two similarly-shaped pits (Hurst and Pearson, 1996). The 156 sherds of Droitwich briquetage with a total weight of 2.825kg seemingly would have had an SSCI approaching 1 judging by the text in the report, but this statistic could not be computed as a vital piece of information, the total weight of the ceramic assemblage, was not provided. Further features making this assemblage significant are the presence of flat salt container bases, a configuration known previously only from Droitwich itself and two further sites, Aston Mill, Kemerton, Worcestershire, and Croft Ambrey, Herefordshire. Madeley Heath is in a region where briquetage is otherwise uncommon, another unique aspect of the site. Had the SSCI been utilised initially, the atypical nature of this site would have been apparent even without knowing these key details.

Confusion between SSCI calculations and other statistical information can occur due to the data presentation in some reports. If the assemblage description refers to the percentage of briquetage in the collection, the statistic could easily be confused with an SSCI value, though if the percentage measurement is based on sherd count rather than on weight as at Throckmorton Airfield (e.g. Griffin et al., 2005), this would not be the case. Especially where the basis for the calculation is not explicitly stated, there exists the possibility of misinterpreting the statistics and attempting to directly compare the two types of statistics. At a minimum, the basis for any percentage calculations in ceramic assemblage descriptions should be clearly provided as guidance for others seeking to use the data.

If the primary purpose of publishing archaeological findings is the preservation of information that would otherwise be lost, work such as this must be regarded as failing to meet that basic standard. It is possible that some of this inadequate treatment of briquetage finds may be due to a bias against the material held by some archaeologists, who prefer to treat briquetage as a type of locally used fired clay industrial container akin to a metal mould, or equivalent to daub or an oven wall sherd; briquetage being, perhaps in their perception, rather a lesser ceramic than a "true" pot. The author would point out that while there is no apparent circulation of partial oven walls or metal objects still in their moulds, the wide exchange networks involving the use of briquetage known in some regions suggests that there was a perceived distinction between the significance of these various classes of ceramic objects in the Iron Age, and that modern efforts to regard them as equivalent types of incidental industrial objects serves only to obscure their original contextual meaning.

A final data problem is the dating of briquetage or VCP, a difficult process generally based upon associated finds. When it is recovered from an extremely small site, or an unstratified site, often little specific information can be provided about its age. Particularly if it is an isolated find, not much can be added other than identifying it as briquetage and saying that it is probably of Iron Age date and represents a link to regional salt distribution if found away from a production site. Even larger assemblages can be hard to date if the briquetage is accompanied only by undecorated Iron Age pottery that was in use for a period of centuries.

4.3 Iron Age salt exchange networks

An inherent assumption in this chapter is that salt production areas had more salt available per capita than areas needing to trade for their salt. Behind this assumption lurks another: that only surplus salt would be made available for trade by the salt producers. This notion assumes, at least in part, that the exchange of salt included some of the same market rationales relative to meeting local demand before expanding the distribution area, as exist in capitalist economic systems, which is surely not a given. It is quite possible, for instance, that salt distribution was controlled by one social segment, such as a warrior or priestly caste, whose interests were not concurrent with those of all members of the social group, and that they could insure that salt export occurred even if there was insufficient salt available to fully meet the demand of all members of the local social group (Sherratt and Sherratt, 1991). Such a system was envisioned by Barrett (1989) who saw that exchange systems came to exist not solely as "...logical responses to demand of consumption, but as the product of discourse between people," and that not all "...cycles of material production, circulation and consumption are driven forward in every instance by the growing demands of consumption." Similarly, Collis (1994) suggested that the alternative logic of pre-capitalist exchange networks was socially based, with kinship and political networks driving the trade pattern of exchanged goods. However, while recognising these social patterns may have existed, and that "salt poverty" near salt sources could have existed, for the sake of simplicity this chapter will work on the assumption of fully adequate salt availability at production areas and lesser availability outside of the production areas, with availability, in general, negatively correlated to distance from the production source.

In considering salt distribution during the Iron Age, the nature of the underlying trade relationship needs to be considered before proceeding further. The quintessential feature of the salt trade was its nature as a demand-led transaction involving a commodity with a geographically limited supply (e.g. Renfrew, 1969, Moore, 2007); people living inland without a local salt spring *needed* salt for a range of uses which are the subject of Chapter 6 of this book. The same was true of other commodities which might not be locally available, such as the raw material for querns or honing stones, iron or bronze ingots, along with other goods less visible archaeologically, such as tanned leather, charcoal or fuel wood and cloth. The essential nature of these commodities ensured not only demand, but widespread demand crossing social boundaries, for these goods. In this they differed from items such as shale, jet, glass beads, imported wine or non-local pottery, whose trade was also demand-led, but which can be seen as *socially desirable,* rather than *necessary,* luxury items (Sherratt and Sherratt, 1991). Such luxury items would have moved in a parallel exchange system where demand was constrained, the trade network peopled only by those who could earn through social means or through trade with appropriate material assets, the desired goods (Pires-Ferreira and Flannery, 1976). While the two exchange systems could have existed simultaneously (Morris, 1983), and both may ultimately have been controlled by the same social elites, they need not have operated under the same social conditions. For example, the exchange of local goods for imported wine and other luxury items in the first century BC at *oppida* may have been controlled by tribal elites (e.g. Strabo, Fulford, 1987, Carver, 2001), while all levels of society may have been involved in local exchange for food.

During the Iron Age, parallel systems of prestige gift giving and commodity exchange of salt may have been in operation, although even the commodity exchange network may have been linked ultimately to established personal or traditional inter-group relationships. Gregory (1982) examined exchange in New Guinea and found that salt, among a range of other goods, could have a dual existence as both a commodity and a gift. Gift exchange was characterised by the creation of a network of personal relationships and social obligations, whereas commodity exchange was based on price-forming and profit maximisation behaviour, even in non-monetary transactions. This ethnographic parallel is relevant as an illustration of the potential dual nature of Iron Age salt procurement, which would also have been conducted on a barter basis in most cases. Money, in the modern sense, would not have been a factor in these exchanges. While it is possible that Kentish Primary potins might date to the very end of the Middle Iron Age (Haselgrove, 2006), even in the Late Iron Age, the period generally held to be the start of coin usage in Britain, coins are not present in all areas (e.g. de Jersey, 2001) and there is an open question as to whether they circulated as a medium of exchange. Their use among the non-elite rural populations may have been limited even after the Roman conquest.

Although Morris identified eight patterns of exchange which might be archaeologically visible (1983), two of these models, borrowed from economics and geography, have since been most frequently suggested for the Iron Age salt trade in Britain. On the one hand, it seemed possible that salt was taken to central places, such as hillforts, and redistributed in smaller amounts to people who gathered there, as shown in Figure 18 (e.g. Cunliffe, 2005). In this model, the central places might be expected to have significantly more briquetage than secondary sites even if at the same distance from the salt source.

In the second model, referred to as down-the-line exchange, as shown in Figure 19, salt passing along the trade network would move from point to point, with some being retained at each point on the distribution network (Renfrew, 1977). In this model, the expectation would be that the greatest amount of briquetage would be found closest to the production site with decreasing amounts farther away, although water transport does perturb the distance-quantity relationship relationship (Ammerman, 1979), a situation discussed below.

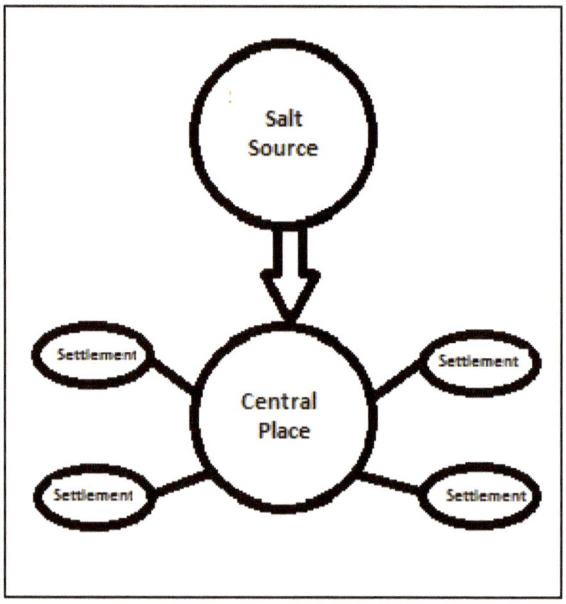

Figure 18. Central places distribution model.
After Cunliffe 2005: 591, Figure 21.3.

Despite the fact that the arrows in these exchange models are shown as moving in only one direction, it is necessary to assume that there was a two-way flow of material, even if only one happens to be archaeologically visible (Renfrew, 1969). Archaeologists studying salt distribution in Britain are not generally as fortunate as those studying the ancient Maya of Central America, whose exchange of coastal solar sea salt for highland volcanic ash intended for use as temper in pottery production is an example of demonstrable two-way trade (Simmons and Brem, 1979). There is also no need to assume salt was moving in socially isolated trade networks. Lambrick and Robinson (2009) point out the similar distribution patterns in the Upper Thames Valley for Droitwich briquetage, Malvernian pottery and May Hill querns, all originating in the West Midlands, suggesting a combined trade system where these bulk commodities "piggy backed" each other's networks along waterways.

Even though both of these models work well when applied to commodities that remain archaeologically visible throughout transport and use, such as obsidian (e.g. Renfrew et al., 1968), there is a significant weakness when they are applied to discussions of salt distribution. In the case of Droitwich or Cheshire salt distribution, the models work only while the salt is continuing to move

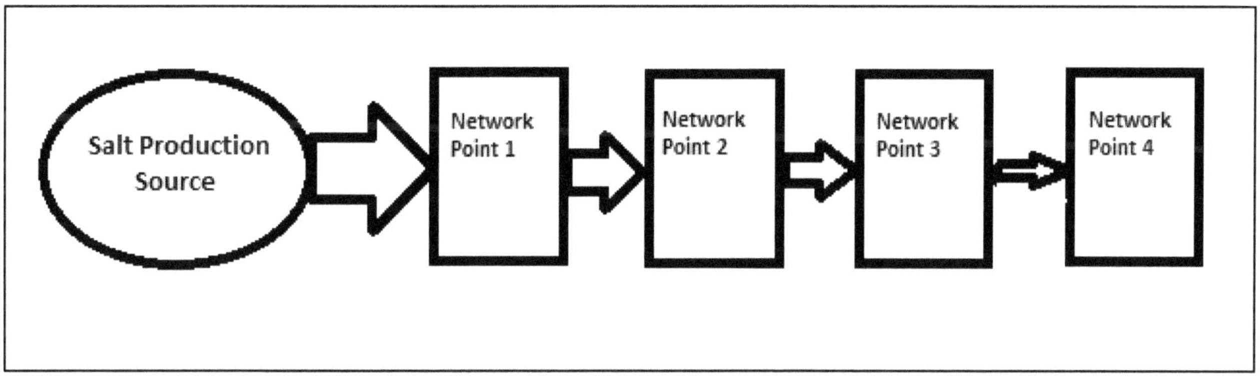

Figure 19. Simplified model of down-the-line exchange.
Source: author.

onwards in archaeologically visible containers, facilitating comparative measurements at various points in the distribution network. For example, while the briquetage evidence from hillforts in Wales seems to support the central place model, there is no way to distinguish such briquetage finds from evidence of on-site consumption (e.g. Pollard et al., 2006). In largely aceramic areas such as this, a strict, if improbable, application of the models could lead to an interpretation which showed that most salt-using activities in the region occurred at the hillforts. A more reasonable explanation, however, is the transfer of salt into the previously discussed archaeologically invisible containers, or possibly no container at all in the case of salt blocks, for further transport within the local distribution network. As Hodder wrote (1982: 203), "Behind' the 'end-result' patterns are strategies and intentions involving the manipulation of material symbols in local contexts." A choice was apparently made to move the salt, but not the briquetage container, onwards; the original container seemed to have changed symbolic meaning at this point or perhaps simply became less desirable within the local aceramic norm. The distribution pattern we observe at the landscape level was not the end goal of the original behaviour which determined the presence of briquetage in these areas, but an artefact of the behaviour and indicative of a change in the human relationships surrounding the movement of salt, such as shifts of kinship alignments through marriages, revision in the military and/or economic linkages between these areas, or possibly the mode of transportation.

The decision whether to use briquetage containers for transport would have been more than a purely functional one based on factors such as the weight of the containers and their fragility; it would have been a choice with deeper social meaning (Hill, 2002: 83). It is likely that participants in these salt distribution networks assigned different values and/or properties to salt depending on the container in which it was received. As Gell (1998: 23) pointed out in his discussion of art and agency, "Any artefact, by virtue of being a manufactured thing, motivates an abduction which specifies the identity of the agent who made or originated it." Feelings and beliefs about the makers of the salt and its providers, not necessarily the same people, and the properties of the salt itself, engendered in part by the salt container, would have been an inescapable aspect of the possession and use of salt in the Iron Age. The Roman author Pliny (*NH*, XXXI, vii, 39 and 40), writing shortly after the British Iron Age ended, distinguished more than 12 types of salt from wide-ranging sources derived from varying production methods. Each of these salts was listed as having unique characteristics and an optimal use. There is little reason to think that Iron Age Britons were less discerning or did not have their own beliefs about the salt at their disposal. The types of container, as well as the unique chemical composition of the salt and the size of the salt crystals within it, were all part of their perceptual experience of salt.

As has been noted earlier, there is also a temporal dimension to the discussion of prehistoric salt distribution, as the availability of salt seemingly increased gradually from the Late Bronze Age to the Middle Iron Age, then began expanding noticeably, with production achieving industrial scale in some areas during the Late Iron Age. Morris (1983, 1985) was the first to demonstrate that the salt exchange networks evolved as well. The link between technical change and the evolution of social relationships was noted by Gell, who wrote "...social relations are themselves emergent characteristics of the technical base on which society rests" (1992: 57). Exchange mechanisms suitable for the sharing of salt produced seemingly at a household level in the Bronze Age would likely bear little resemblance to the networks in place in the Late Iron Age of Essex in the early first century AD when briquetage-based salt production was at its peak, approximating an industrial level of production. We can envision small scale, local networks for sharing salt in the earlier periods, with more formalised long distance exchange networks in the later period. What is far less certain are the detailed social factors related to these networks, for instance, the gender and status of individuals involved in salt exchange, or the relative values placed upon goods being exchanged, i.e., was a kilo of salt exchanged for a kilo of quern stone or ten times its weight? We do not know whether the salt exchange networks were operated by individuals, collective social groups or social elites, or whether exchanges were conducted in the realm of the ritual or ordinary barter, or a combination of both. While historical or ethnographic parallels might be cited to suggest certain practices in the salt trade of Iron Age Britain, recovery of these details through archaeological practice alone is unlikely.

With this long litany of cautions and caveats in mind, a review of the briquetage distribution patterns from Droitwich and Cheshire incorporating some of the methodological approaches used in Morris' PhD research follows.

4.4 Droitwich and Cheshire distribution patterns: updating the early work of E. Morris

The pioneering research of Elaine Morris in the late 1970s and early 1980s (e.g. Morris, 1983, Morris, 1985) demonstrated conclusively that Iron Age briquetage from Droitwich and Cheshire was used in the distribution of salt. Morris selected Iron Age sites in the Severn and Avon watershed area for which large ceramic assemblages existed. She studied the forms and fabrics of both pottery and a ceramic material that was then often identified as "Very Coarse Pottery" or "VCP", of uncertain function (e.g. Gelling and Stanford, 1965) when she began her studies, and is now classified as briquetage and Stony VCP. Although the material was not present within all of the assemblages she studied, she determined through petrological analysis of inclusions in the ceramics viewed as thin sections under polarising light, that at 33 sites the material was briquetage that had come from Droitwich, and at 35 sites the sherds could be identified as Stony Very Coarse Pottery which originated in the Middlewich or Nantwich area. Her work demonstrated that the material had been used for salt distribution from these sites. Noting that briquetage or VCP could be present or absent in an assemblage at different points in time, she found that the distribution patterns were dynamic, with material of both types becoming more widespread in the later Iron Age. She illustrated this with time-phased maps of the distribution patterns for both Droitwich and Cheshire briquetage shown below as Figures 20 and 21. Morris also demonstrated a statistical correlation between the distribution of Malvernian Iron Age pottery and Droitwich briquetage (1983, 1985). Morris concluded, citing models used by Collis in 1971 for Late Iron Age coin production and distribution (1983: 378), that Cheshire VCP patterns reflected decentralised production and distribution, but a centralised and controlled exchange network after 500 BC for Droitwich salt.

Beyond considering distribution patterns, Morris utilised a variety of statistical tools to consider in greater depth the ceramic assemblages she was studying and the nature of the trade networks. To help compare briquetage find sites, Morris developed the Specific Salt Container Index or SSCI (1983: 307-8) which considered briquetage as a proportion of the recovered ceramics found at a site. The index values range from 0 to 1.0. An index value close to 1 does not necessarily signify that a large quantity of briquetage was recovered; only that it was a large

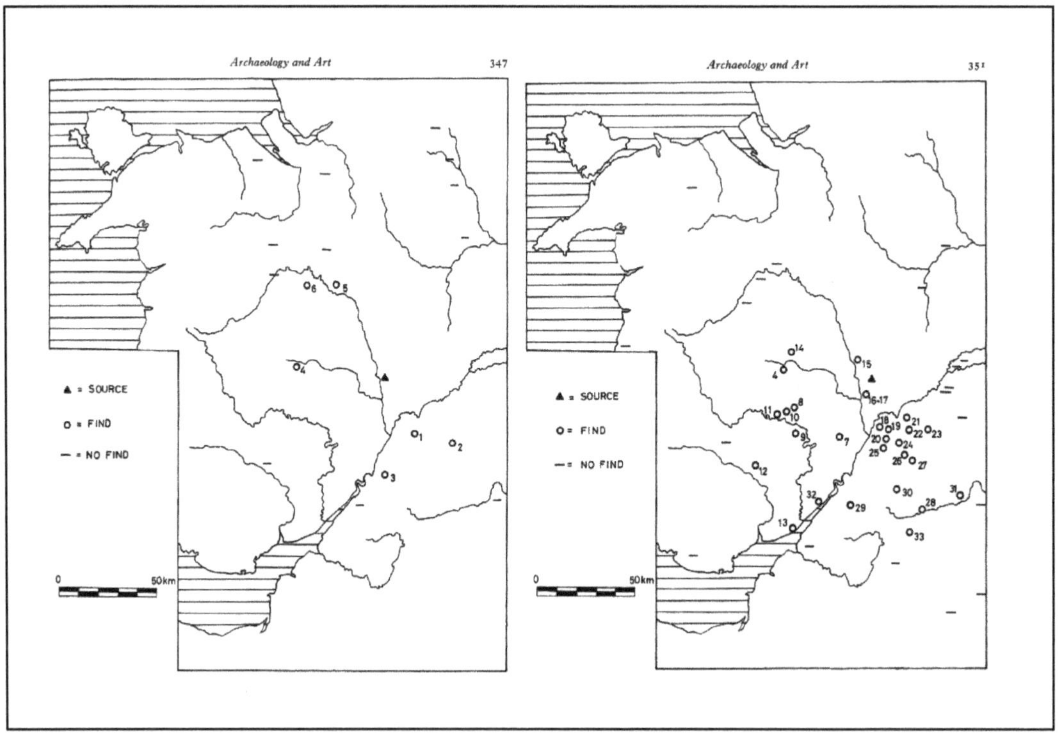

Figure 20. Early (left) and later (right) Droitwich briquetage distributions.
From Morris, 1985: 347 & 351.

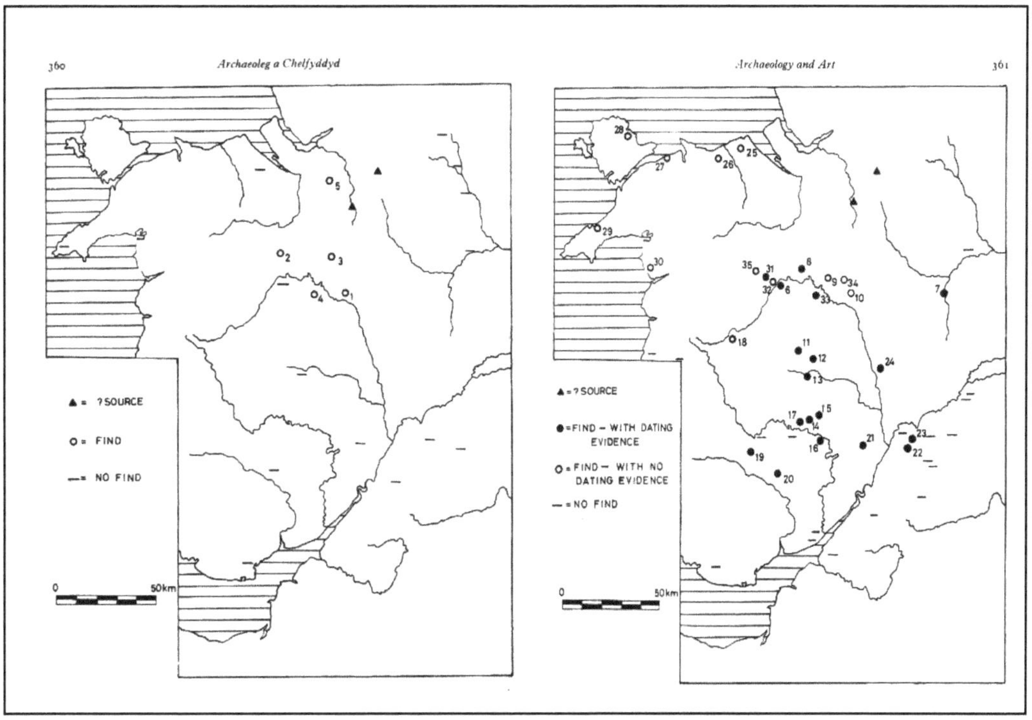

Figure 21. Early (left) and later (right) distributions of Cheshire Stony VCP.
From Morris, 1985: 360- 1.

proportion of the ceramic assemblage. VCP or briquetage dominated the ceramic assemblage at sites in aceramic areas, such as Wales, giving them particularly high indices. This statistical tool allowed sites to be compared regardless of the absolute quantity of briquetage that had been recovered. However Morris (1983: 307) found that the indices were only comparable within a dataset associated with a single production site, not between the two production sources due to the preponderance of high index values in the largely aceramic Cheshire distribution area and lower values in the largely ceramic-using Droitwich distribution area.

Morris' findings with respect to briquetage were presented to a wider audience in a paper which appeared in *The Bulletin of the Board of Celtic Studies* (1985). Since that date, she has been a consistent voice in British archaeology, identifying briquetage and VCP within ceramic assemblages and encouraging others in the study of these materials. Morris' work is the starting point for most briquetage research which has followed, making the inland salt trade networks from the Droitwich and Cheshire salt production areas by far the best known to British archaeologists. However, while the intervening years have seen the identification of many new sites where briquetage or VCP were found, often supported with the review of the suspected briquetage by Morris, to date there has been no comprehensive reconsideration of her initial work and conclusions in light of this new evidence.

Morris' initial models and her conclusions about the dynamic nature of the Cheshire and Droitwich distribution patterns, supported by evidence of changing pottery styles found with the briquetage (1983, 1985), have been widely accepted and have held up well in the face of subsequent discoveries (e.g. Matthews, 1999), therefore the reconsideration undertaken here will focus on integrating information from briquetage find sites located subsequent to her thesis into her models and methodology. Accepting as secure Morris' conclusion about the changing nature of the distribution patterns over time has allowed the author to pool her own data and present the finds on unphased maps. For the reasons discussed above relative to movement of salt without the use of briquetage containers and taking into account the scale used on the distribution maps in this chapter, the author has elected not to indicate investigated sites where briquetage was not located as Morris had done on the maps in this chapter. In Chapter 5, where detailed case studies of salt distribution within small areas are presented, sites without briquetage will be included on the maps.

Where Morris utilised direct examination of selected ceramic assemblages to develop her distribution models, the current research has been primarily a desk-based literature review aimed at compiling updated information about the range of sites where briquetage and VCP have been found. An examination of books, articles in archaeological journals, published excavation reports with Iron Age phases, "grey" literature reports held by county archaeological archives or available on the internet or through commercial archaeology firms, archaeology websites and HER records was conducted with the results being compiled as an Access database.

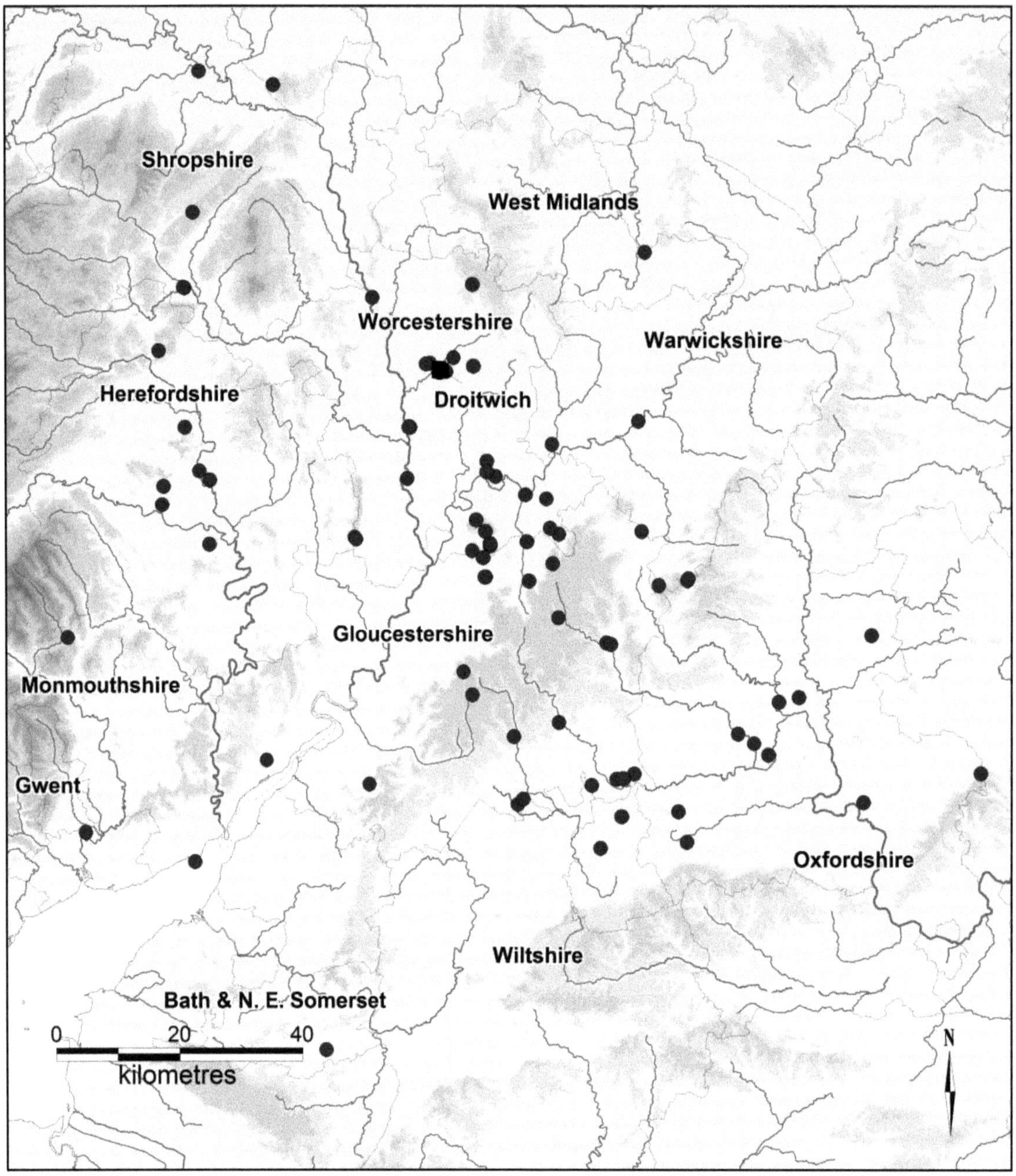

Figure 22. Unphased Droitwich briquetage find sites map.
Site names listed in Appendices 3 and 8. Underlying map © Crown Copyright Ordinance Survey.

These reviewed materials covered counties identified by Morris as containing finds, counties geographically adjacent to those find sites and any county mentioned in any of these sources as having a find not otherwise part of the initial geographic search parameters. Despite the limitations noted while compiling the data that were discussed above, this search methodology has allowed the author to compile records of 85 Droitwich briquetage find spots unrelated to salt production, including those initially identified by Morris, as detailed in Appendix 3 and displayed graphically in Figure 22. This is more than 10 times the number of production sites known within Droitwich, and more than twice the total number of finds known when Morris wrote her PhD thesis.

The same literature search methodology was applied with respect to reports of Cheshire Stony VCP finds resulting in the compilation of data from 67 sites, including the 35 sites used by Morris. The details of these sites are contained in Appendix 4, with the site locations shown in Figure 23. No claim is made here that these lists represent all new finds from the recent past; other find site names which were encountered during research but for which critical information, such as county, National Grid Reference, dating or source of the briquetage could not be confirmed, are excluded from these datasets.

The character of many archaeological sites where briquetage has been found since 1983 differs considerably from those available to Morris at the time of her work. The list of Droitwich briquetage and Cheshire VCP find sites compiled since the implementation of

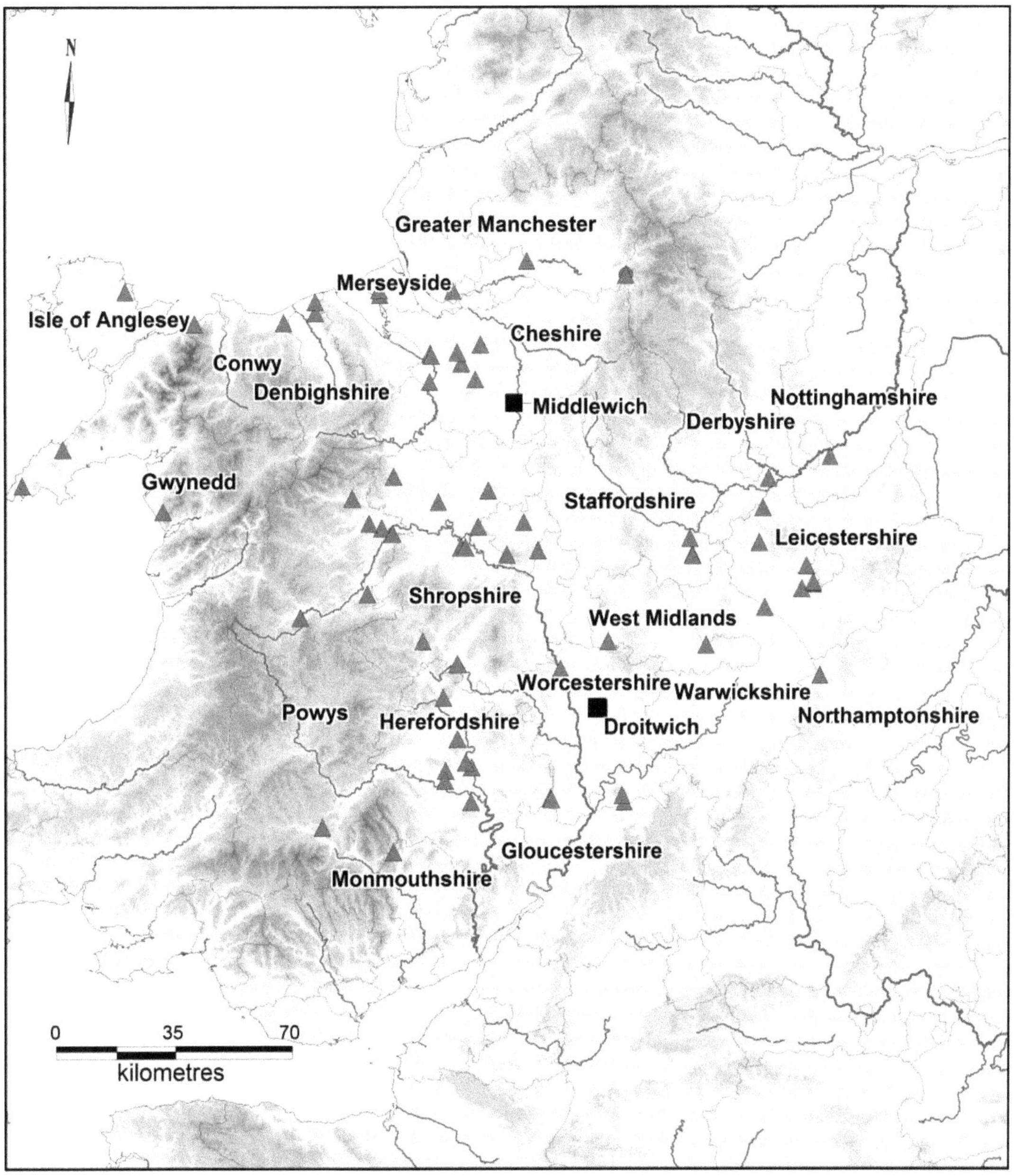

Figure 23. Unphased Cheshire VCP finds map.
Site names listed in Appendix 4. Underlying map © Crown Copyright Ordinance Survey.

PPG 16 in 1990 is dominated by sites other than hillforts, as shown in Tables 3 and 4 below. For comparison, almost 40% of Morris' Droitwich briquetage samples came from hillforts, popular excavation sites during the years prior to her early research, but they make up approximately 20% of the dataset utilised here, comprised of the original sites used by Morris and the subsequent finds. For the combined salt distribution networks of Droitwich and Cheshire, hillforts now constitute less than one-third of the finds dataset. This has not occurred through a concerted effort to improve the understanding of Iron Age settlement patterns but as an unintended consequence of the archaeological funding process, with most investment coming from developers focused on specific sites rather than on research agendas.

In compiling Tables 3, 4, 5 and 6 the archaeological debate about what, precisely, defines a hillfort has been placed to one side (e.g. Cunliffe, 2005) and the author has chosen to accept that, if it has been published as a hillfort site, even if on a hillslope or valley bottom, it is considered a "hillfort" for the purpose of this analysis. The tables clearly show that salt, in briquetage containers, was reaching not just hillforts but many other types of sites as well, some of which were at a 100km distance from the salt sources. Non-hillfort find sites are particularly common to the southeast of Droitwich in contrast to the west, suggesting that there may have been differences, such as the decanting of salt from briquetage at hillforts before wider distribution, occurring in that region. These differences will be discussed further below and in greater detail in Chapter 5.

The dating of briquetage finds, as mentioned above, is often inexact. Table 5 summarises the site dating information for the Droitwich find sites referenced

Production source	Hillfort find sites	% of total find sites	Non-hillfort find sites	% of total find sites	Total number of find sites
Droitwich	13	39.4	20	60.6	33
Cheshire	24	68.6	11	31.4	35
Total	37	54.4	31	45.6	68

Table 3. Types of find sites used by E. Morris in her PhD thesis and 1985 article.

Production source	Hillfort find sites	% of total find sites	Non-hillfort find sites	% of total find sites	Total number of find sites
Droitwich	17	20.0	68	80.0	85
Cheshire	29	43.3	38	56.7	67
Total	46	30.3	106	69.7	152

Table 4. Droitwich briquetage or VCP find sites by location type used in this book.

in Table 4 as provided in the source documentation. As was highlighted in Chapter 3's discussion of Bronze Age salt production, in some cases it was necessary to adjust the originally published dating to match changes in Iron Age chronology since the initial find report, with the Early Iron Age now considered to extend back to *c.*800 BC. These changes are reflected in Table 5. The identification of Early Iron Age sites exhibiting Droitwich briquetage used for salt distribution is somewhat surprising, as this evidence has not yet been paralleled by evidence at Droitwich of salt production utilising briquetage from that period. There was no clear evidence of an Early Iron Age salt production phase involving briquetage at the Droitwich sites at the Old Bowling Green or Friar Street (Woodiwiss, 1992), Upwich (Hurst, 1997), 1A High Street (Sworn and Griffin, 2004), Dodderhill Roman Fort (McAvoy, 2006) or Bays Meadow Roman Villa (Barfield, 2006). The absence of finds from hillforts dated to the Late Iron Age or later is consistent with information on the general waning of significance of this type of site during that period (e.g. Cunliffe, 2005). The finds categorised in Table 5 as generic or those which are very broadly phased often come from small assemblages with limited amounts of easily dated material or those from sites with poorly defined stratigraphy. The general impression that more salt in briquetage and VCP was moving around the landscape in the Middle and Late Iron Ages is supported by the larger number of sites in these categories, although, alternatively, this could be a reflection of the differences in site visibility between the often unenclosed sites of the Early Iron Age and the more common enclosed ones of the Middle and Late Iron Age.

As shown above in Figure 22 the expanded corpus of sites now in evidence confirms Morris' earlier conclusion that Droitwich briquetage was to be found mainly to the south, southeast and southwest of Droitwich, but far less frequently to the north or northeast. A possible explanation for the shape of this distribution pattern is found in the work of Bradley (1971) who discussed the fact that trade distribution patterns tended to be distorted by areas of especially severe competition. If this observation also held true in the Iron Age for salt distribution, then the pattern may indicate that there was direct competition between the Cheshire and Droitwich salt sources north and northeast of Droitwich. Hurst (2001) suggests that this competition may be reflective of tribal loyalties, as discussed at greater length below. However, it is equally plausible that the archaeologically visible pattern simply reflects a preference for salt distribution in archaeologically invisible containers in this region.

The number of find sites to the southeast of Droitwich, clustered roughly along the course of the modern A44, suggests that this route was in use for salt transport long before its utilisation as a saltway at the beginning of the early medieval period (Hooke, 1985) or even the arrival of the Romans, who built a road along the route (McAvoy, 2006). Similarly, the map of the medieval saltway running from Droitwich to Uffington, which passes through a gap in the Iron Age earthwork known as Grim's Ditch, previously referred to as Grim's Dyke (Case et al., 1964/1965), has proven to be a good predictor of briquetage finds in western Oxfordshire. The new array of sites also suggests strongly that the northern reaches of the Upper Thames watershed, across the Oxford clay vale, south to the Chiltern Hills and Berkshire Downs, was an area supplied primarily with salt from Droitwich, the closest large supplier. At the time of Morris' initial work, the maximum straight line distribution distance was thought to be 60km in the early phase and 75km in the later period (1985). With the additional sites considered here, the maximum distancefor the Late Iron Age distribution is now known to slightly exceed 100km, although only 8 sites (9.4%) are 80 or more kilometres from Droitwich. A

Droitwich finds by period	Hillfort find sites by period	% of total find sites	Non-hillfort find sites by period	% of total find sites
Early Iron Age (EIA)	5	29.4	6	8.8
Early to Middle Iron Age	0	0	2	2.9
EIA to Romano-British (RB)	0	0	1	1.5
Middle Iron Age (MIA)	2	11.8	20	29.4
Middle to Late Iron Age	9	52.9	22	32.4
MIA to Romano-British (RB)	0	0	1	1.5
MIA and RB	0	0	1	1.5
Late Iron Age (LIA)	0	0	2	2.9
LIA to RB	0	0	1	1.5
LIA or RB	0	0	1	1.5
Generic Iron Age (IA)	1	5.9	6	8.8
Generic IA to RB	0	0	2	2.9
IA and/or RB	0	0	3	4.4
Droitwich totals	17	100	68	100

Table 5. Droitwich briquetage finds as reported by time period and site type.

Cheshire VCP finds by period	Hillfort find sites by period	% of total find sites	Non-hillfort find sites by period	% of total find sites
Late Bronze Age and M-LIA	0	0	1	2.6
Early Iron Age (EIA)	3	10.4	2	5.3
E-LIA	1	3.4	0	0
EIA to Romano-British (RB)	0	0	1	2.6
Middle Iron Age (MIA)	3	10.4	6	15.8
M-LIA	16	55.2	11	29.0
MIA-RB	1	3.4	3	7.9
Late Iron Age (LIA)	1	3.4	6	15.8
LIA-RB	0	0	3	7.9
Generic IA	4	13.8	4	10.5
Generic IA- RB	0	0	1	2.6
Cheshire totals	29	100.0	38	100.0

Table 6. Cheshire briquetage finds as reported by time period and site type.

further confirmation of Morris' work (1983) concerns the overlapping distribution patterns of Malvernian "A" and "B" pottery and Droitwich briquetage, and these dual find sites have continued to steadily increase in number in the Upper Thames Valley (e.g. Morris and Crosby, 2007, Lambrick and Robinson, 2009).

Information on Cheshire VCP find site phasing for the expanded data set is contained in Table 6. When writing in 1983 (Morris, op. cit.: 246), 47km was the farthest VCP was known to have travelled in its early phase of distribution, but that distance was known to increase to 140km during the later Iron Age. The major change now visible in the enlarged data set, as shown in Figure 23, is not a greater absolute distribution distance but the number of sites in the east and central Midlands regions where VCP has been found, arriving apparently as the result of a transport route that moved salt either directly across, or to the south of, the Peak District. Several of the sites, including Crick Covert Farm (Woodward and Hughes, 2007), Coventry Road (Chapman, 2004) and Meriden Quarry (Stevens, 2005), are physically closer to Droitwich than to the Cheshire salt springs. All of the Leicestershire sites, including Breedon-on-the Hill (Morris, 1999), Coventry Road (Chapman, op. cit.), Enderby I (Elsdon, 1991) and II (Marsden and Morris, 2004), Grove Farm (Clay, 1992), Huncote (Marsden, 2004), Kirby Muxloe (Cooper, 1994), and Normanton-le-Heath (Thorpe et al., 1994), are actually physically closer to the Fenlands, another major Iron Age salt producing area, than to the Cheshire salt source (Marsden and Morris, 2004), thought to be at Middlewich (Morris,

2007a). Even by the standards of briquetage studies, only small absolute amounts of Cheshire VCP have been found in this region, suggesting that salt from this source may never have been present in large quantities there, or generally arrived in archaeologically invisible containers. Possible explanations for the presence of VCP in this region will be explored below.

Although the finds at Fox Covert Farm, Derbyshire (Morris, 1999), and Kirby Muxloe (Cooper, 1994), could not be dated more precisely than "Iron Age", of the remaining 11 sites in the Midlands geographic cluster, 2 date to the Middle Iron Age, 2 are Middle to Late Iron Age, 6 are Late Iron Age and 1 seems to span the Late Iron Age-Early Romano-British period. These dates coincide with the expansion of Cheshire VCP distribution across north Wales and the Severn watershed, illustrated by Morris (1985: 360 and 61) and shown in Figure 21. As distribution of Cheshire salt in VCP began in the Early Iron Age (e.g. Morris, 2001b), this expanded distribution area reflects a significant change in the pattern that had initially been established, an alteration perhaps reflecting changing social relationships driven by the expansion of sea trade through the port of Meols, connected to the Cheshire salt springs by the rivers Weaver and Mersey, beginning *c.*500 BC (Matthews, 1999, Philpott, 2007).

It is quite probable that the majority of the salt used in the Midlands region during the Middle and Late Iron Ages came from the nearer sources in the Fens (Morris, 2001b), from producers who do not appear to have used briquetage for salt movement (Lane, 2001), possibly exchanged on some sort of a commodity basis while in parallel small quantities of Cheshire salt in VCP containers arrived as prestigious gifts. This would be akin to some of the sites studied by Morris (1983, 1985) where dual sourcing of salt from Droitwich and Cheshire is suggested by the presence of both briquetage and VCP. The potential value of such dual sourcing was noted by Morris (1983) who wrote that such activity can act as a form of social storage and buffer against periodic fluctuations in availability. However, recognition of potential duality between Cheshire and Lincolnshire salt sources had not been predicted. Now, with the benefit of hindsight and 11 VCP finds in the central and east Midlands, the existence of such duality can be hypothesised.

The Midlands VCP finds can serve as potential examples of the range of social factors involved in long distance trade networks as discussed above. All salt used in the Midlands had to be acquired through some form of exchange, and if efficiency was the only source selection criterion, distance to source and ease of movement would be the main determinants; but human behaviour is not that simple. The presence of VCP serves to emphasize the point that the logic driving the movement of salt in briquetage could be governed by factors quite distinct from notions of economic efficiency, for example, as was mentioned above, the use of multiple sources to strengthen social links and buffer against regional shortages. Haselgrove (1982) defined prestige goods as any items "... only available from outside the local system ...," a category which clearly describes salt in this region in general, and Cheshire salt east of the Peak District in particular. This would have been equally true at the other extension of the VCP distribution region into south Wales. Salt in VCP, probably transported by boat through the port of Meols (Philpott, 2007) would also be unique in north Wales and the Severn watershed where it appears during the Middle and Late Iron Ages. By virtue of its rarity, using Haselgrove's definition, it would have transformed into a prestige good.

The Collfryn Hillslope Enclosure

An example of the importance of the salt trade between the Cheshire salt makers and the occupants of the Severn Valley watershed is clearly in evidence at Collfryn, shown in Figure 24, excavated between 1980 and 1982 (Britnell, 1989). The heavily defended farmstead, dating to the fourth or third century BC, contained multiple roundhouses and four- and six-post structures thought to be used for various types of storage. The scale of the site's defences, along with the artefactual evidence recovered there, has led to its interpretation as a farmstead belonging to a powerful chieftain. This site yielded the largest assemblage of VCP or briquetage not associated with a salt production source that has been found in Britain: 2,839 sherds with a total weight of 13,285gm, all of which was Cheshire VCP. The finds are particularly noteworthy as the site was otherwise essentially aceramic. Occupation at the site may have been continuous until the fourth century AD. The VCP finds at the site span the entire period from the mid-fourth century BC in the Iron Age to the early Romano-British period when the use of VCP ended (ibid.: 124 and microfiche 30), indicative of sustained trade in salt at this site.

As illustrated in Figure 25, VCP was found in many locations at the site, often as isolated sherds (ibid.: microfiche 30). The largest concentration came from the north terminal of Enclosure Ditch 2, where the suspected remains of a single vessel were recovered. Post slot 3703, a component of Post Structure 29 shown in Figure 26, also contained a larger than average quantity of VCP. Drainage ditches and several of the roundhouses indicated above, particularly numbers 3, 4 and 8, were also found to have more VCP. The absence of VCP and other artefacts from either roundhouse 1 or 7 caused the excavator to believe that these structures had been used for a different purpose than the other roundhouses despite their similar construction.

Most of the detailed information about the distribution of VCP finds at this site comes from information provided in the microfiches rather than the main text of the article although even this resource fails to provide context level information for most of the finds. The major exceptions to this statement relate to the illustrated VCP rim and base sherds (ibid.: 124-25) which do include context

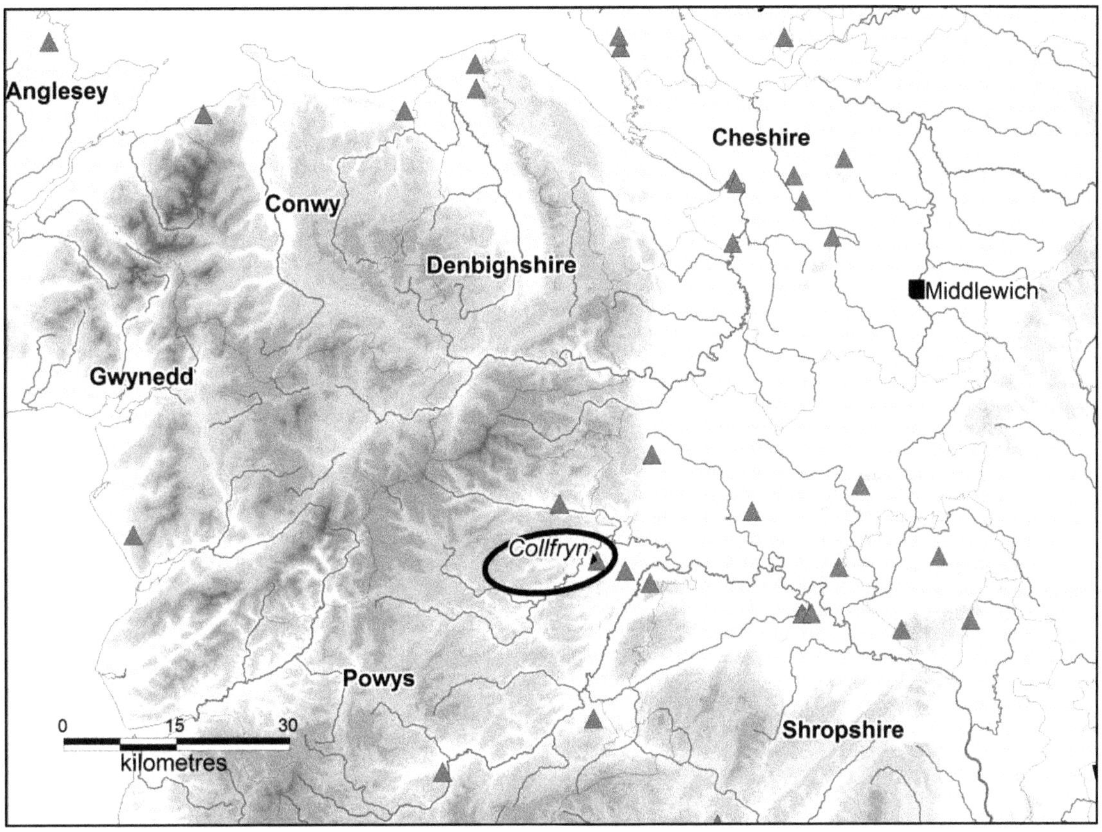

Figure 24. Location of the Collfryn Hillslope enclosure.

Key: Triangles mark VCP find sites. despite their similar construction. Underlying map © Crown Copyright Ordinance Survey.

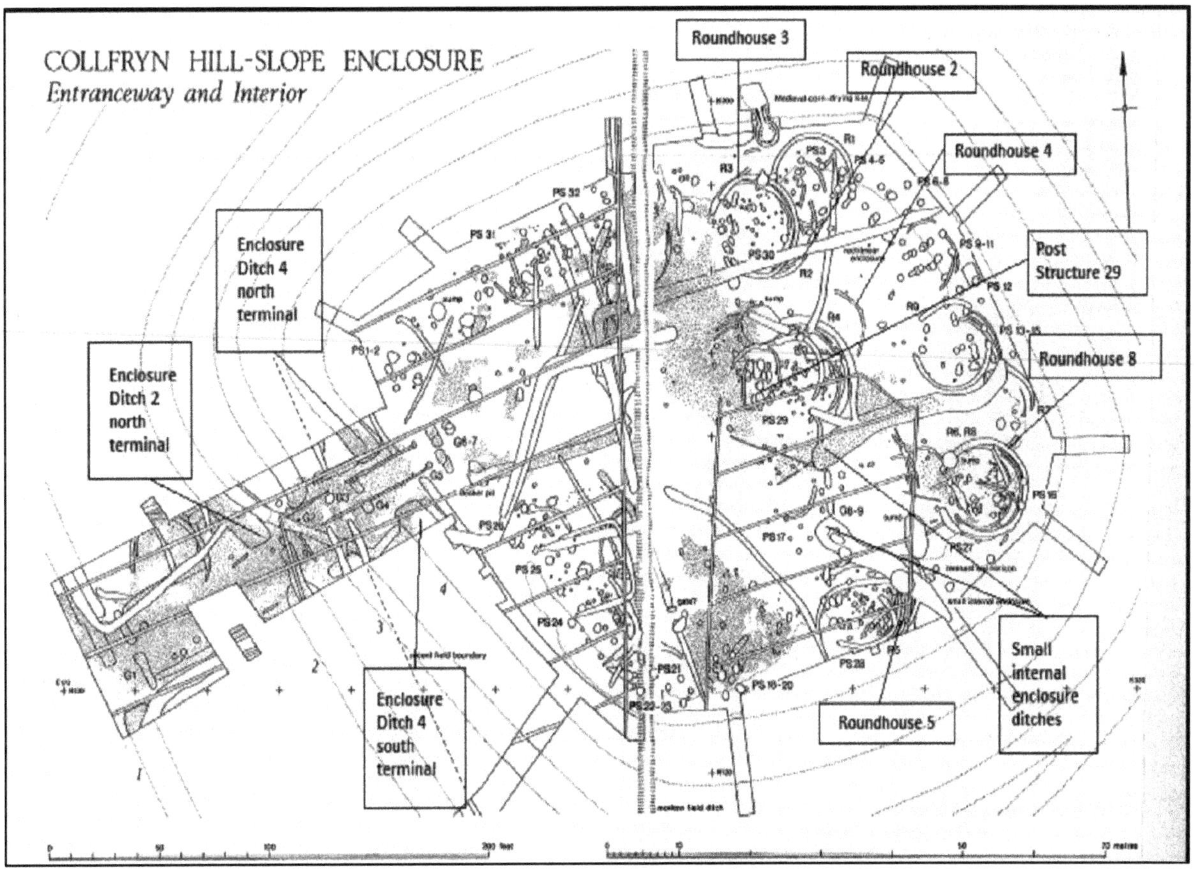

Figure 25. Plan of Collfryn excavation trench with major VCP-bearing structures noted.

After Britnell, 1989: 132, Fig. 37. Scales equal 200 feet (right) and 70 metres (left).

level information in their descriptions. Figure 27 illustrates three contexts from the fills of Ditches 2 and 4 which contained either rims or bases. Figures 28 and 29 show the plans of two roundhouses associated with finds of either rim or base sherds along with the contexts which contained VCP. In the case of Roundhouse 8, context 3524, Britnell advised that the VCP came from layer 2967 (ibid.: 124) but no section drawing has been provided for this feature so it is not available for use here. Other contexts with illustrated finds not shown here include context 3006, layer 2008, of the north arm of the small internal Iron Age enclosure ditch which held a base; context 3652, a rubbish pit near Roundhouse 3 that also contained a base; and a base thought to be residual in context 3531, layer 2959, a fill in a late third century-early fourth century Romano-British pit in the south arm of the small internal enclosure ditch.

Despite the presence of such a large volume of VCP at the site, there is no evidence that the material was treated as anything other than occupation debris during the hundreds of years the site was in use as there are no examples of special deposition of VCP at this site. As this region was largely aceramic in the Iron Age, it is likely that the VCP assemblage represents both the residue from local use of the salt and the refuse from the removal of salt from its transport container and possibly its transfer into organic containers for distribution locally.

The large assemblage of VCP at Collfryn might be interpreted as representing social exchange as much as the movement of a vital commodity. Coin evidence from the Late Iron Age (e.g. de Jersey, 2001, Cunliffe, 2005) is one type of information that has been used to define approximate British tribal areas, shown here in Figure 30, identifying the groups with names used by the Romans. If one assumes that the coin evidence represents residency patterns that actually extend back to the Middle Iron Age, as suggested by ceramic style evidence (e.g. Cunliffe, 2005), one could interpret the possible gifting of prestigious salt in VCP from the Cornovii area to the Deceangli, Ordovices, Silures and Catuvellauni, the apparently non-salt producing tribal groups surrounding the Droitwich-controlling Dobunni, either as an exercise in measured diplomacy or possibly as a more sinister effort to encircle the Dobunni within a network of competing alliances established and built up over time. In this same fashion, finds of Cheshire VCP within supposed Dobunnic territory itself might reflect the residue of diplomatic missions or surreptitious forays to find allies within an opponent's sphere of influence. If this was the case, the asymmetric distribution of Droitwich briquetage to the south and southeast may suggest an alliance network focused in that direction. An equally plausible, if less dramatic, alternative, however, is that salt in VCP could have arrived through a range of other social transactions, such as inclusion in bridewealth (Morris, 2007a) or dowry payments, tribute payments or use as a form of currency, personal gift exchange or perhaps even as prestigious Iron Age souvenirs acquired on a journey.

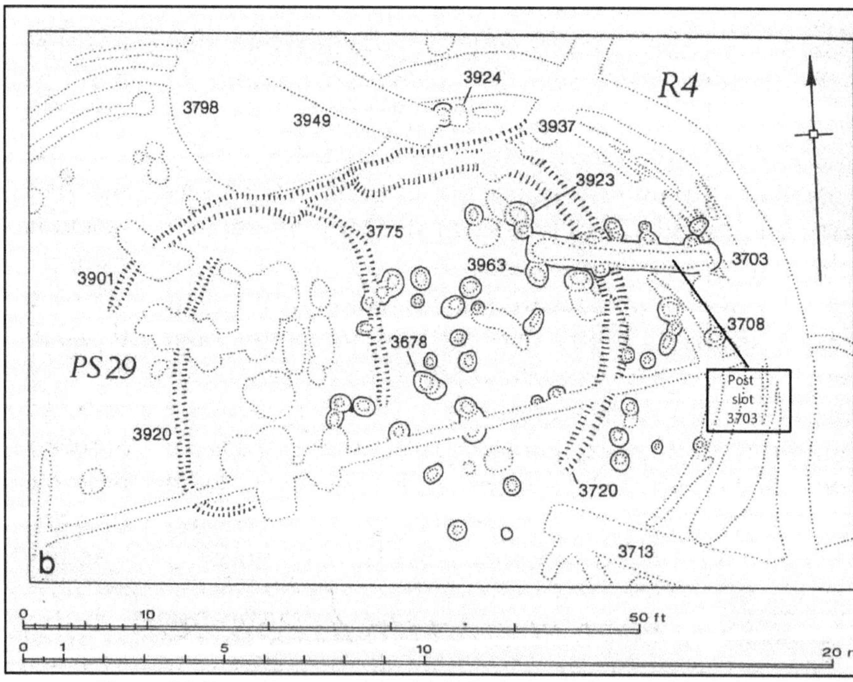

Figure 26. Post Structure 29 with VCP-bearing feature identified.
After Britnell, 1989: 118, Fig. 21. Scales equal 50 feet (upper) and 20 metres (lower).

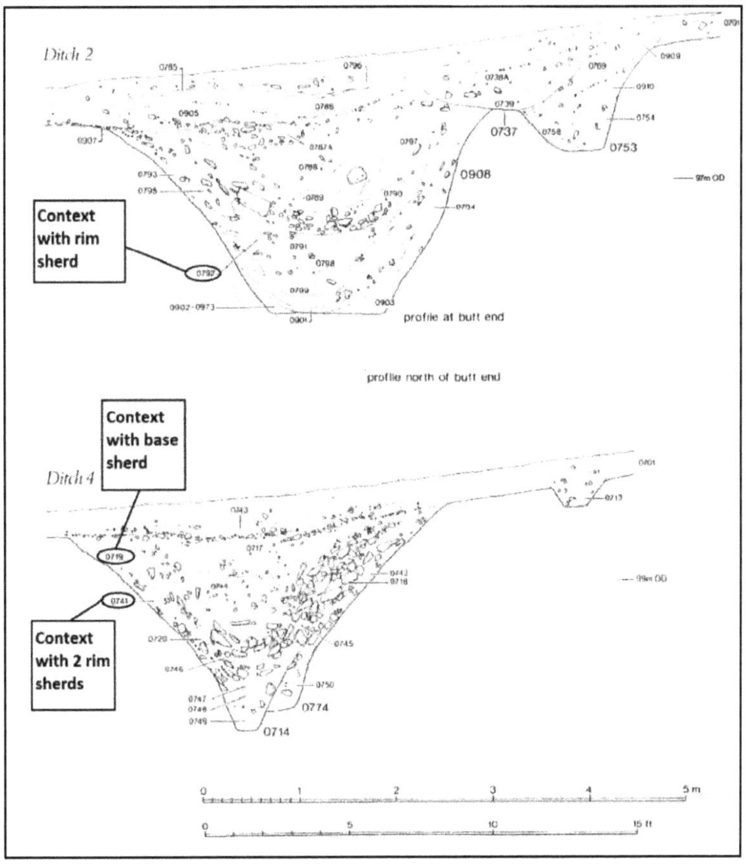

Figure 27. Section views, Ditches 2 and 4, north side of entranceway
After Britnell, 1989: 132, Fig. 38. Scales equal 15 feet (upper) and 5 metres (lower).

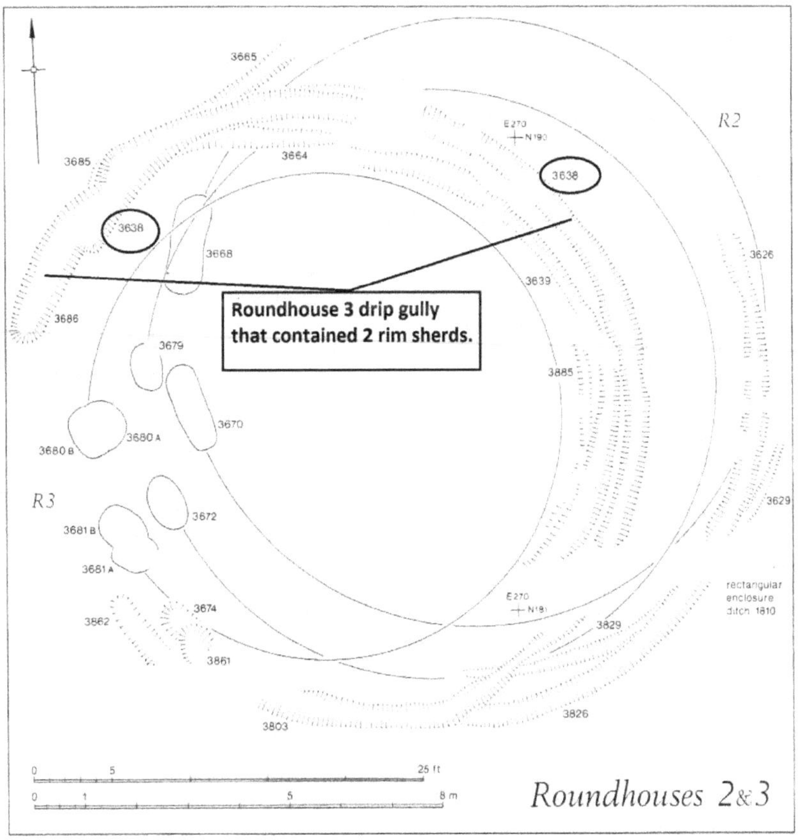

Figure 28. Plan of Roundhouse 3, context 3638.
After Britnell, 1989: 99, Fig. 7. Scales equal 25 feet (upper) and 8 metres (lower).

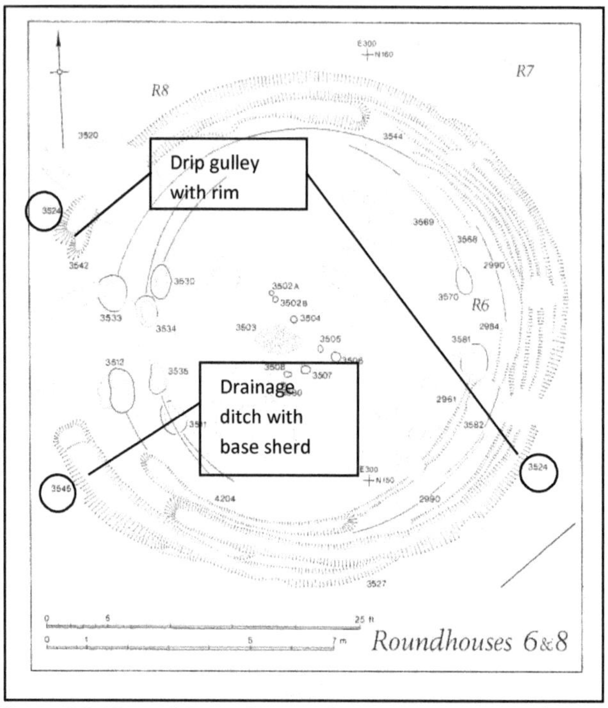

Figure 29. Plan of Roundhouse 8, contexts 3524 and 3545.
After Britnell, 1989: 105, Fig. 11. Scales equal 25 feet (upper) and 7 metres (lower).

Figure 30. Map of estimated Late Iron Age tribal areas in Britain.
Drawn by Peter Bridgewater. From Scullard, 1999: 24.

4.5 Using the Specific Salt Container Index (SSCI)

Site Name	Site Description	Distance from source (km)	SSCI
Kenchester	enclosure	50	0.003
Mingies Ditch, Hardwick-with-Yelford	enclosed settlement	75	0.004
Shenberrow Hill Camp	hillfort	35	0.004
Ireley Farm, Hailes	find site	37	0.005
Spratsgate Lane, Somerford Keynes	partially enclosed settlement	50	0.007
Claydon Pike, Lechlade	enclosure	71	0.011
Chastleton Hill Camp	hillfort	50	0.013
Gravelly Guy, Stanton Harcourt	EIA-MIA settlement	80	0.014
Gravelly Guy, Stanton Harcourt	LIA-ERB settlement	80	0.015
Hampton Lovett and Westwood	settlement	2	0.016
Dinedor	hillfort	46	0.032
Watkins Farm, Northmoor	enclosed settlement	80	0.032
Bredon Hill	hillfort	24	0.033
Shorncote Quarry, Somerford Keynes	roundhouse gulley	68	0.041
Wormington Farm, Aston Somerville	trackways	30	0.046
Guiting Power	settlement	42	0.046 (1974)
Guiting Power	settlement	42	0.047 (1997)
Beckford I	settlement	28	0.054
Sharpstones Hill site A, Weeping Cross	enclosed farmstead	63	0.063
The Knolls, Oxenton	unenclosed settlement	34	0.065
The Wrekin	hillfort	52	0.069
Broadway	enclosure	32	0.070
Uley Bury	hillfort	65	0.094
Throckmorton Airfield	enclosed settlement	15	0.096
Meriden Quarry, Solihull	enclosure	37	0.102
Wellington Quarry	ditched enclosure	43	0.104
Salmonsbury	large enclosure	50	0.106
Elm Farm, Beckford	settlement	30	0.108
Evesham, 93-97 High Street	domestic site	24	0.116
Fladbury, Evesham Road	settlement	19	0.117
Bromfield	enclosure	43	0.119
Conderton Camp	hillfort	27	0.119
Clifton Quarry, Severn Stokes	unenclosed settlement	18	0.161
Blackstone	enclosure	17	0.164
Croft Ambrey	hillfort	45	0.164
Stonebridge Cross, Westwood	farmstead	2	0.325
Lodge Hill Camp	hillfort	116	0.355
Midsummer Hill	hillfort	29	0.471
Midsummer Hill site 2	hillfort	29	0.471
Credenhill Camp	hillfort	48	0.708
Twyn-y-Gaer	hillfort	74	0.715
Droitwich, 1A High Street	find site	0	1.000

Table 7. SSCI for selected Droitwich briquetage find sites ordered by increasing index values.

NB: Distance from source measured in a straight line; actual distance travelled between source and find spot may have been considerably greater.

Following the methods current at the time (e.g. Renfrew et al., 1968, Hodder and Orton, 1976), Morris (1983: 308) utilised mathematical modelling in an effort to better understand the nature of the trade networks revealed through the distribution of briquetage finds. She developed a standardised measurement, the Specific Salt Container Index (SSCI) (1983: 307) to facilitate comparison of sites. As discussed above, this index represents the proportion of briquetage within a ceramic assemblage as measured by weight. Further, she utilised regression analysis to model the distance fall-off patterns within the Droitwich and Cheshire trade networks (ibid.: 308-09). Although exponential models had been used successfully on some long-distance trade networks, Morris found that a logarithmic model of the distance-decay pattern was needed for her network models (1983: 308-9), potentially a side-effect of her datasets, that were relatively small for detailed statistical analysis. By pooling all data points by supply source, rather than dividing them into two phased datasets, there are now more Droitwich and Cheshire sites to evaluate statistically and exponential models can be applied.

Despite the data problems discussed earlier, it was possible to enlarge the data sets used by Morris using published information. Tables 7 and 8 expand the number of assemblages for which Morris' SSCI has been calculated from 22 to 42 assemblages from 39 sites for Droitwich briquetage and from 17 to 24 assemblages from 23 locations for Cheshire VCP finds. Note that 14 sites appear on both tables due to the presence of briquetage from both sources. Based on its relatively high SSCI as shown in Table 7, Morris speculated that the Droitwich distribution network centred on Midsummer Hill (1983). As shown in Figures 22 and 31, the Droitwich find site locations discovered after her thesis was published suggest that the geographic centre of the distribution network is closer to other hillforts, Bredon Hill or Conderton Camp, formerly called Danes' Camp, to the south of the River Severn. The low SSCIs at these sites, however, would suggest that either of them was an unlikely option for the centre of the distribution network. Hurst (1992b) has suggested the little-known Hanbury hillfort, now partially beneath St Mary's Church, as the central control and distribution point for the springs. Had it been possible to calculate the SSCI for this site, the case for this claim could have been more rigorously evaluated here. However, an alternate explanation is also possible: if salt was routinely decanted from briquetage at Midsummer Hill for further transport, providing the taphonomic event leading to the briquetage presence detected by modern archaeologists, but generally passed onwards from Bredon Hill or Conderton Camp in its original container leaving far fewer sherds behind for archaeologists to discover, the actual relationship between the hillforts and Droitwich may be obscured by the available archaeological evidence. Figure 32 graphically represents the relationship between the distance and SSCI values for the Droitwich find sites.

As mentioned above, the SSCI of Stonebridge Cross distinguishes it from other Droitwich finds sites in ceramic-using areas. It is a Middle-to Late Iron Age site with an unusually high SSCI of 0.325 located only 2km from Droitwich near a ford of the River Salwarpe, as shown in Figure 34. Welsh quernstones, Malvernian pottery and 824gm of Droitwich briquetage were all found at this site, suggesting that its location gave it access to a range of goods that were being traded at Droitwich, though it was otherwise considered a fairly ordinary site: a "...middling sized farmstead with a mixed agricultural base" (Miller et al., 2003). It can be compared to the nearby site of Hampton Lovett and Westwood (Napthan et al., 1997), of similar date and position north of the river that also had quernstones, Malvernian ware and more than a kilo of briquetage suggesting a role in the salt exchange network as well. However, the ceramic assemblage was larger overall at this site than at Stonebridge Cross and the SSCI was only 0.016, far lower than at its western neighbour. Although the sites may have functioned differently, with Stonebridge Cross perhaps more focused as a salt exchange point, both may have been dealing with persons prevented for some reason from travelling to the salt springs after journeying from the north and west of Droitwich, perhaps by a ritual taboo or something more prosaic like a flooded river. This possibility seems more likely when these assemblages are compared to that from the Middle Iron Age double ditched enclosure at Stoke Lane, Wychbold, only 3km east of Droitwich but south of the river, where only a single sherd of briquetage was recovered (Jones and Evans, 2006). Unfortunately, the data presentation in the Wychbold report does not allow the calculation of the SSCI for this site for direct comparison.

Lodge Hill Camp, a hillfort with a triple banked enclosure adjacent to the Caerleon Roman legionary fort, has a relatively high SSCI that is calculated based on an assemblage of only 161gm of briquetage and 292gm of pottery, virtually all of which was imported Malvernian ware (Pollard and Morris, 2006, Peterson et al., 2006), suggesting that it may have be part of an exchange network which differed from other hillforts in the region, none of which have yielded Droitwich briquetage. Two further outliers, Credenhill Camp and Twyn-y-Gaer, are also sites from largely aceramic areas, causing them to differ dramatically from the rest of the dataset, which frustrated Morris in her attempts to calculate distance-decay correlations using the SSCI for VCP distribution (1983: 309-310) in the combined ceramic-using and aceramic areas. These three points will be removed in the next analytical iteration to restrict the model to ceramic-using sites. The next iteration will also exclude the find from 1A High Street in Droitwich, where the SSCI of 1.0 might allow for its interpretation as a type of production site, possibly a location where salt was repackaged, rather than a site to which salt was distributed. The final outlier is Stonebridge Cross, Worcestershire, the site discussed above, whose SSCI is so high as to flag it as distinctly different from other sites in the dataset. It will, however, be used in the next calculation. When the four atypical

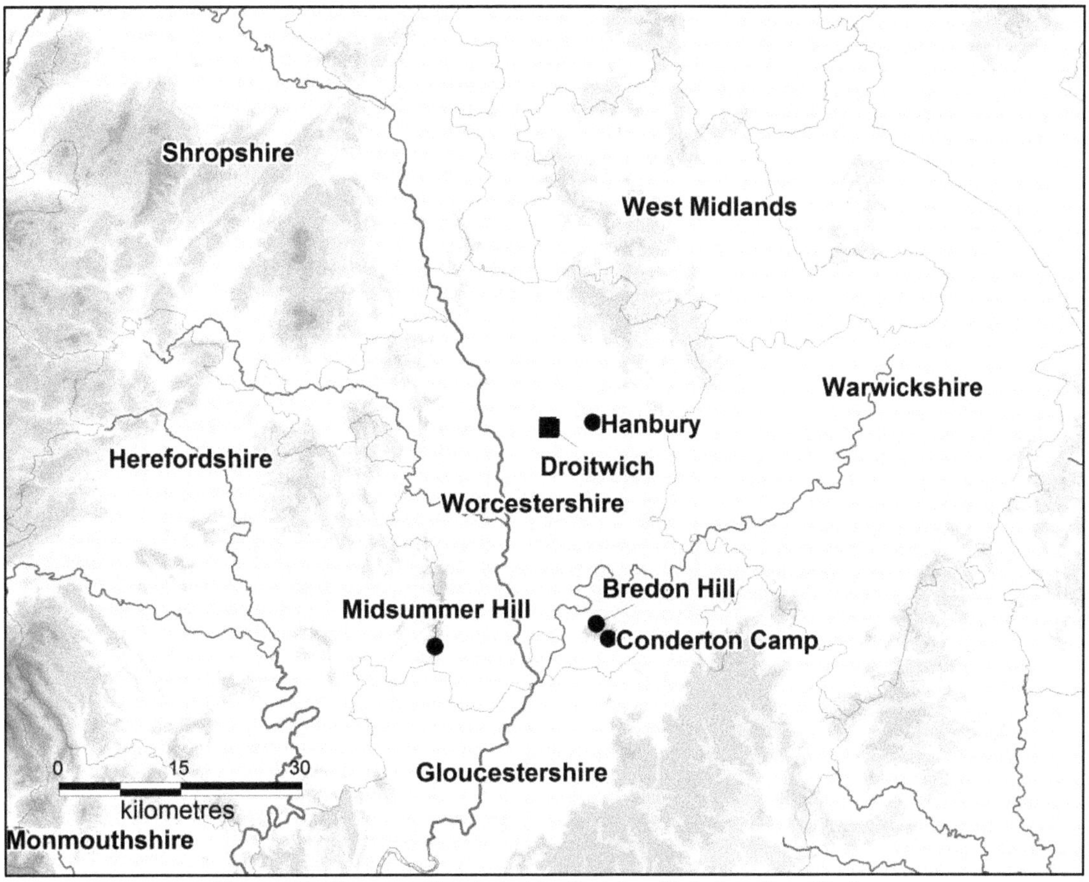

Figure 31. Closest hillforts to Droitwich with briquetage finds.
Underlying map © Crown Copyright Ordinance Survey.

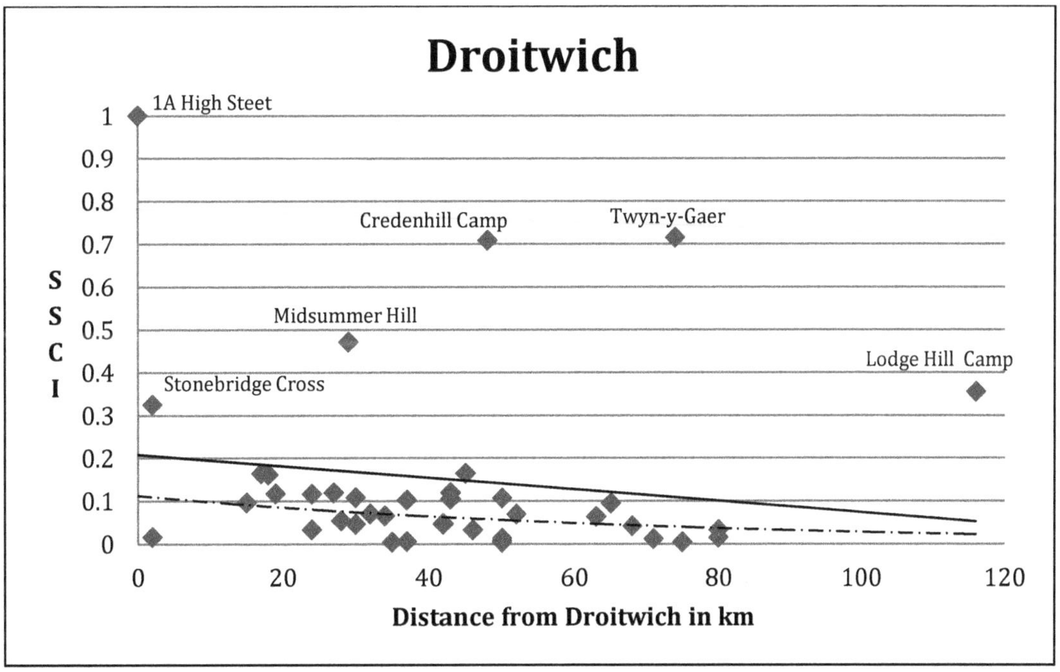

Figure 32. SSCI and find distance with exponential (dashed) and straight (solid) trendlines.

sites are removed from the dataset, the remaining 37 data points generate the graph shown in Figure 33. With the exponential best fit line now more closely approximating a straight line model of the distance-decay function normally used to describe long-distance trade, it can be shown that this model is applicable to Droitwich briquetage finds in ceramic-using areas.

Even with the additional VCP find sites for which the SSCI has been calculated, the sample size remains problematically small for robust statistical analysis. Figure 35 below shows the SSCI dataset for 24 Cheshire VCP finds, graphically illustrating the problems encountered when attempting to blend the data from aceramic and ceramic-using areas into a single chart. Some of the plotted sites fall far off both the exponential and straight best fit trendlines and a large gap exists between the two lines. Morris used only 16 sites in her regression analysis of Cheshire VCP distribution (1983: 325-328). However, if the dataset is segregated into ceramic-using and aceramic sites, as shown in Figures 36 and 37, the exponential trendline in both cases becomes similar to the trends shown for Droitwich, illustrative of down-the-line trade, suggesting that this model may be applicable to at least some of the distribution of Cheshire salt as well as that from Droitwich. The limited size of the separated datasets, however, particularly with respect to aceramic areas with VCP, makes the results far less reliable than those obtained for Droitwich.

In a simple graphical presentation of the down-the-line trade model, there would be a straight line negative relationship between distance and the amount of briquetage found, which appears to be the relationship between the SSCI and distance shown in the charts above for the revised Droitwich dataset in Figure 33 and for the Cheshire VCP distribution in ceramic-using areas in Figure 36. The VCP distribution dataset from aceramic areas in Figure 37 is also trending towards a negative relationship, although the number of sites is too small for a robust conclusion using this method of analysis.

Although the charts suggest that down-the-line models may fit both Droitwich and Cheshire salt distribution patterns, the trade areas differ in size and shape as there is some distortion of the trade distance relationship caused by the apparent mixture of both land and water transport in varying proportions between the two production centres. A study of the maps in Figures 22, 23 and 38 strongly suggests that both water and land transport routes were in use as part of the salt distribution process, a point made by Morris (1983: 297) and since restated by others (e.g. Matthews, 2002).

The use of water to transport bulk commodities during the Iron Age should not be a surprise in a period when evidence for paved roads and bridges is lacking and when there is earlier archaeological evidence for the use of boats. A number of Bronze Age boats suitable for river transport have been found in Britain (McGrail, 1990), suggesting that a well developed riverine transport network already existed in that period. Finds such as the Hasholme logboat (Millett and McGrail, 1987), radiocarbon and dendrochronologically dated to the Middle Iron Age, demonstrate that substantial cargo carrying vessels existed in this period as well. At 12.58m in length, 1.5m wide at the stern and with an estimated side height of 1m, it was estimated that it could carry a crew of 5 and 8,602kg of cargo (ibid.: 134). This, however, was probably an extraordinary watercraft, having been made from a single 600 to 820 year old oak tree, and most boats were probably smaller, carrying substantially less cargo. Most of the boats found (McGrail, 1990) seem comparable to twentieth century South American dugout canoes, ranging from 8m to 10m in length and 1m to 1.2m wide, which were reported as being able to transport loads weighing up to a ton (approximately 907kg) a distance of 20km upstream or 40km downstream in a single day (Drennan, 1984), so even these smaller craft would facilitate rapid movement of substantial amounts of bulk commodities. For movement along the ocean coast, the coracle was a better-suited vessel than a logboat. Citing early historic sources, Henderson (2007) claims that this type of craft could achieve a speed of 5 knots and travel about 35 miles, approximately 65km, on a favourable tide with a load of several tons.

Although water transport had advantages for bulk commodities, such routes have obvious limitations: at some point, the salt would need to move across the landscape. Land transport could have been achieved through human labour, pack animals or with carts. Late Bronze Age leather pouches on wooden frames found at Hallstatt (Wells, 1981) suggest a possible means for an individual to transport a quantity of salt across the landscape. Drennan (1984: 105) calculated that a 30kg load could be moved 36km over even terrain in 8 hours. Roughly 45% of the Droitwich briquetage finds are within this theoretical 8 hour movement distance, while approximately 22% of the Cheshire sites are this close to Middlewich, statistics which imply a greater use of water transport for Cheshire salt, a theory supported by the distribution of the find locations in this dataset. The use of pack animals and/or horse or oxen-drawn carts would have increased the volume of material per trip, although not necessarily the speed of movement. Considering the state of the road network and the need to ford streams and rivers, overland trips would have been easiest during the drier summer and early autumn months, although for most people the trips would have needed to be coordinated with agricultural work and possibly salt production cycles. Alternatively, it is possible that some Iron Age individuals earned a living as itinerant salt traders during the Iron Age, freeing them from concerns with such agricultural tasks.

The probability of two-way exchange for salt and the presence of imported quernstones and pottery in the Droitwich and Cheshire salt distribution areas have already been mentioned. Archaeologists know about these particular imports because they are archaeologically

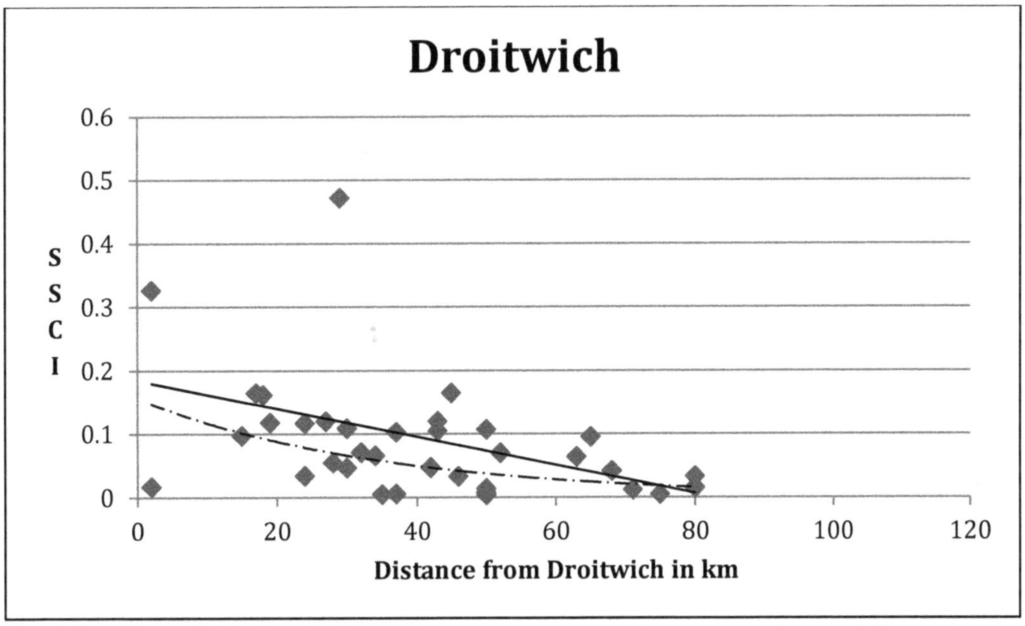

Figure 33. Purged dataset with exponential (dashed) trendline and straight (solid) trendlines.

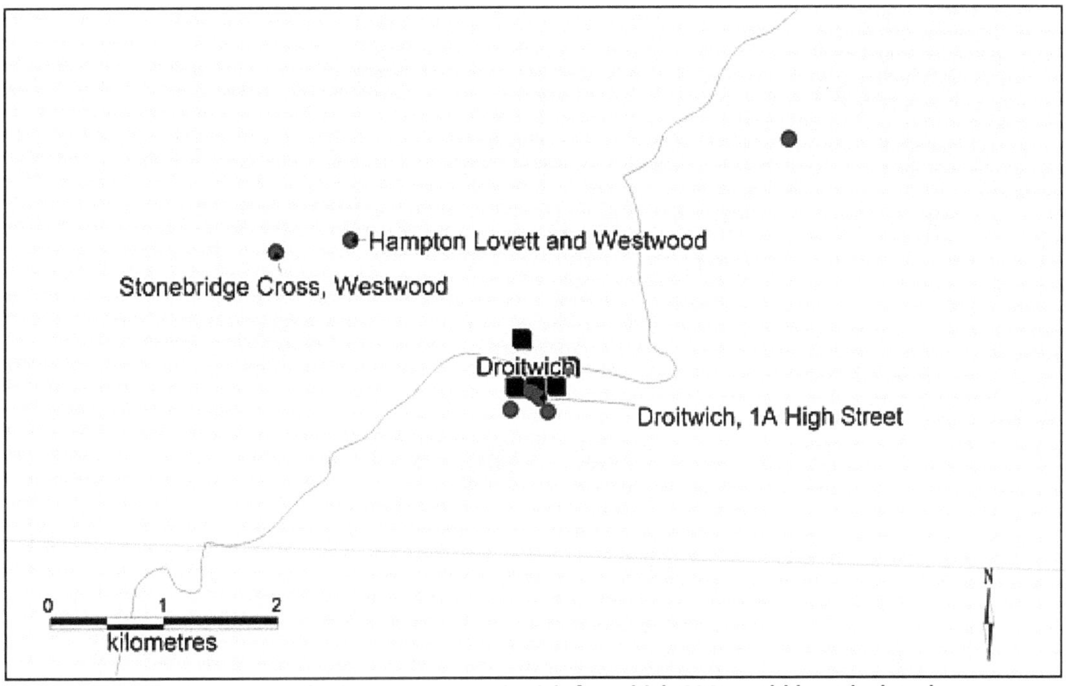

Figure 34. Briquetage find sites near Droitwich for which SSCI could be calculated.
Squares are Iron Age salt production sites, while circles are briquetage find sites. Underlying map © Crown Copyright Ordinance Survey.

visible materials whose origins can be determined through detailed study. These artefacts undoubtedly circulated alongside many other types of archaeologically invisible goods, including grain, other foodstuffs, raw hides, finished leather, cloth and live animals. For example, the Hasholme log boat was found to be carrying joints of beef, a sheep skull and three horse bones, which were originally interpreted as food for the trip (Millett and McGrail, 1987) though it is possible that some of this could have been salt preserved meat intended for trade. Fuels, including charcoal and firewood, may have been traded into areas with depleted forests, a type of trade documented in the Anglo-Saxon period at Droitwich, when cartloads of firewood were traded for salt (Hooke, 1985). Charcoal would have been a particularly desirable bulk commodity as a long distance trade item since it is relatively light weight and provides as much energy as wood weighing 6-7 times as much (Gale, 2001). With just these few examples, it becomes apparent that the exchange networks in the Iron Age were

Site Name	Site Description	Distance from Source in km	SSCI
Blackstone	enclosure	78	0.0006
Conderton Camp	hillfort	135	<0.001
Midsummer Hill	hillfort	113	0.001
Midsummer Hill site 2	hillfort	113	0.001
Kenchester	enclosure	112	0.003
The Wrekin	hillfort	44	0.009
Croft Ambrey	hillfort	87	0.018
Foxcovert Farm, Aston-on-Trent	enclosures	74	0.018
Credenhill Camp	hillfort	110	0.019
Whitemoor Haye Quarry	pit alignment	103	0.024
Dinedor	hillfort	116	0.044
Bury Walls	hillfort	27	0.048
Bromfield	enclosure	78	0.059
Coventry Road, Hinckley	enclosure	89	0.060
Twyn-y-Gaer	hillfort	135	0.063
Meriden Quarry, Solihull	enclosure	87	0.083
Beeston Castle	hillfort	14	0.088
Wellington Quarry	ditched enclosure	97	0.186
Old Oswestry	hillfort	42	0.284
Sharpstones Hill site A, Weeping Cross	enclosed farmstead	44	0.419
Fisherwick 1974-1975	enclosures	68	0.560
Great Woolden Hall Farm, Glazebrook	enclosure	50	0.667
The Breiddin	hillfort	52	0.787
The Berth	hillfort	36	0.845

Table 8. Specific Salt Container Index for Cheshire VCP find sites ordered by increasing SSCI. Distance from source measured as a straight line; actual distance travelled between source and find spot may have been considerably longer.

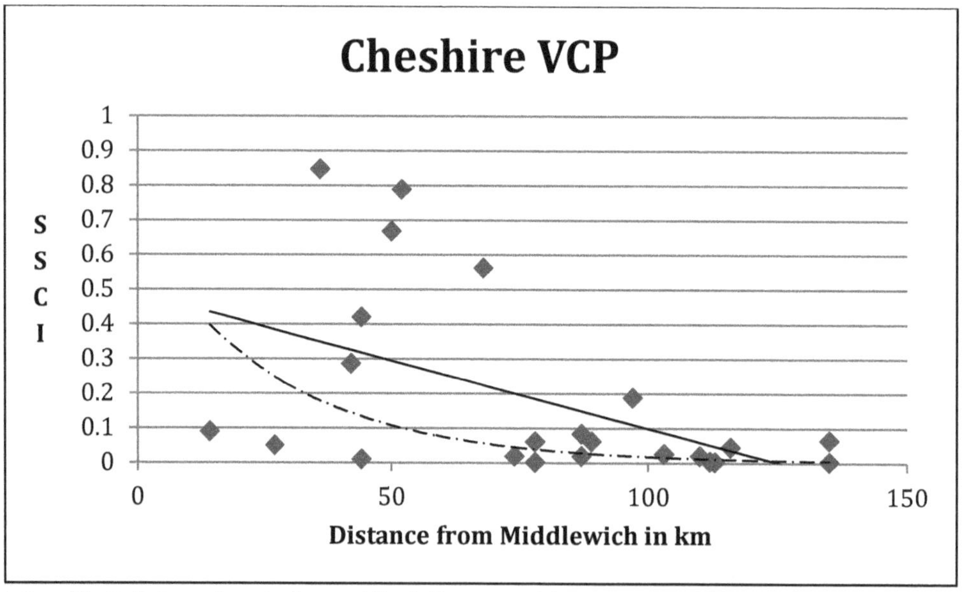

Figure 35. Specific Salt Container Index and find distance with exponential (dashed) and straight trendlines.

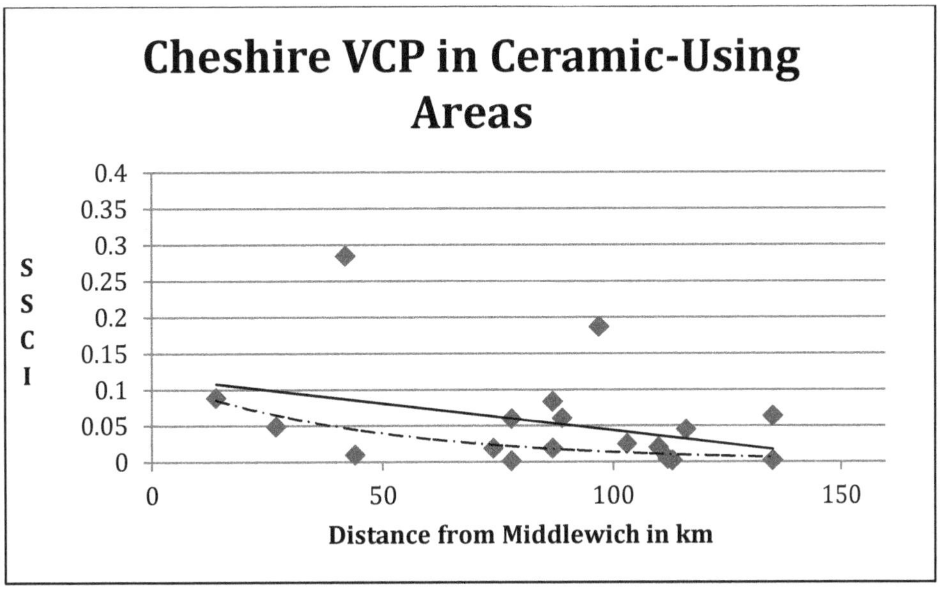

Figure 36. Selected dataset with SSCI and distance with exponential (dashed) trendline approximating straight best fit line.

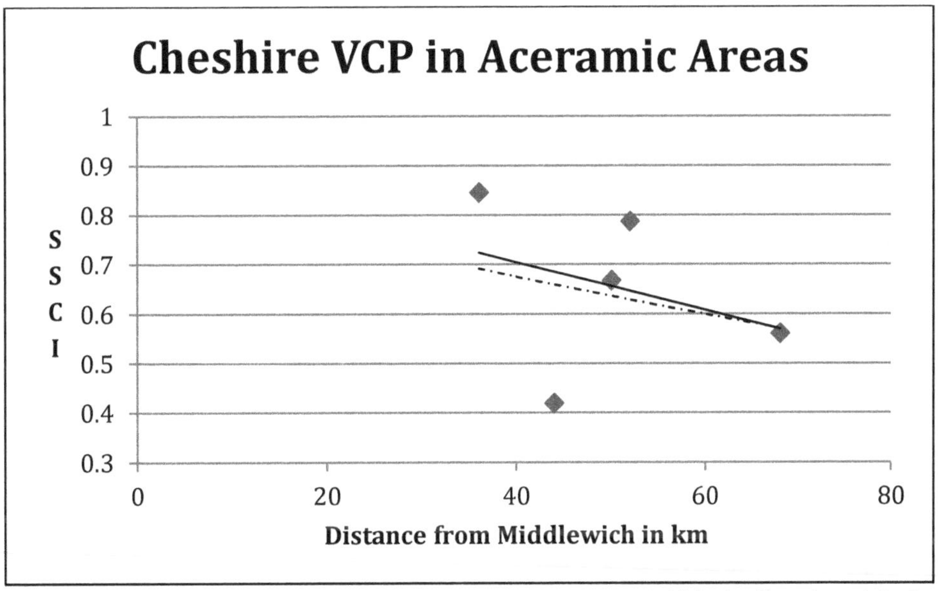

Figure 37. Selected data subset with SSCI and distance with exponential (dashed) and straight best fit trendlines.

undoubtedly far more complex than the quern/pot/salt evidence demonstrates.

4.6 Distribution and the absence of evidence: Essex and the south coast

The accumulated information about the distribution of salt from Droitwich and Cheshire tells us that most Iron Age Britons in the Upper Thames Valley, the Welsh Marches, the Midlands and the northwest of England had access to salt, and sometimes a choice of suppliers. They may have had a choice of salt container as well, choosing between briquetage and an alternative container, depending upon the source of supply and the intended usage of the salt. Evidence of long distance trade for vital commodities and even some luxury goods, such as Malvernian pottery, appears repeatedly in these distribution areas. This section draws a comparison between the Droitwich and Cheshire information just presented and the evidence for Iron Age salt distribution from sea salt supply areas, focusing particularly on Essex and the south coast.

As discussed above, there is a substantial body of evidence to demonstrate the distribution of salt in briquetage from both Droitwich and Cheshire inland salt springs. The case is reversed, however, for the coastal sea salt production areas as there is little inland Iron Age evidence of salt distribution in briquetage from any of the coastal production sites. The large number of coastal production sites stands in sharp contrast to the limited

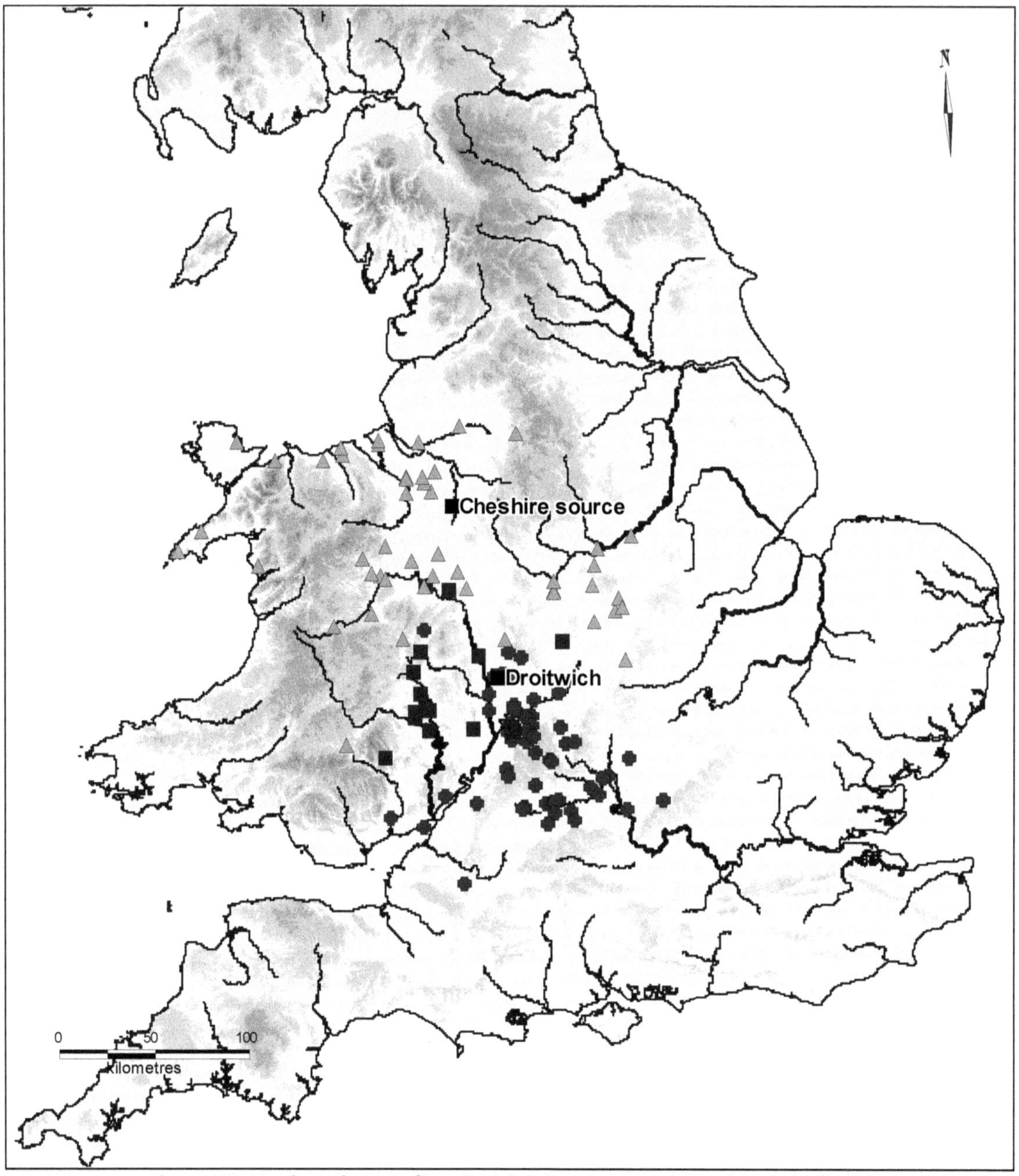

Figure 38. Unphased map of Droitwich briquetage and Cheshire VCP finds.
Dots indicate Droitwich finds, triangles Cheshire VCP finds and squares are sites with both materials present. Underlying map © Crown Copyright Ordinance Survey.

number of inland finds. This evidentiary dichotomy will be made clear as the nature and extent of the salt distribution evidence from Essex is considered.

Figure 39 demonstrates that although more than 250 Red Hill salt production sites containing large quantities of briquetage covering the time span through the Bronze Age to the early Romano-British period have been documented, there have been only 31 prehistoric finds of Essex briquetage since the early 1970s at sites that were not thought to be associated with salt production. Basic information on these non-production find sites is listed in Appendix 5 with further information available in the Appendix 8 database. The list of well-documented but undated inland briquetage finds adds only a further handful of sites to the list. One indisputable fact is that at least one piece of briquetage did manage to reach Baldock, Hertfordshire (Stead and Rigby, 1986), more than 50km from the nearest Essex salt production source, a distance well within the range known for the distributions of Cheshire and Droitwich briquetage.

Each of these inland finds has been subject to intense study as they are atypical of the Essex region and their existence has caused much debate (e.g. Sealey, 1995, Barford, 2000). However, the debate has seemed to hinge in large part on attempts to find a single explanation for all of the finds, such as the use of briquetage for salt

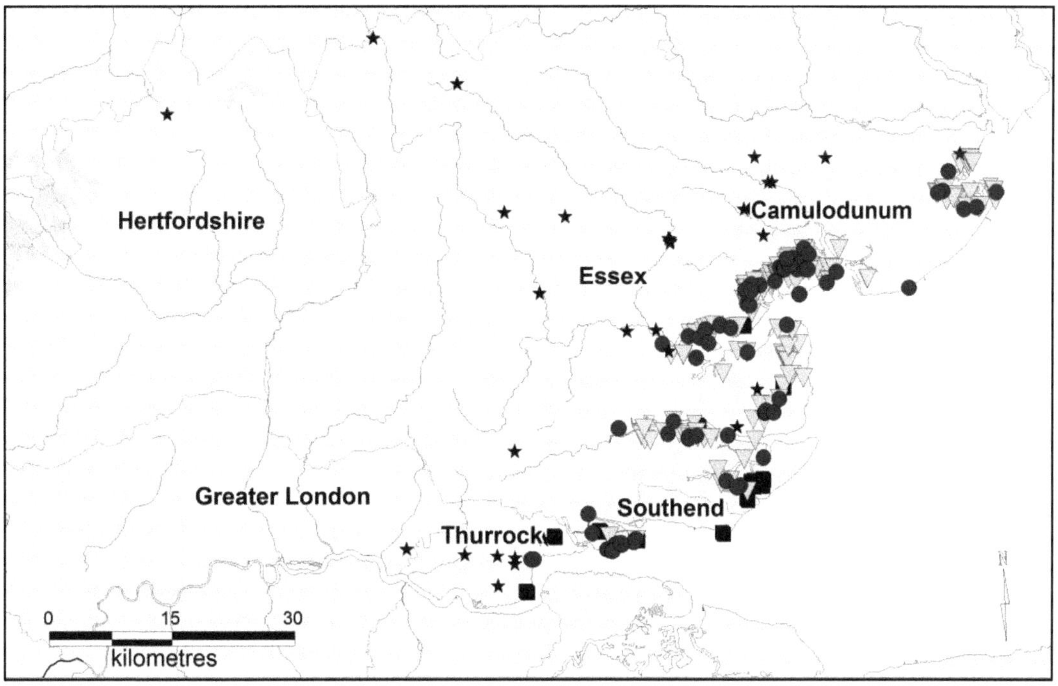

Figure 39. Essex Bronze Age, Iron Age and Romano-British salt production sites using briquetage and inland briquetage finds.

Bronze and Iron Age sites shown as dots, squares represent Romano-British production, triangles are undated Red Hills or briquetage finds, stars are prehistoric inland briquetage find sites. Inland find site details are shown in Appendix 5. Underlying map © Crown Copyright Ordinance Survey.

transport (Rodwell, 1988); accidental inclusion of briquetage with salt during shipment (Sealey, 1995) or the use of briquetage as salt licks for domestic animals (Barford, 2000). Each of these explanations seems to have significant problems. Rodwell (1988: 81), suggests that the Kelvedon finds might occur because the occupants of that site were salt producers returning with the fruit of their labour, an interpretation applied earlier to the Mucking inland briquetage finds (Jones, 1977), an explanation that works for this site but would be difficult to apply to all inland Essex briquetage finds. Sealey's explanation, though sounding plausible in general, suggests that the separation of salt crystals from briquetage in Essex, as a process, reached a quality level that would leave the greatest modern quality control experts in awe (e.g. Ishikawa, 1983). Barford's explanation seems particularly unlikely to be sufficient to this author, as it suggests that briquetage should be ubiquitous at prehistoric Essex sites where evidence for domestic animal keeping exists, which is certainly not the case. All of these explanations also fail to elucidate how briquetage can reach Baldock, Hertfordshire, while missing all the intervening settlements between that *oppidum* and the coastal salt production areas. Ultimately, it is the narrowness of these approaches which makes them problematic, when perhaps multiple explanations are more appropriate given the long date range covered by the finds, a period in excess of 1,000 years, and the variety of site types included in the group.

Some examples will demonstrate why a single explanation is inappropriate for this body of evidence. The Late Bronze Age Corringham finds were recovered among spoil from a 60cm square hole along with 10 pot sherds (Barford, 1984-85). The briquetage and pot forms and fabrics were comparable to those found nearby at Mucking North Ring, another Late Bronze Age site where salt production briquetage had been found, with both sites being on gravel terraces above the Thames. Mucking was considered to be a possible salt drying site (Jones, 1978), but the nature of activity at Corringham could not be discerned due to the extremely limited excavation carried out at the site. The similarities between the sites suggest the possibility that a larger excavation might have found evidence to suggest that Corringham was also a salt drying site. Contrast this to the presence of briquetage at Early Iron Age Asheldham Camp hillfort (Sealey, 1995), which might be interpreted as evidence it once served as the heart of a central place distribution system, as discussed above with respect to Droitwich and Cheshire salt. Neither of these sites is directly comparable to the briquetage find from two mortuary enclosure ditches at Late Iron Age or Early Romano-British Stanway (Crummy, 2007), where the briquetage is interpreted as relating to the use of salt in some fashion during the funerary rituals at the site. Most of the possible explanations given for inland Essex briquetage put forward to date, other than that proposed for the Stanway find, have been functionalist in nature, studiously avoiding any phenomenological aspects of human behaviour as part of the discourse. The Stanway find, however, serves to remind archaeologists that the social and ritual aspects of behaviours should be considered as applicable to the other sites as well.

One possible explanation which could be put forward to

explain much of the atypical limited distribution of Essex briquetage derives from the work of John Chapman (2000), who has considered the relationship between the distribution of objects within trade networks and the reinforcement of social linkages between the trading partners. He reasoned that if such a link could be based upon the exchange of complete objects, it would be a small intellectual leap to see the same relationship existing as well in partial objects (ibid.: 37). Applying his model to this case, there would be no accidental inclusion of briquetage with Essex salt; its presence would be intentional and symbolic, reminding both the provider of the salt and the recipient of these meaningful social linkages. The map in Figure 39 shows that many of the inland finds occur on or near rivers that could have been the primary transport arteries for salt produced on the coast, allowing for the possibility that many of the finds are indicative of a direct social link between the salt producers and the people at the find site, with the salt possibly being decanted before it moved elsewhere. It is also possible that no decanting was necessary, as the symbolic sherd could have arrived in an archaeologically invisible container of salt. This would be consistent with the very small quantities of material found at some of these sites, making them more likely to reflect a deliberate activity to enhance relationships within a symbolic context rather than representing residual evidence of transport of large quantities of salt. This model also addresses the problem of how accidental contamination of salt with briquetage would have occurred when the salt crystals would have been removed from the briquetage brine heating pan to a presumably clean, possibly organic, container for drying.

The author believes a combination of several of these models is needed to explain the range of inland Essex briquetage finds that occur at sites active during the first millennium BC and/or early first century AD. In some instances, such as the Corringham case, briquetage may have been used at the site in some later stage of the salt production process, such as salt crystal drying, or the site may itself have been the location of briquetage production, an interpretation proposed by Jones for Mucking North Ring (1977), although transport of fragile, heavy, empty briquetage containers any significant distance seems improbable if they could be made closer to the saltern. While impossible to discount human fallibility leading to accidental contamination of salt with briquetage, the author thinks this explanation unlikely, given that salt would have been a highly valued commodity and hence great care would have been taken with its preparation. Further, the likelihood of archaeologists finding such rarely occurring "error" briquetage inclusions would also be quite low, as the combined probability of two events, each of which has a low probability on its own, is exceedingly low. It would, however, be difficult to tease this specific explanation, rather than any other alternatives, out of the archaeological evidence from any site if such a processing error had occurred and been found during an excavation. The author would suggest, therefore, that the majority of these inland finds may relate to the multiple types of social linkages envisioned by Chapman, which would have been symbolised by briquetage sherds moving across the landscape.

The south coast

The scarcity of Essex inland briquetage find sites is similar to the limited number of inland south coast briquetage finds, yet there are distinct differences to be noted as well. As in Essex, some of the finds near the coast may have been associated with secondary processing to further purify the fresh salt crystals formed near the seashore or remove dried salt cakes to other containers for shipment. More than sixty years ago Calkin (1948) asserted that a large briquetage find at Hobarrow, 100 feet (30.48m) above sea level, was associated with such an activity. A similar explanation of secondary processing was also offered by Bradley (1975), though he also discussed the potential for trade of salt in briquetage containers. As shown in the map in Figure 40, while there is only very limited evidence of inland movement of salt in briquetage in Dorset or Sussex, a route possibly involving both land and water transport methods having some similarities to the path of the modern A34, connects coastal Hampshire and Oxfordshire. Salt was probably not the only link between the coast and these inland areas; for example, objects fashioned of Kimmeridge shale, found near the Purbeck area, that were produced between the sixth century BC and the Roman period, are found widely within 80km of that location (Cunliffe, 2005) and there are distinct similarities between Figure 40 and Cunliffe's maps of Late Iron Age Atrebatic coin and pottery distributions (Cunliffe, 2005). The similarity with the Late Iron Age coin distribution map suggests a possibility that coins may have been involved in the salt exchange process although there is no firm evidence that monetary exchange was part of the salt distribution process in this period and their presence may have related to wealth displays or use as personal adornment (Van Arsdell, 1989). The pottery distribution correspondence raises the possibility of a regional trade network where Atrebatic pottery might have served as salt containers, as has been speculated for Malvernian pottery and Droitwich salt (Morris, 2008 pers. comm.).

The existence of a south coast salt transport route echoing a modern roadway is reminiscent of the link between Droitwich and Oxfordshire briquetage finds occurring along the route of the modern A44 noted earlier. It is possible that Oxfordshire was a region where salt packed in briquetage was a preferred choice over that received in other containers, making it worth the extra effort to transport the bulky briquetage containers these distances. In one instance, south coast briquetage has been found beyond even distant Oxfordshire, at Bourton-on-the-Water, Gloucestershire (Nichols, 2001). Referring again to Chapman's theory (2000), the single piece of Dorset briquetage found nearly 153km from Southampton via the modern road network, can reasonably be interpreted as having acted as a material reflection of a social connection marked with a gift of salt rather than as evidence the existence of an otherwise unknown major long-distance salt exchange network between the south coast and southern Gloucestershire.

A further distinction between the south coast and Essex inland briquetage finds comes from differences in the nature of the sites where the finds have been made. In the Essex case, of the 34 sites listed in Appendix 5, only Asheldham Camp hillfort, the Baldock *oppidum* and the Stanway cemetery would typically be defined as types of central places, though they have vastly different functions. Whether Stanway would have served as a location for distribution of salt when it was an active cemetery is a matter for speculation. Such an activity, had it occurred, would imply a view of the place in the Late Iron Age and the Early Roman period quite dissonant from the modern view of cemeteries as well as being at variance with the current archaeological understanding of these sites during that period. A more traditional view of the Stanway evidence would be supportive of an interpretation of the site as a final destination for salt rather than its use as a site destined to pass salt onwards. In comparison, as shown in Appendix 6, 9 of the 34 inland south coast finds (26%) come from either hillforts or an *oppidum*, all archetypical central places from which salt could be distributed (Cunliffe, 1995). It could be argued that the difference derives purely from the general absence of hillforts in Essex (e.g. Cunliffe, 2005: Figure 15.1, 348) , or that it is an artefact of the large number of hillfort excavations which have been carried out in Wessex, though this latter suggestion does not explain the general absence of briquetage from southern hillforts outside of modern Hampshire. It seems more likely that the difference in the find site pattern is representative of an authentic difference in the social mechanism for salt distribution, with centralised distribution of salt occurring at some hillforts and possibly at the *oppidum*. Other explanations, such as the presence of briquetage at these central places as part of the residue of the conduct of periodic rituals involving salt there are equally plausible, though again they suggest a difference between the rituals in Wessex and those in Essex, which, had they required salt, did not generally require that it come packaged in a briquetage container.

Another possible distinction noted between the inland Essex finds and the south coast material is the distribution overlap between south coast briquetage and Droitwich briquetage which occurs in the area south of the Thames and to the north of the Berkshire Downs, as shown in Figure 41. Knight's Farm (Bradley et al., 1980) is a Late Bronze Age site, with Basingstoke (Morris, 1994a) producing evidence of a long-term trade in salt with the south coast starting in the Middle Iron Age; the Vineyard and Segsbury Camp are late Middle Iron Age sites (Lambrick and Robinson, 2009), while the Abingdon Town Centre site is Late Iron Age or Early Romano-British and seems to relate to the *oppidum* known to exist under central Abingdon (Allen, 2000). Collectively these sites suggest long term movement of south coast salt into this region in parallel with that arriving from Droitwich, though perhaps in lesser amounts. This perceived difference with the Essex evidence, however, should not be taken to imply that there is no potential for dual sourcing of salt from either Kent to the south or the Fens to the north within what one may construe as the likely Essex salt distribution area, only that there is no briquetage evidence to support such a theory of multiple sourcing at this time.

4.7 Theoretical salt: the blank spaces on the Iron Age distribution maps

To this point, the discussion has focused upon the sites where briquetage has been found, either in conjunction with salt production or distribution. However, as has been pointed out above, these finds are not universally distributed across all Iron Age sites in Britain, and large swathes of the country, as shown in the map in Figure 42 and detailed in the Figure 43 bar chart, have, as yet, failed to produce any briquetage finds whatsoever. Even if an attempt is made to develop hypothetical distribution ranges for the various production sites, as attempted by Morris (2001b) for the later Iron Age, large areas of Britain appear to have no salt available to them. The southwest of England, north Sussex, and western Kent, the middle Thames Valley, and most of Hertfordshire, south western Wales, the east Midlands, north western England and all of Scotland would appear, on the briquetage evidence alone, to have been salt-free zones.

Yet if salt were vital for food preservation, a key assertion used by archaeologists to explain briquetage distribution patterns (e.g. Haselgrove and Pope, 2007), then it is difficult to imagine such large swathes of Iron Age Britain going without it. There are at least two possibilities, the first of which is that our knowledge of Iron Age salt production and distribution with briquetage is still quite incomplete. For example, recent surveys of the Somerset Levels have now located 5 salterns which may have begun production using briquetage in the Late Iron Age (Rippon, 2000, Rippon, 2006, Moore, 2006) and further finds may change the map sufficiently to make the Levels a more important Iron Age salt source than it appears in the map in Figure 42. The second possibility is that these regions were involved in networks producing and distributing salt in the Iron Age in archaeologically invisible ways. For example, the find in the Mendips of an Iron Age wide-mouthed carinated bowl of a type known previously only from Lincolnshire and Northamptonshire (Todd, 2001) raises the possibility of its use for salt distribution from Lincolnshire to Northamptonshire and beyond. This would be a scenario similar to the speculation that Malvernian ware might have been filled with Droitwich salt as it travelled south of this salt production source. Cornwall, a Roman salt production area, has produced no evidence of Iron Age salt production, yet there is no physical barrier to its production there at the earlier date. Production on any of the suitable coastal areas, in fact, using archaeologically invisible means should not be discounted out of hand during prehistory.

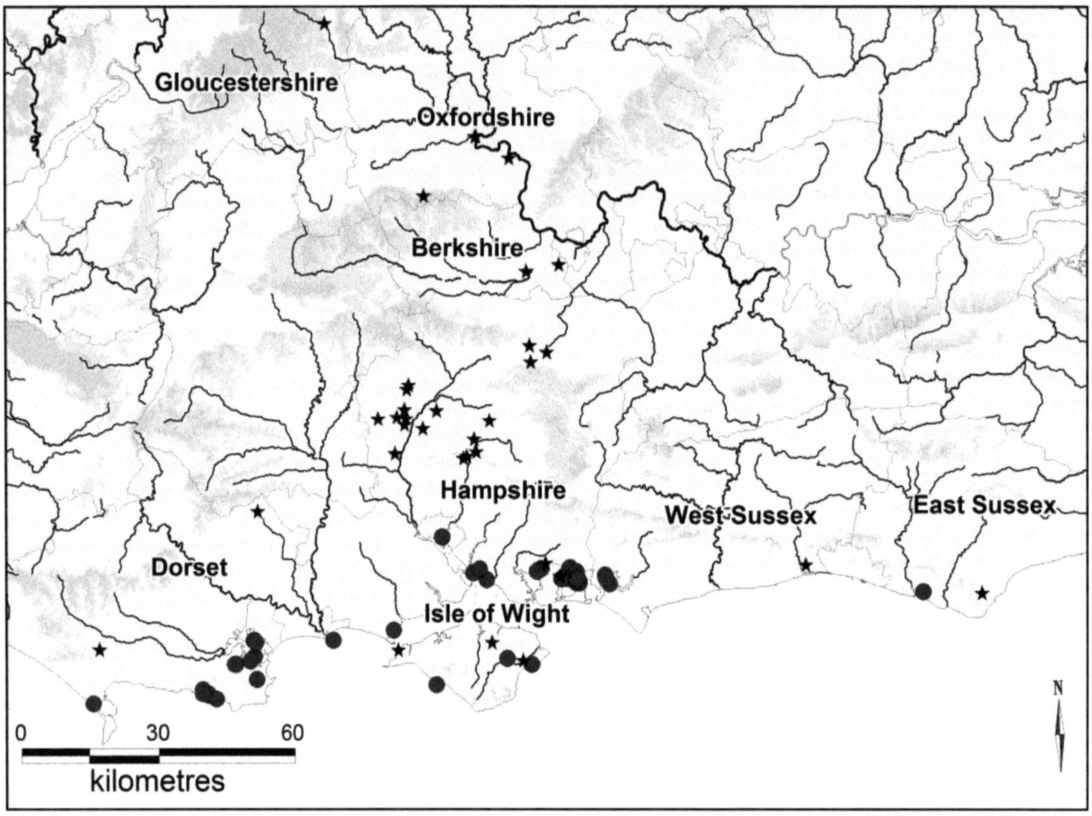

Figure 40. Inland briquetage finds from south coast production sites.
Production sites are shown as dots, inland finds as stars. Underlying map © Crown Copyright Ordinance Survey.

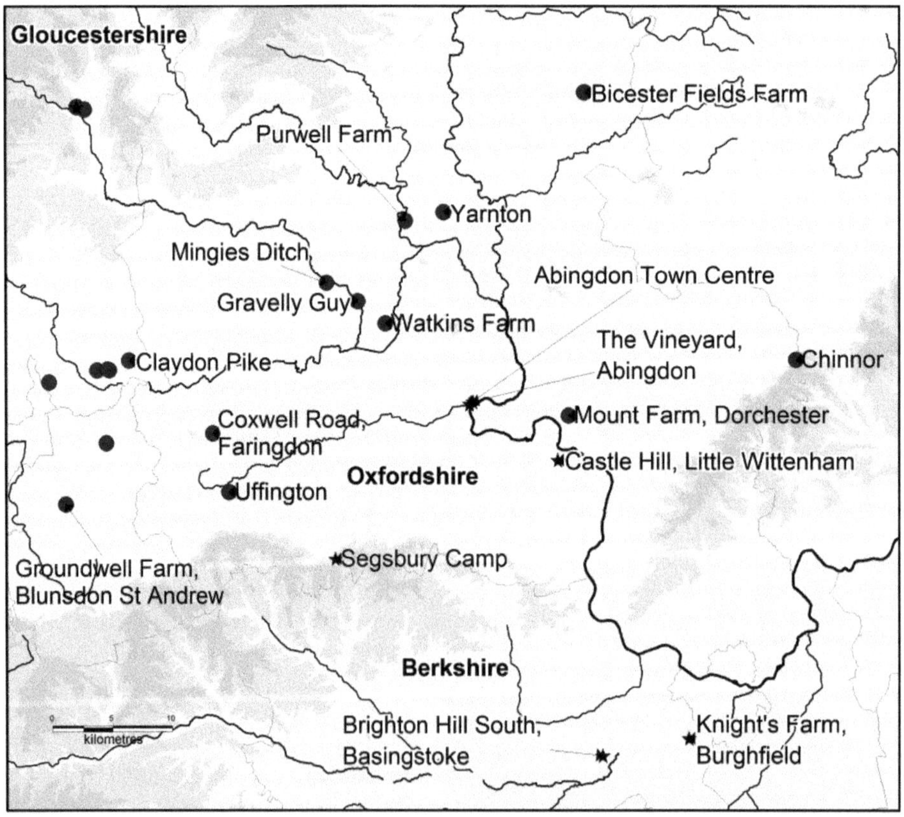

Figure 41. Map of Droitwich and south coast briquetage finds in Oxfordshire and adjacent counties.
Droitwich finds shown as dots, south coast briquetage shown as stars. Underlying map © Crown Copyright Ordinance Survey.

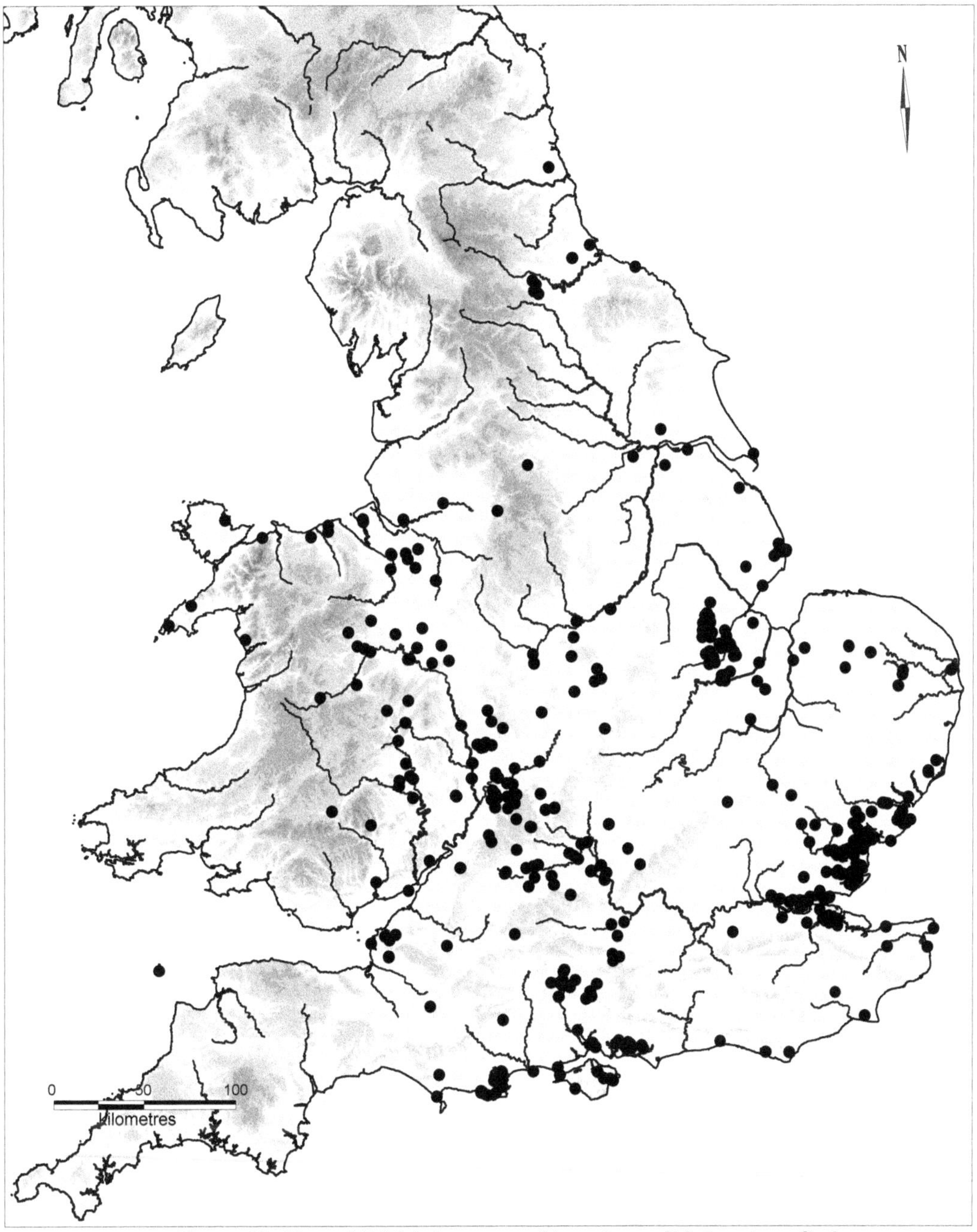

Figure 42. Map of British Bronze and Iron Age salt briquetage finds.
Underlying map © Crown Copyright Ordinance Survey.

A third possible solution to some of the apparent problem created by the lack of an archaeologically visible salt supply network might be solved by looking beyond Britain and positing that south western England, south western Wales, north western England and Scotland met their salt requirements through trade with Armorica (Cunliffe, 2004) along the coastal seaways. Salt production is known from Morbihan and Nalliers starting in the Neolithic period (Nenquin, 1961, Gouletquer and Daire, 1994), suggesting that Bay of Biscay salt could have been used as a trade good in exchange for the British products, such as those enumerated by Strabo, "grain, cattle, gold, silver and iron....also hides, and slaves and dogs that are by nature suited to the purpose of the chase" (*Geog.*, IV, v, 2). The links between the south-west peninsula and Armorica are clearly evidenced between the Middle Iron Age and the first century BC through pottery styles (Cunliffe, 2005). The availability of salt through such trade networks would have tended to suppress the need for local salt production and distribution, potentially contributing to the lack of evidence now observed archaeologically.

On the other side of the country, the London region and the northern Home Counties probably received salt from Essex and Kent via the River Thames and its tributaries in the previously discussed archaeologically invisible

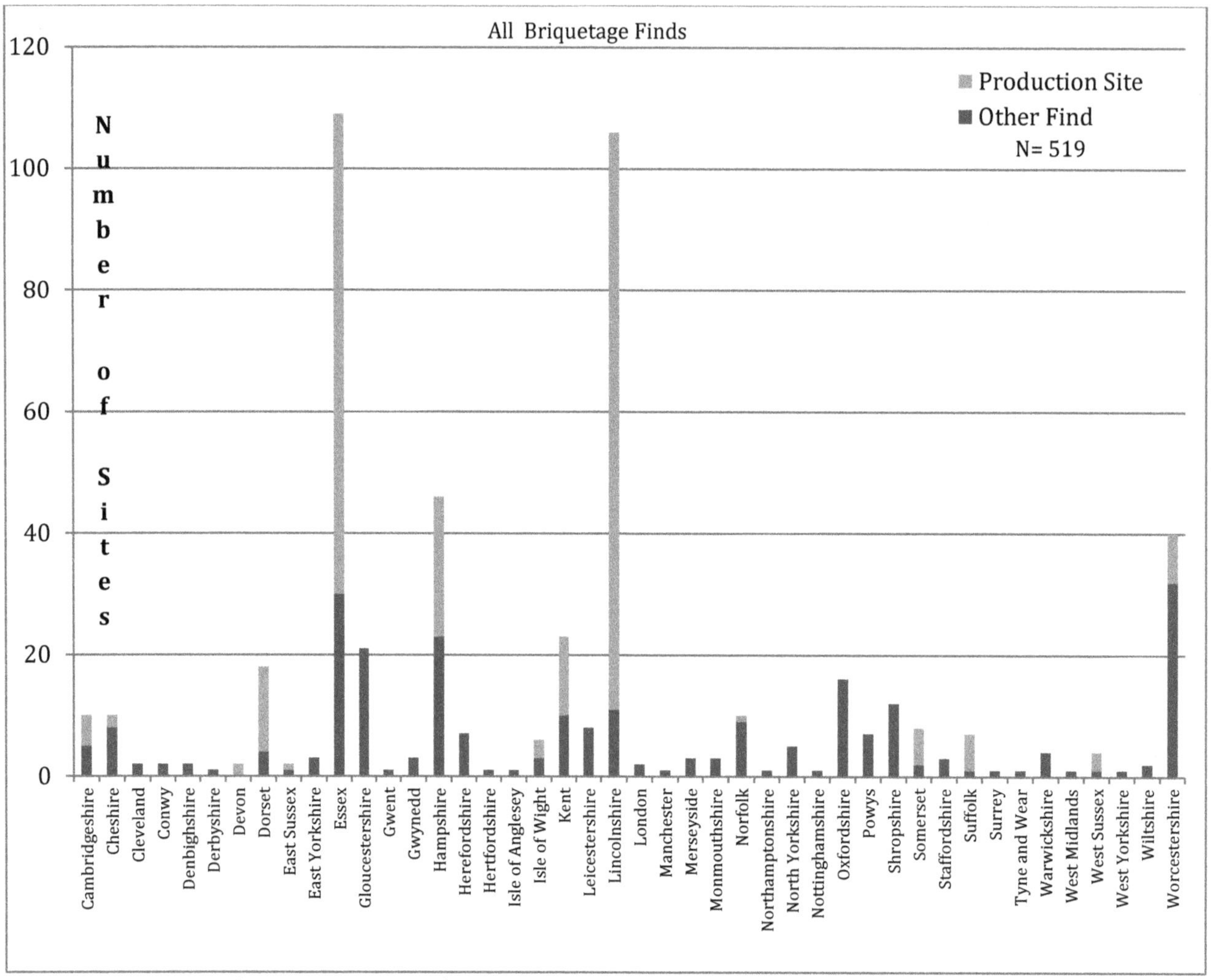

Figure 43. Bronze and Iron Age sites by county as shown in Figure 42 map.

containers, while the southern Home Counties had these potential sources and the south coast to draw upon. The limited number of finds of Cheshire briquetage in the east Midlands discussed above now suggests that region may have traded for salt with Cheshire, Lincolnshire or even Essex. Similarly, in north eastern England, as predicted by Haselgrove in 1982 (per Fitts et al., 1999), work by Willis (1994, 1995) demonstrated a prehistoric salt production industry existed in that region and recent finds of briquetage in the area such as that from Bronze Age Easington (e.g. Faulkner, 2008) suggested it had an early start. There is little of Britain, therefore, for which a reasonable Iron Age salt source cannot be imagined, even if there is limited archaeological evidence for this trade.

4.8 Iron Age salt distribution in Britain

This chapter has focused on some of the briquetage evidence for a domestic salt trade in Britain during the Iron Age. Evidence to date suggests that the use of briquetage for distribution was most common for the inland salt springs at Droitwich and Cheshire and infrequent for the coastal salt producers discussed here and those at the major salt production area in the Lincolnshire Fens (Lane and Morris, 2001). Relatively rare finds of briquetage inland from the Essex and south coast production areas are likely to show the existence of social links and salt trading networks which would otherwise be archaeologically invisible, and are suggestive of a much larger volume of exchange than that which is solely demonstrated by this limited evidence. Salt was only one commodity moving in regional trade networks that included quernstones, iron, ceramics and other bulk goods, as well as luxury items, transported along water and land routes which were already ancient by the start of the Iron Age. Archaeological and historical evidence suggest that these trade routes often continued through the Roman, Saxon and medieval periods and may even have laid the basis for some of the modern road network. In the following chapter, the perspective on salt distribution will shift from a focus at the total distribution network level to one of detailed distribution case studies for Droitwich, Essex and the south coast.

Chapter 5
Another view of salt distribution within the Iron Age

5.1 Introduction

In the prior chapters, the discussion had focused largely on consideration of briquetage finds at a large, regional scale, but it is desirable to look more closely at the finds and put the information into the wider context of archaeologists' ideas of the Iron Age. This chapter, therefore, will focus on consideration of a different range of Iron Age sites, especially those where there is no evidence directly associated with salt, as well as those where it has been found. To achieve this within a reasonable scope, four 400km^2 areas will be used as case studies in which sites with briquetage finds and/or salt production evidence will be compared and contrasted to other, excavated Iron Age sites where no evidence of salt production or use was recovered. Within the study areas, a "site" is defined as an excavated location of archaeological interest dated to the Iron Age which has been publicly described and has a unique British Ordinance Survey National Grid Reference (NGR). Thus a location which is discussed in both a county Historic Environment Record and in a commercial archaeology firm's report is treated as a single site despite multiple records of its existence, if the reports have a common NGR. Using this method, a large scale location, such as a gravel quarry worked over a multi-year period, may appear as multiple sites in this analysis if records or reports exist where it has different NGRs. Effort has been taken to avoid the inflation of site numbers during the reconciliation of records when the NGR has been reported incorrectly through human error such as transposition of digits, making a single site initially appear as two separate ones, with site locations verified on maps. Where multiple records exist for a single site, the location has been reported using the highest level of precision possible, e.g. using an 8 character NGR rather than a 6 character one where a choice was possible.

Although each of the study areas possesses many suspected Iron Age cropmarks and land features that have been identified from aerial photographs and field surveys, only sites that have shown through excavation, even if only of a limited nature, to actually have an Iron Age phase will be part of the information used in the case studies. Sites where Iron Age finds have been termed "residual" have also been excluded, as the available evidence is too limited to provide clear guidance on the nature of Iron Age activity at the site and whether or not the site had any relationship to salt production, distribution or use. This partition of the available data used here is equivalent to levels 1 and 2 plus the excavated evaluations included in level 3 of Moore's information quality scale (2006: 12). While it varies by study area, an example of the difference these chosen parameters make to the studies is shown in Figures 44A and 44B, where the Oxfordshire study area with all sites thought to be of Iron Age date are compared with the

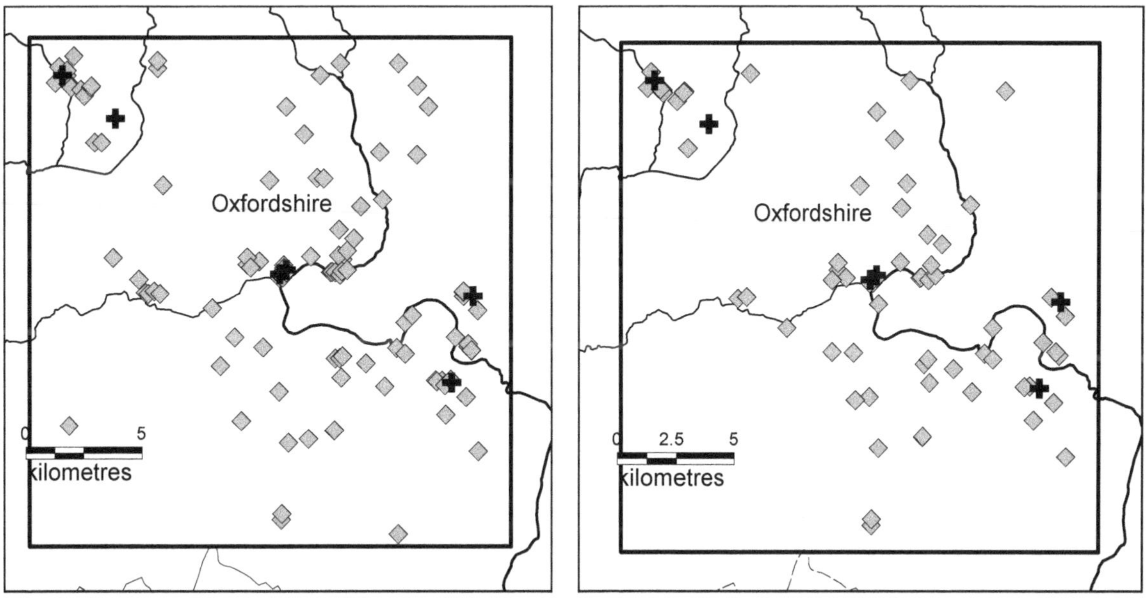

44A 44B

Figure 44. Maps of the Oxfordshire study area: (44A) with all suspected or excavated Iron Age sites and (44B) excavated sites with proven Iron Age phase.

Key: Sites with briquetage finds are shown as red crosses, while black diamonds are sites without briquetage. Underlying maps © Crown Copyright Ordinance Survey.

smaller number of excavated sites in the study area with proven Iron Age phases. For the sake of clarity, sites that lie beyond the bounds of the study areas are not shown on the maps. A full list of briquetage-negative sites and their locations used in the case studies can be found in Appendix 7.

The maps in Figure 44 also illustrate the point that the distribution of excavated sites largely follows the pattern of areas subject to disturbance or development during the past 20 years that were favoured with archaeological attention under the recently replaced Planning Policy Guidance 16 (PPG 16), while less work has been done in the wider countryside. In this specific example, more excavation has occurred in the Thames watershed gravel terraces and urban centres than on archaeologically interesting cropmarks in rural areas away from main roads, a pattern that will be repeated elsewhere.

Sources used to compile the information on the location of Iron Age sites include county HER reports, English Heritage records, the information held online by the Archaeology Data Service including their library of archaeological grey literature, the annual excavation summaries published by the Council for British Archaeology, national and regional archaeology journals, fully published excavation reports and the websites of commercial archaeology firms. It must be acknowledged that a meta-analytical approach as utilised here can be flawed by underlying methodological issues which have been touched on elsewhere in this work. For example, there is a risk that briquetage has been misidentified in assemblages as daub, fired clay or completely overlooked in post excavation work, creating an impression of difference between "negative" locations and verified find sites which is completely artificial. The variability of the information in published sources will also be discussed in respect to efforts to reconstruct detailed briquetage distribution patterns at find sites using the published excavation reports for a number of sites in or near study areas.

A further issue arises from the calculation of the number of "sites", either positive or negative, identified here and how they relate to the Iron Age purpose of the site. As the redevelopment of central towns generally took place on relatively small parcels of land, each of which is considered as an individual find site here as it was reported separately, the original Iron Age activity which occurred there may have operated over several of these sites at once. Conversely, some negative sites in farm fields can cover far larger areas but are also considered a single site due to the way they have been reported. As a result of the inherent non-standardised approach in archaeological meta-data sets drawn from excavations performed over a wide time span by numerous practitioners of varying skill levels covering a large geographic area and divergent research objectives, discovering crucial details such as the context of individual finds is not always possible.

The study areas will centre on Droitwich, Worcestershire; a portion of the Thames Valley within Oxfordshire with finds of both Droitwich and south coast briquetage; the Colchester, Essex, area and Danebury hillfort in Hampshire. This pattern allows a pairing of two production regions, Droitwich and Essex, and two salt consuming areas, central Oxfordshire and inland Hampshire. Within each study area only some of the excavated sites will have been fully published, and these sites will therefore be the most important for the detailed discussions, though they may not have been the most important locations in the study region during the portion of the Iron Age when they were in use.

The negative sites in the case studies do not objectively demonstrate the absence of salt at these locations during the Iron Age. As discussed at length in Chapter 4, distribution of salt in organic containers which do not survive in the archaeological record may have been more common than the use of briquetage containers even in areas where such evidence exists.

5.2 Worcestershire

This study area centres on the town of Droitwich and is used as the initial case study as the distribution of Droitwich briquetage is well documented, as was discussed in Chapter 4. The study area in Worcestershire contains both the urbanised area of Worcester, shown by the cluster of sites in the southwest of the map in Figure 45, and the town of Droitwich, but is otherwise largely rural in character. Of all the study areas, only in this region does the number of briquetage find sites exceed the number of briquetage-negative Iron Age sites, with seventeen finds in the study area compared to only thirteen Iron Age sites where briquetage was not found. However, ten of the finds occurred in Droitwich and seem to be associated directly with salt production or possibly its distribution in briquetage. These further sites are listed in Table 9, where the term "active period" relates only to the phase(s) where briquetage has been found. Of the finds shown in Table 9, the potential roles of Stoke Lane, Stonebridge Cross and Hampton Lovett and Westwood in regional exchange networks were discussed at length in Chapter 4, so this information will not be repeated here other than to say that the salt trade network may have had marshalling points beyond the modern corporate limits of Droitwich. Although little information was available about the briquetage find in Sidbury, of the Late Iron Age Worcester finds, one came from a site with a roundhouse and associated pit and the second from within a bank and ditch complex which may have been contemporary. Due to the nature of the urban excavations, it is not possible to report with full confidence how the sites may have related to one another, though it is thought likely that some of the sites in Droitwich functioned as a single larger activity area in the Iron Age (Woodiwiss, 1992).

The limited size of the briquetage-negative sites dataset here precludes much detailed analysis. Dating of these seven sites lacks great precision, with four of the seven sites broadly dated to the Iron Age. Of this group, one has Middle to Late Iron Age pottery, another is considered Late Iron Age and the final one is classified as Late Iron Age. At least the last three of these sites would, therefore, span the Middle to Late Iron Age period of well documented salt production using briquetage at Droitwich (e.g. Woodiwiss, 1992), although there are at least two reports of Droitwich briquetage having been found at sites with Early Iron Age pottery (Morris, 2003b) making it possible that all these sites were contemporary with salt production at Droitwich. Thus site date is not sufficient in itself to explain the absence of briquetage.

Given this dating information, the question arises as to why some sites did not yield briquetage in what was apparently a salt rich area where briquetage was commonly used for transportation of salt, allowing for speculation that the activity at the site may not have required salt. But does this theory seem to work with the data? The two negative sites to the southwest of Droitwich at Hindlip were associated with an Iron Age pit alignment thought to be near an unexcavated settlement (Wainwright, 2010). Such pit alignments are now generally interpreted as linear boundary markers (e.g. Lambrick and Robinson, 2009: 57, 60-61).

As is typical of such features, there were few material

Name	Description	Location	Active Period
Hanbury, St Mary's Church	hillfort	SO 9540 6430	IA
Stoke Lane, Wychbold	enclosure	SO 9215 6560	MIA
Sidbury	find site	SO 848 547	M-LIA
Stonebridge Cross, Westwood	farmstead	SO 87648 64647	M-LIA
Newdix Court, Worcester	bank & ditch	SO 851 547	LIA
Deansway, Worcester	roundhouse & pit; R-B farm	SO 849 548	LIA, RB
Hampton Lovett & Westwood	settlement	SO 8830 6475	IA-RB

Table 9. Briquetage find sites within the Worcestershire study area from non-production locations.

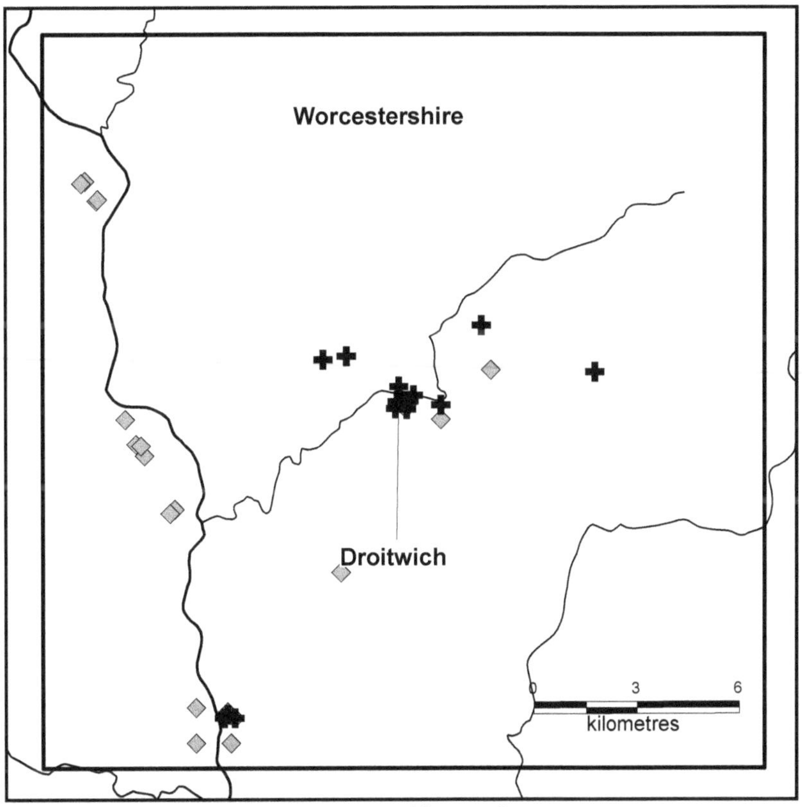

Figure 45. Worcestershire study area.

Key: Briquetage finds (crosses) and excavated Iron Age sites without briquetage (diamonds). Underlying map © Crown Copyright Ordinance Survey.

culture finds in the pits, which were dated based upon the presence of iron slag and hammerscale and absence of Roman material to the Iron Age. One might say, therefore, that the presence of briquetage in a pit alignment would have been unexpected, and that its absence is not surprising even in this region. Yet the majority of the negative sites here seem to have been associated with domestic activity. All but the pit alignments can be broadly classified as representing domestic activity, and the site descriptions refer to enclosure ditches filled with domestic material (e.g. Hunt et al., 1986) and suspected hearths (e.g. Walker, 1958). If one assumes most Iron Age salt usage was associated with food preservation, which would have been a significant domestic concern, then the absence of briquetage at these sites cannot be explained unless one considers explanations beyond site function.

Intriguingly, seven of the negative sites, 53.8% of the total, and none of the finds, occur to the west of the River Severn in the study area. Unfortunately this is a small subset of what is itself a statistically marginal dataset, preventing meaningful conclusions being drawn about the pattern. As was illustrated in the total distribution map shown as Figure 42 in Chapter 4, the westward gap from Droitwich in the briquetage finds map shown above extends as far as the River Wye watershed. It is possible to suggest that if salt was taken across the Severn in the study area, it was not done so in briquetage. A potential interpretation of this pattern may be found in the view of the Severn as a Dobunnic tribal boundary (Scullard, 1999) suggesting that a possible group identity characteristic reflected in material culture – the rejection of briquetage as an appropriate salt container – led to the observed archaeological pattern (e.g. Hodder, 1977, Jones, 1997). However, both Dobunnic coins and Droitwich briquetage are found west of the Severn north and south of the study area (Hurst, 2001), so the rejection of briquetage containers, if it existed as a group identifier, may have been a highly localised phenomenon.

Clearly a lesson to be taken from the study area surrounding Droitwich is the fact that geographic proximity to the salt production source does not inherently equate to the reported presence of briquetage at an Iron Age site. It would appear that there were some places in the Iron Age landscape where salt in a briquetage container was not needed, or perhaps desired, for the business at hand.

Conderton Camp

The Conderton Camp hillfort, although it lies outside the study area 27km southeast of Droitwich, is worthy of further discussion here. As shown in Table 10, it is the source of the fifth largest assemblage of briquetage in Britain (by weight) that has been found away from a production site. Although the site was excavated in 1958 and 1959, it was not fully published until 2005. While such a time lag is obviously not desirable, the delay did allow for modern methods to be applied to evaluating the site and the finds. One obvious benefit is the correct interpretation of the briquetage, which would likely have been misidentified as daub, as happened initially at sites such as the Breiddin and Ffridd Faldwyn hillforts (Musson et al., 1991, O'Neil, 1942), or as crude Iron Age pottery, as at Dinorben (Gardner and Savory, 1964). It should be noted that the weight of the Conderton Camp assemblage as shown here is taken from the more detailed Table 53, page 308, of the site monograph, and is 24gm higher than an alternate weight which appears in summary Table 19, page 279, of the same report (Morris et al., 2005). There is also a variance of one sherd in the total sherd count between these two tables, although Morris repeats the number 451, as shown in Table 19, in a discussion of deposition at this site which appears on page 144 of the monograph, and may therefore be correct. All but two sherds in the assemblage, with a total weight of 38gm, came from Droitwich, though the presence of any Cheshire VCP so close to Droitwich was somewhat unexpected (Morris et al., 2005).

Although there was no discernable patterning to the briquetage distribution within pits at Conderton (Thomas, 2005), the level of detail provided in the monograph supports the generation of plans and section drawings which highlight the contexts in which briquetage finds were made, and these will be presented below. As shown in Figure 46, briquetage finds were widely distributed within the enclosure.

A total of 44 pits were excavated at Conderton, 15 of which (34.1%) contained briquetage (Thomas, 2005). The location of these pits is shown in Figure 47 and the range of finds from the pits with briquetage is detailed in Table 11. By weight, 33.4% of all the briquetage recovered at this site came from these pits with the remainder located in or near domestic structures, working hollows and trenches. Thomas stated that there was "...no clear pattern..." in the presence of briquetage, pottery and fired clay when the pits were analysed, though he noted that he had not considered the pits in the same manner as Hill's (1995) analysis of Wessex pit contents. A further 1,196gm of briquetage (17.8%) was recovered from the trenches surrounding pits A, B, C, D, E, V/Vi, W, X and Y. Both sandy temper and organic temper Droitwich fabrics were present in the pits and their associated trenches.

As shown in Table 11, the contents of Pit A are significantly different from the other pits that yielded briquetage, and not just with respect to this material. It is the only pit at Conderton in which a complete decorated vessel was deposited (in fill layer 7) (Morris et al., 2005). It is also the only pit at the site where the amount of sandy Droitwich briquetage (556gm) exceeds the amount of organic fabric (538gm) (ibid.: 145), consistent with the pit's dating on the basis of pottery typology to the earlier part of the Middle Iron Age. The mean briquetage sherd weight, 21gm for the 52 sherds found in Pit A, "...is at least 3.5 times greater than that of any other significant pit assemblage... " (Morris et al, 2005.: 144). The combination of goods in this pit as well as the volume of material there led Morris to suggest that this may have been the only pit at Conderton where special deposition was clearly in evidence and to suggest that the volume of briquetage may have been connected with a ritual role for salt at this site (ibid: 145). A plan and section drawing of Pit A appear in Figure 48.

Site Name	Total Briquetage Found (gm)	Source(s)
Collfryn Hillfort, Powys	13,285	Cheshire
Twyn-y-Gaer Hillfort, Monmouthshire	9,856	Droitwich and Cheshire
Gravelly Guy, Oxfordshire	8,457	Droitwich
Danebury Hillfort, Hampshire	7,596	South Coast
Conderton Camp Hillfort, Worcestershire	6,736	Droitwich and Cheshire
Elm Park House, Essex	5,891	Essex
Beckford II, Worcestershire	5,180	Droitwich and Cheshire

Table 10. Weight comparisons of selected non-production briquetage assemblages.

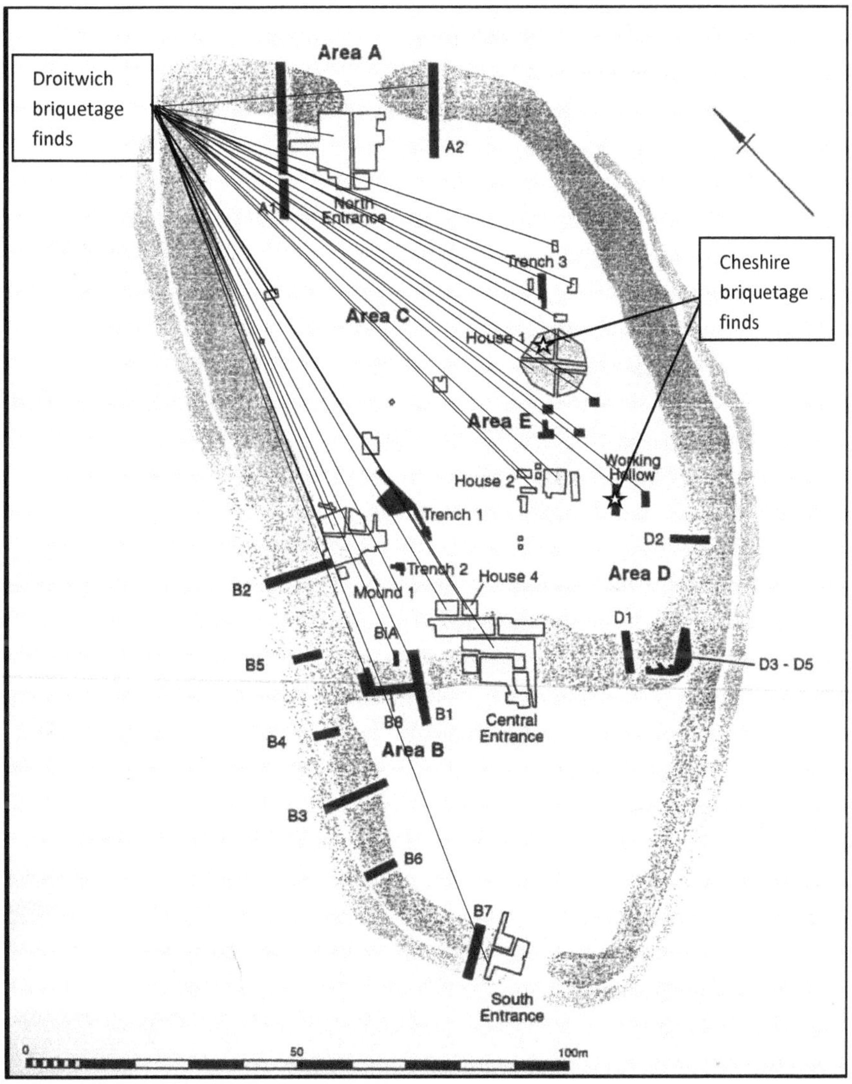

Figure 46. Conderton Camp showing 1958 and 1959 trenches with briquetage finds.
After Thomas, 2005: 16, Figure 6. Reproduced by permission of Nicholas Thomas and the Council for British Archaeology. Scale equals 100 metres.

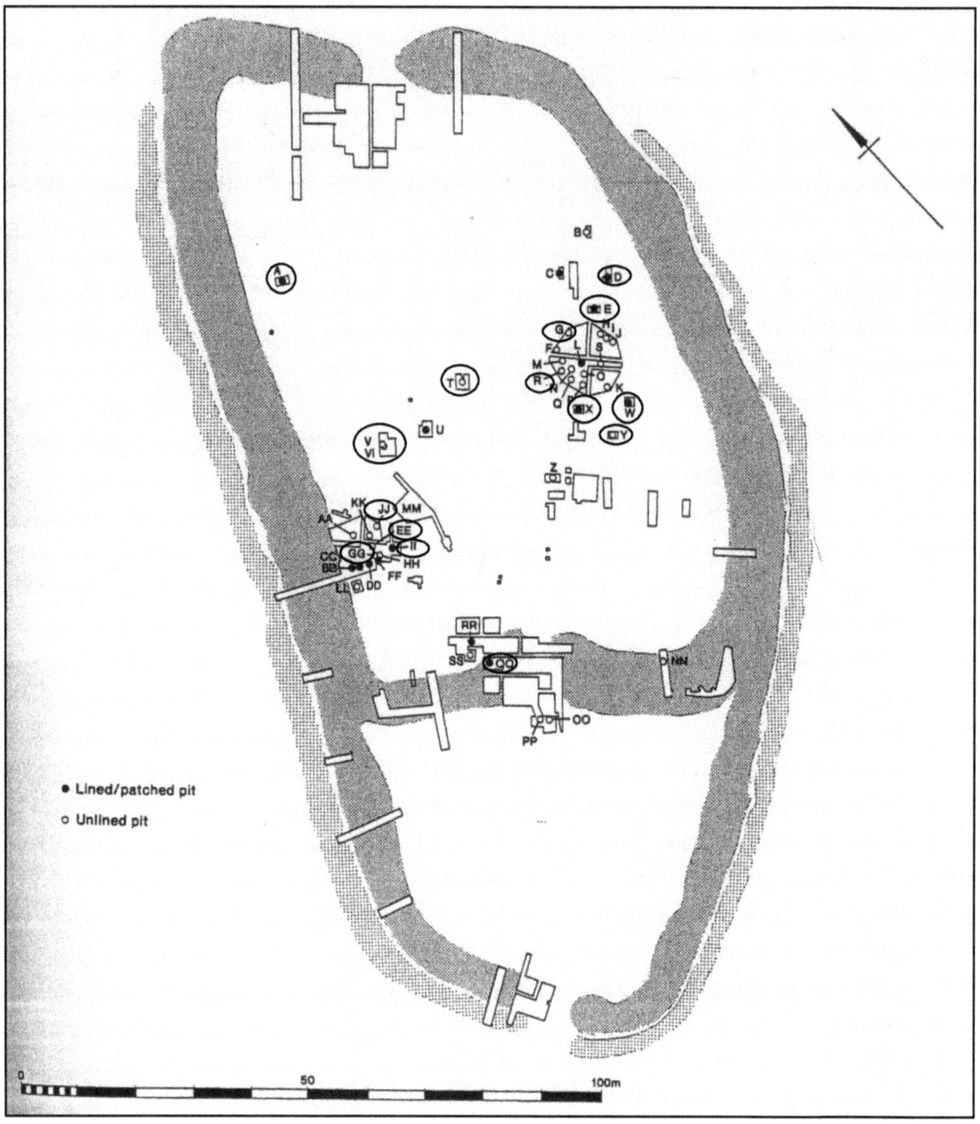

Figure 47. Distribution of excavated pits with briquetage at Conderton Camp.

After Thomas, 2005: 97, Figure 30. Reproduced by permission of Nicholas Thomas and the Council for British Archaeology. Scale equals 100 metres.

Detailed plans and section drawings of the remainder of the pits, other than Pit T, from which briquetage was recovered, appear below in Figures 49 through 59. Although not illustrated here, briquetage was recovered from fill layers 3, 4 and 5 of Pit T (Morris et al., 2005).

If the briquetage found in some of the excavated pits appeared to have reached these locations as inclusions with domestic rubbish, it should not be surprising that, after the pits, the next largest quantity of briquetage was found in and around Houses 1, 2 and 4 and the multiple houses found in the area identified as Mound 1, presumably the ultimate source of much of the occupation debris. A plan of House 1 appears below in Figure 50, House 4 is shown in Figure 59 and the plan of Mound 1 shown in Figure 54. The excavated portion of House 2 where briquetage was found is shown in Figure 60. Morris et al. (2005) provided detailed information on the location and metrics of the finds, also differentiating those made in occupation or use layers from those coming from layers associated with the abandonment of the structures. This information suggests that briquetage was usually removed from the houses while they were in use, with only 202gm being associated with the occupation layers of houses (ibid.: 289) and 179gm being found in contemporary occupation layers on their periphery (ibid.: 291), far less than came from the pits as shown in Table 11 and in sharp contrast to the 2,092gm which was retrieved from layers associated with the abandonment of these houses. The pattern suggests that the houses were regularly cleaned of briquetage when occupied although dumping refuse on the house sites once abandoned was an acceptable practice. Only a small amount of briquetage, totalling 298gm, was recovered

Pit	Location	Briquetage (gm)	Pottery (gm)	Fired Clay (gm)	Ceramic Objects (gm)	Copper Alloy (gm)	Iron (gm)	Worked Bone or Antler (gm)	Worked Stone (gm)	Animal Bone (gm)
A	nw corner of hillfort	1,094	6,660	2,640	1	2	1	2	3	2,829
D	ne area of hillfort, se of Pits B & C	24	384	870	0	0	2	3	4	2,700
E	outside House 1 (H1), to the ne	2	328	20	0	0	0	0	0	1,610
G	outside the entrance of H1	3	63	80	0	0	0	0	0	100
R	outside H1	11	102	160	0	0	0	0	0	580
T	Area C, nw of H1	242	749	60	0	0	0	0	0	1,828
V	Area C, ne of Mound 1 (M1)	21	572	0	0	0	1	0	0	680
W	Area E, s of H1, Trench 11	61	287	0	0	0	1*	0	0	340
X	Area E, outside & sw of H1	299	314	0	0	3	0	0	0	1,100
Y	Area E, between H1 & H2, Trench 9	105	541	0	0	0	0	0	0	30
EE	M1, s quadrant, n of Pit DD	14	53	20	0	0	0	0	0	0
GG	M1, s quadrant extension, under eastern arc of Pit FF	75	494	20	0	0	0	0	1	270
II	M1, eastern edge	214	987	45	0	1	0	0	0	240
JJ	M1, outside H3, east quadrant	39	127	20	0	0	0	0	0	160
QQ	central entrance, inside H4, in sw corner of Cut 2, nw corner of Cut 3	44	417	25	0	0	2	0	0	460
	Totals	2,248	12,078	3,960	1	6	7	5	8	12,927

* smithing debris

Table 11. Other finds in Conderton Camp pits with briquetage.
After Thomas 2005, Table 14, page 107, and Appendix 2, pages 265-278.

from the hollow at House 1, as shown in Figure 50, and the Working Hollow cuttings shown in Figure 61 (ibid.: 292).

Rather unexpectedly, small amounts of Cheshire VCP were found at Conderton Camp 135km from the centre of Cheshire. Of the two sherds of Cheshire briquetage found at Conderton, 1 sherd, weighing 8gm, came from an abandonment layer inside the northern quadrant of House 1, shown in Figure 50, and the second, a larger sherd weighing 30gm, came from layer 4 of Working Hollow Cutting 1, as shown in Figure 61. Morris et al. (2005: 144) suggest this material reached the site as a result of down-the-line trade. Other possible explanations for the presence of Cheshire briquetage so close to the Droitwich production source were explored in Chapter 4.

Although multiple trenches were cut across the entrances and the ramparts as shown in Figures 46 and 47, only a total of 31 sherds of briquetage, 398gm in total, less than 6% of the site briquetage assemblage by weight, came from these areas. This is not an unexpected occurrence; Morris et al. (2005: 144) note that finds of briquetage within ramparts are infrequent. Sherds found in these areas were small, with a mean sherd weight (msw) of 12.8gm, or 14% below the average msw for the whole site (14.9gm). None of these trenches had significant concentrations of material, as shown in Table 12. This pattern would suggest that relatively little briquetage reached the site before the construction of the ramparts, and that little or no salt was decanted at the entrances.

Although many hillforts' ceramic assemblages have included briquetage, as was shown in Table 10 above, no others have the combination of proximity to Droitwich and a briquetage assemblage of the size known from Conderton Camp, suggesting that the site may have had a significant role in controlling the distribution of salt being produced at Droitwich while it was active, which would have overlapped with the beginnings of archaeologically evidenced salt production there in the sixth century BC (Woodiwiss, 1992).

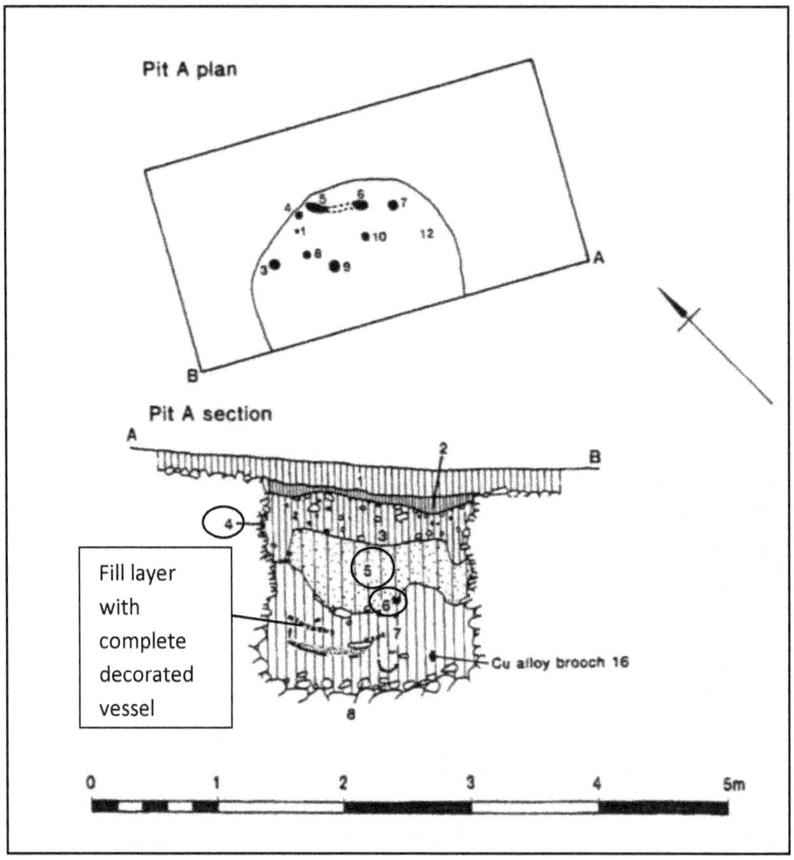

Figure 48. Conderton Pit A with briquetage-bearing fill layer numbers circled.

After Thomas, 2005: 98, Figure 31. Reproduced by permission of Nicholas Thomas and the Council for British Archaeology. Scale and north arrow apply to pits A, D and E.

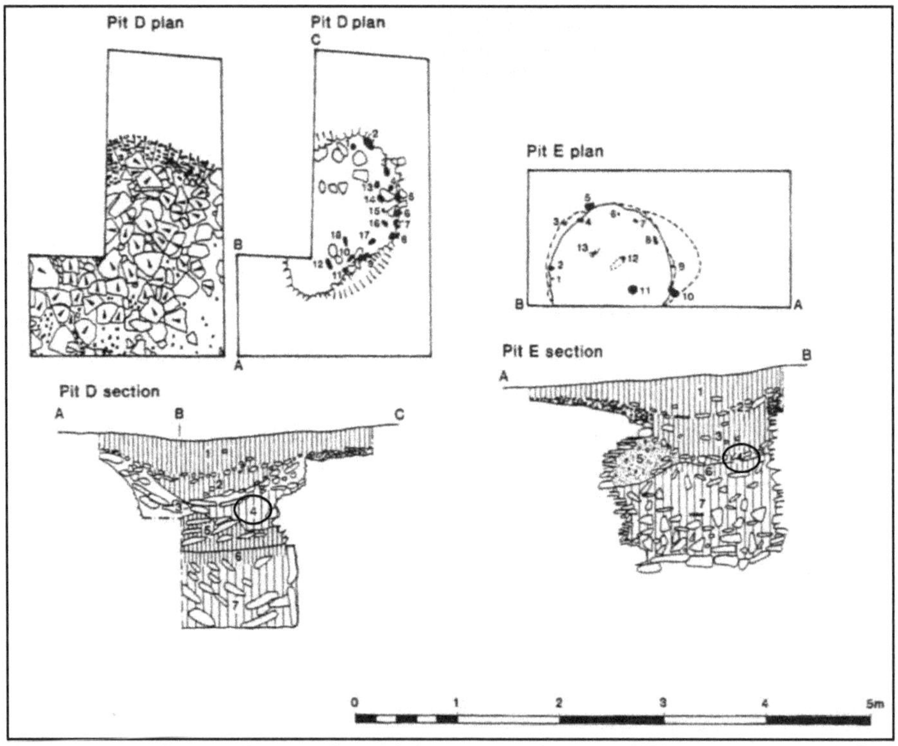

Figure 49. Conderton Pits D and E with briquetage-bearing fill layers number circled.

After Thomas, 2005: 98, Figure 31. Reproduced by permission of Nicholas Thomas and the Council for British Archaeology. Scale shows 5 metres.

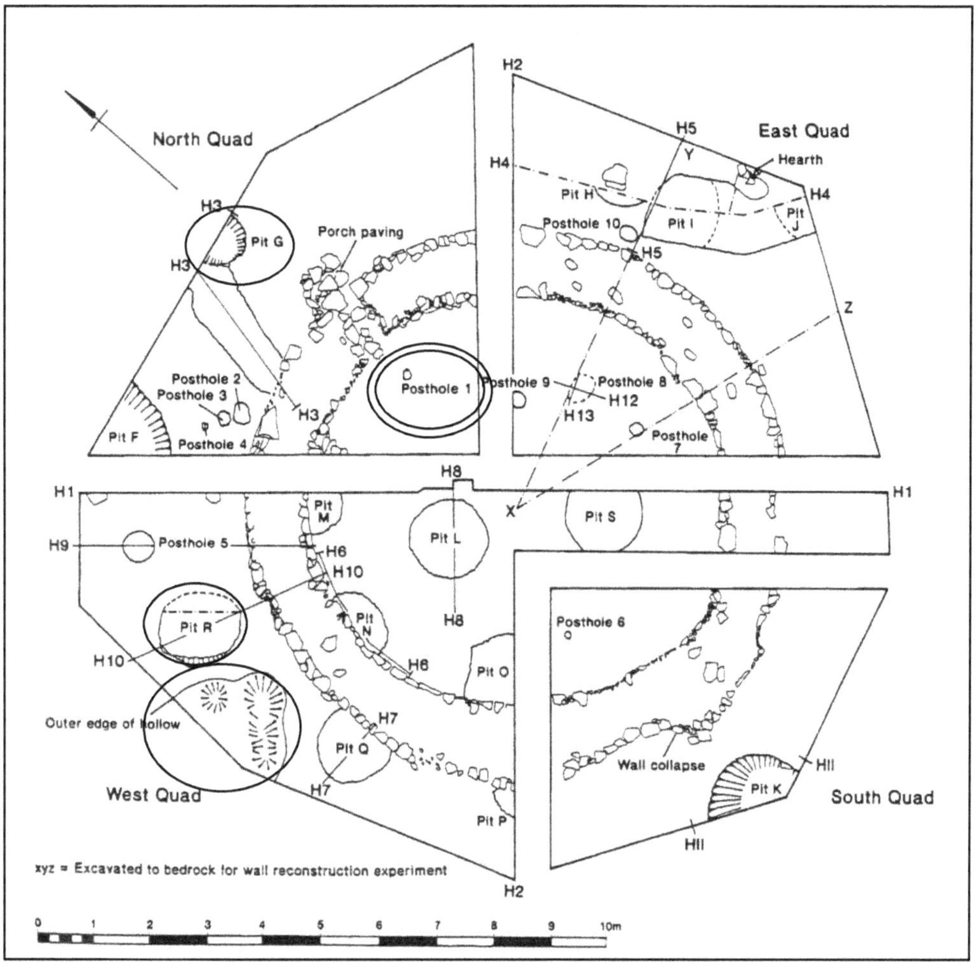

Figure 50. Plan of House 1 with Droitwich briquetage bearing West Quad hollow and Pits G and R circled; general location of Cheshire briquetage find marked with double circle.

After Thomas, 2005: 59, Figure 20. Reproduced by permission of Nicholas Thomas and the Council for British Archaeology. Scale shows 10 metres.

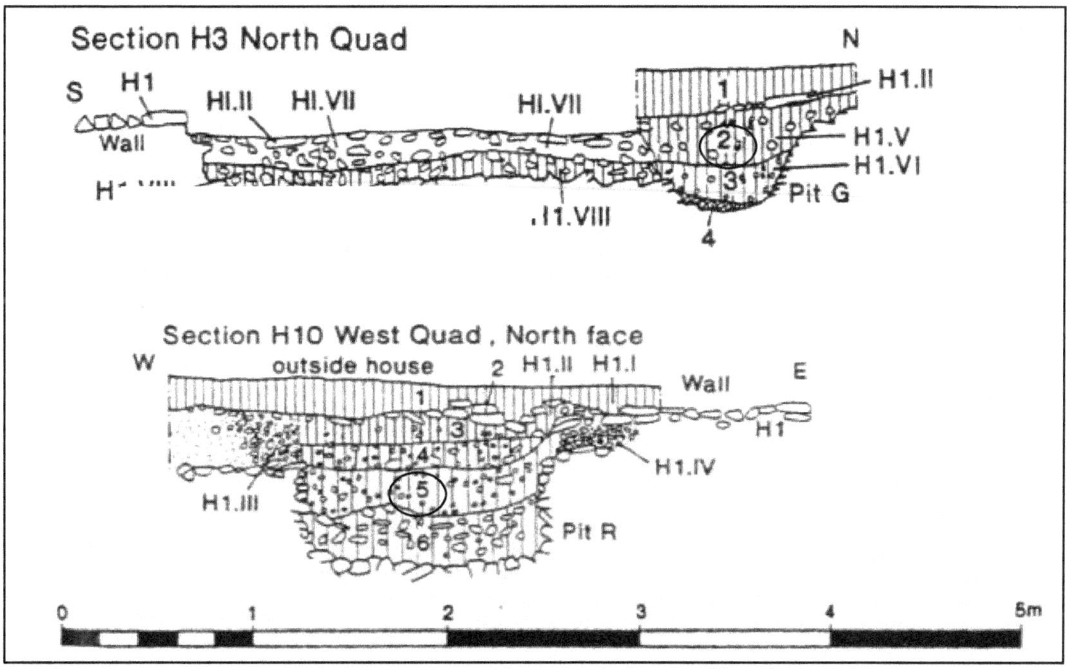

Figure 51. Section views of Pits G and R, with circles indicating fills which yielded briquetage.

After Thomas, 2005: 60, Figure 21. Reproduced by permission of Nicholas Thomas and the Council for British Archaeology.

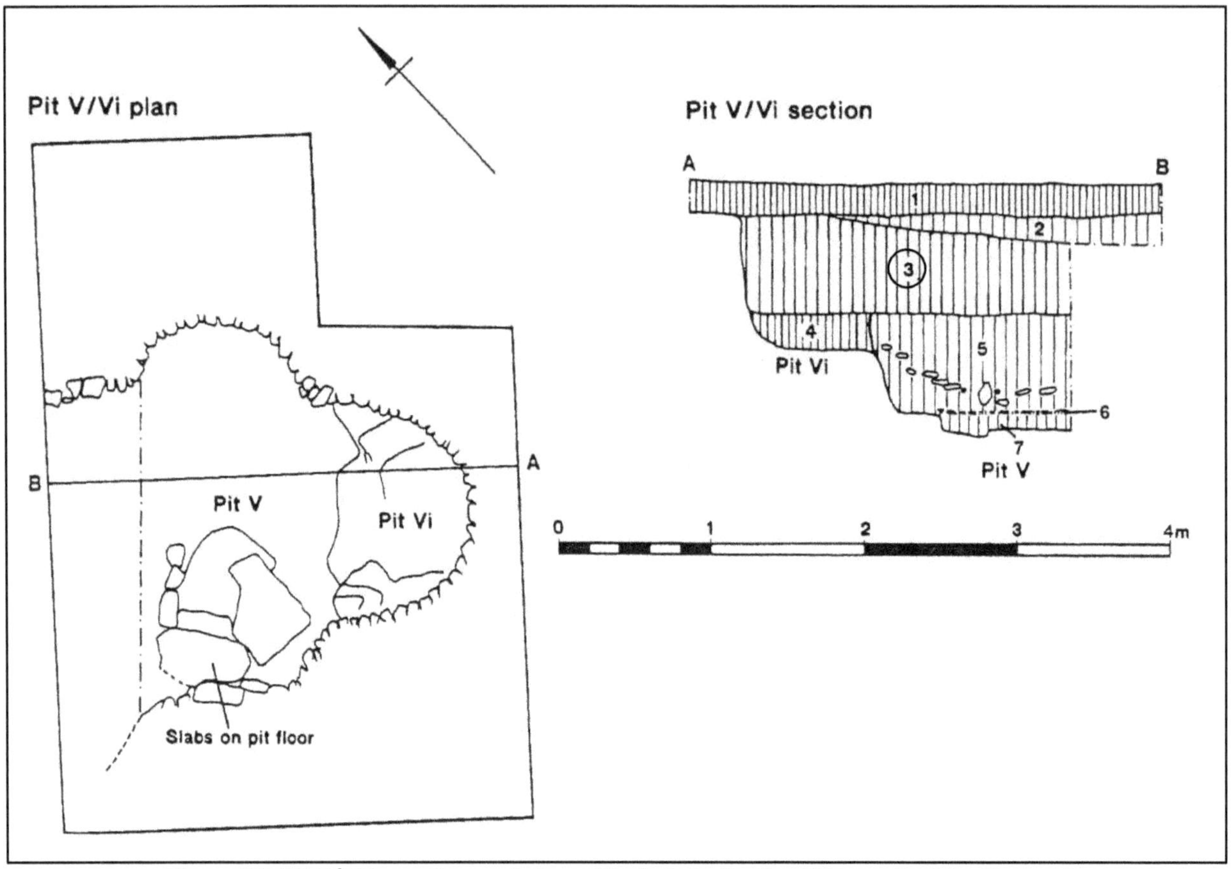

Figure 52. Pit V/Vi in section and plan with circle indicating fill layer that had briquetage.

After Thomas, 2005: 109, Figure 33. Reproduced by permission of Nicholas Thomas and the Council for British Archaeology.

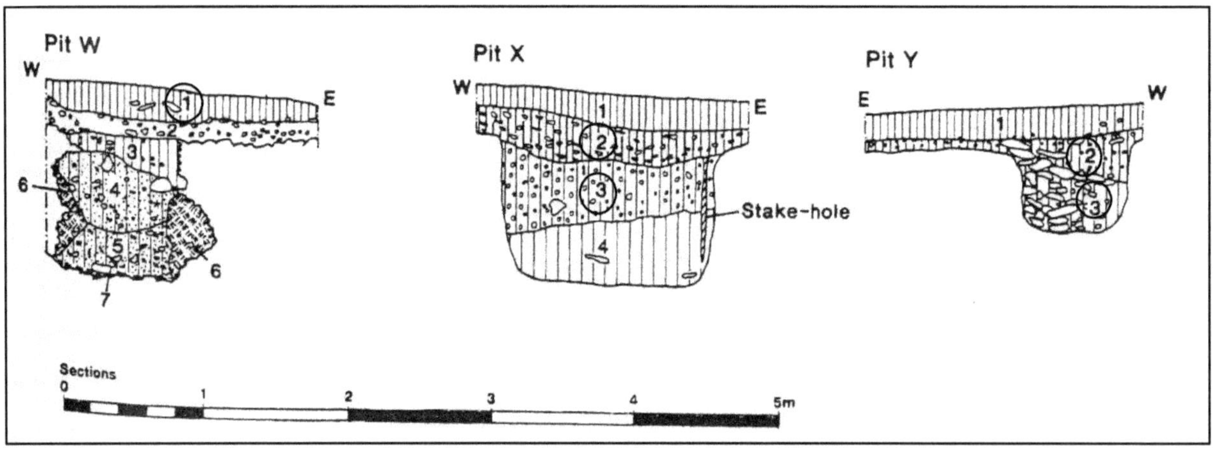

Figure 53. Pits W, X and Y with circles indicating fill layers with briquetage finds.

After Thomas, 2005: 111, Figure 34. Reproduced by permission of Nicholas Thomas and the Council for British Archaeology.

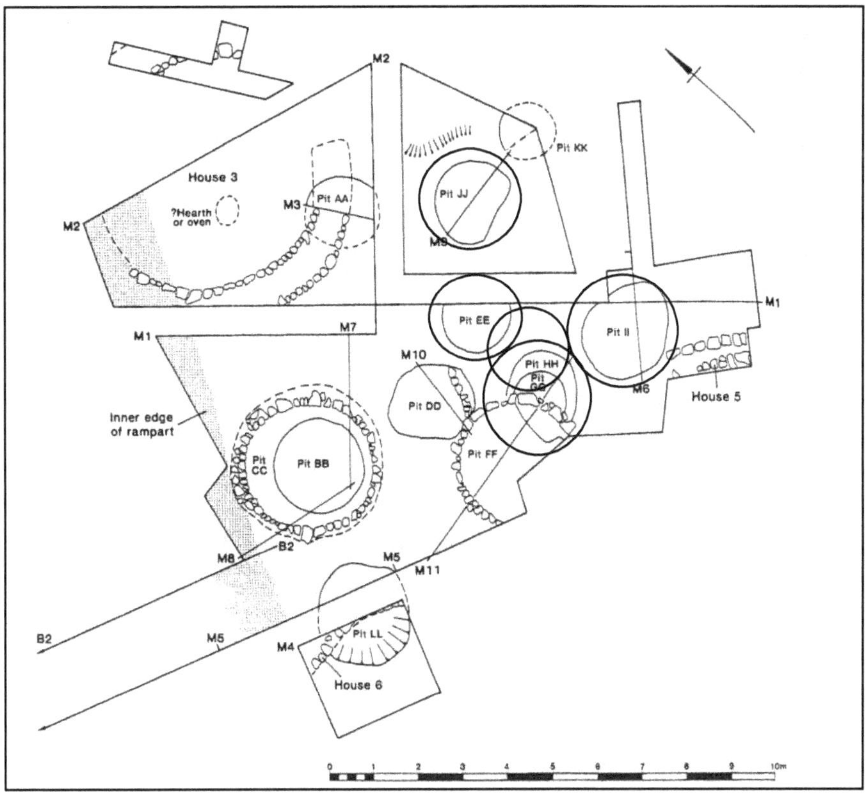

Figure 54. Plan of Mound 1, showing briquetage-bearing Pits EE, GG/HH, II and JJ.

After Thomas, 2005: 70, Figure 25. Reproduced by permission of Nicholas Thomas and the Council for British Archaeology. Scale shows 10 metres.

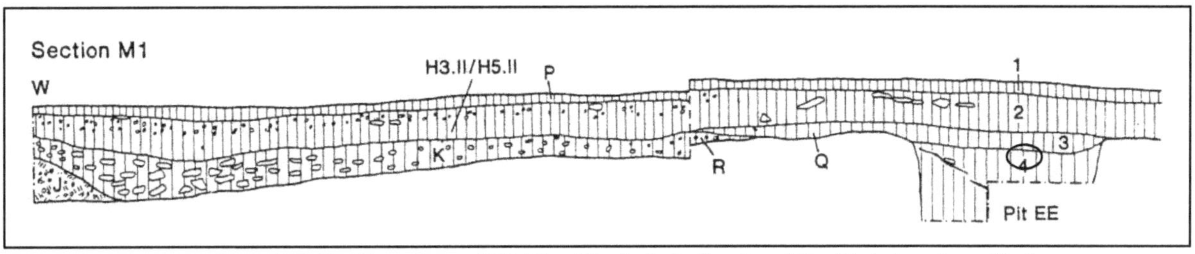

Figure 55. Section view of Pit EE with circle indicating fill with briquetage find.

After Thomas, 2005: 71, Figure 26. Reproduced by permission of Nicholas Thomas and the Council for British Archaeology. Same scale as Figures 51, 52 and 53.

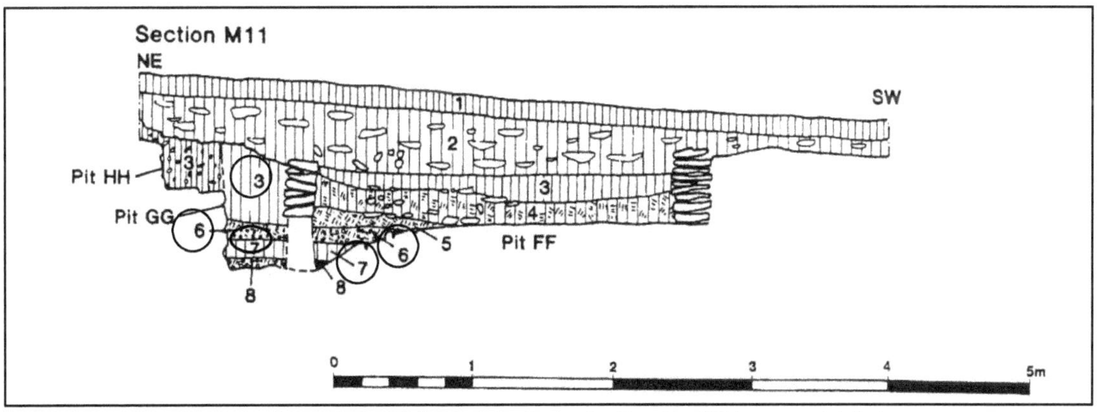

Figure 56. Section of Pit GG/HH with circles indicating fill layers with briquetage.

After Thomas, 2005: 73, Figure 27. Reproduced by permission of Nicholas Thomas and the Council for British Archaeology. Same scale as Figures 51, 52, 53 and 55. A single briquetage sherd was found on the boundary between fill layers 6 and 7.

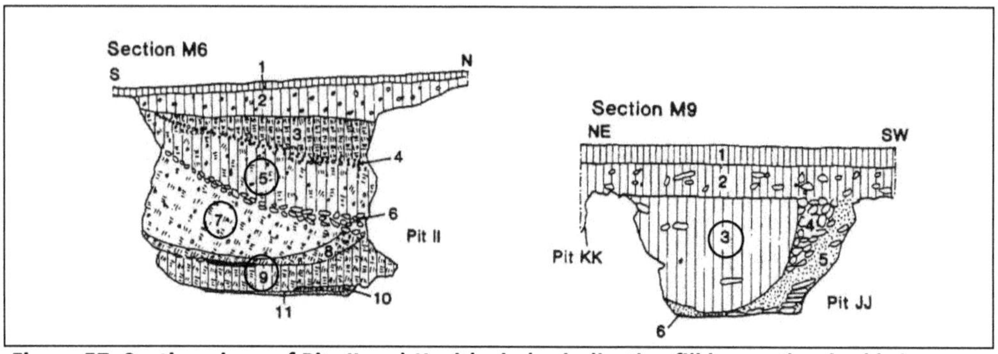

Figure 57. Section views of Pits II and JJ with circles indicating fill layers that had briquetage.
After Thomas, 2005: 73, Figure 27. Reproduced by permission of Nicholas Thomas and the Council for British Archaeology. Same scale as Figures 51, 52, 53, 55 and 56.

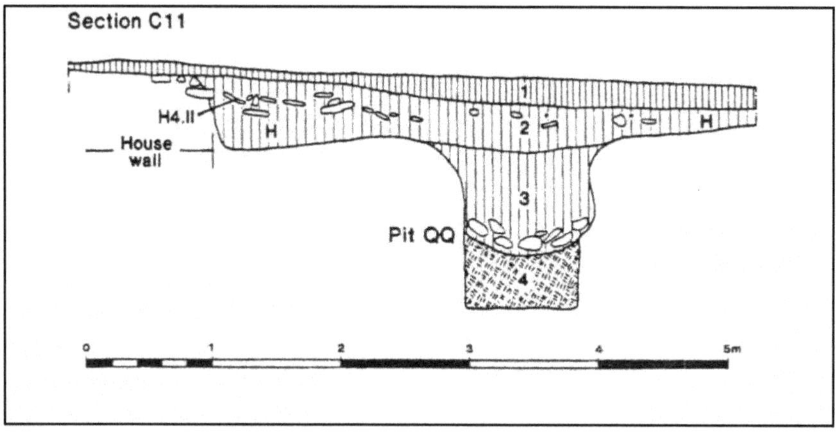

Figure 58. Section view of Pit QQ where briquetage was recovered from an unspecified fill layer.
After Thomas, 2005: 80, Figure 29. Reproduced by permission of Nicholas Thomas and the Council for British Archaeology. Scale shows 5 metres.

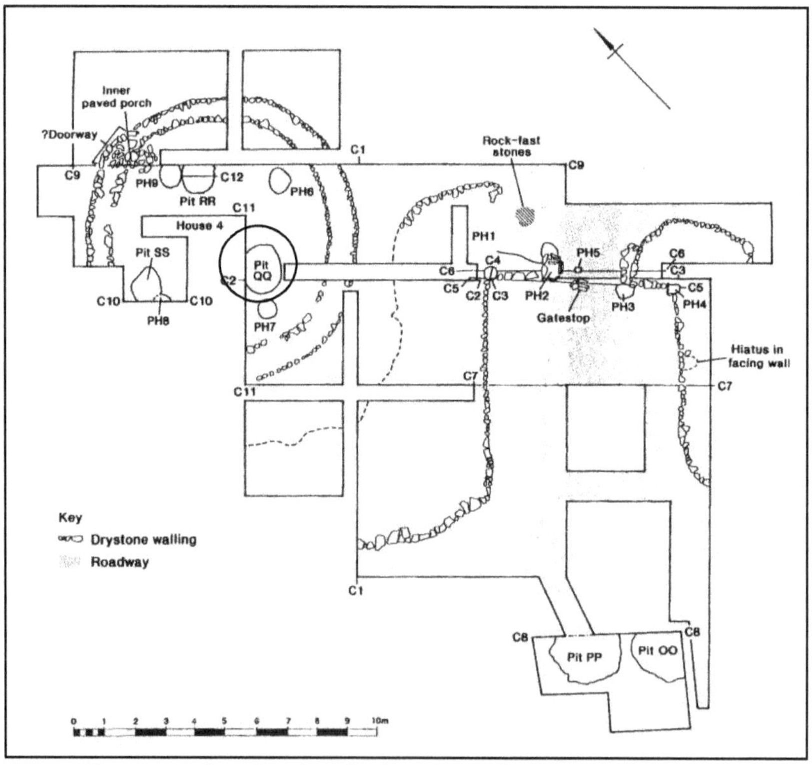

Figure 59. Plan showing the locations of Pit QQ and House 4.
After Thomas, 2005: 43, Figure 16. Reproduced by permission of Nicholas Thomas and the Council for British Archaeology. Scale shows 10 metres.

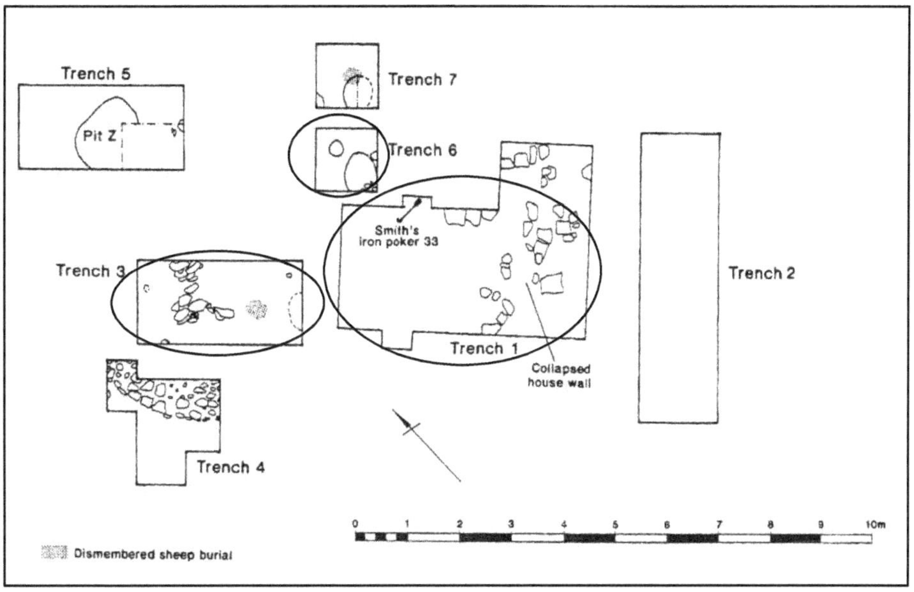

Figure 60. Partial plan of House 2 trenches with circles indicating briquetage find areas.

After Thomas, 2005: 66, Figure 23. Reproduced by permission of Nicholas Thomas and the Council for British Archaeology. Scale shows 10 metres.

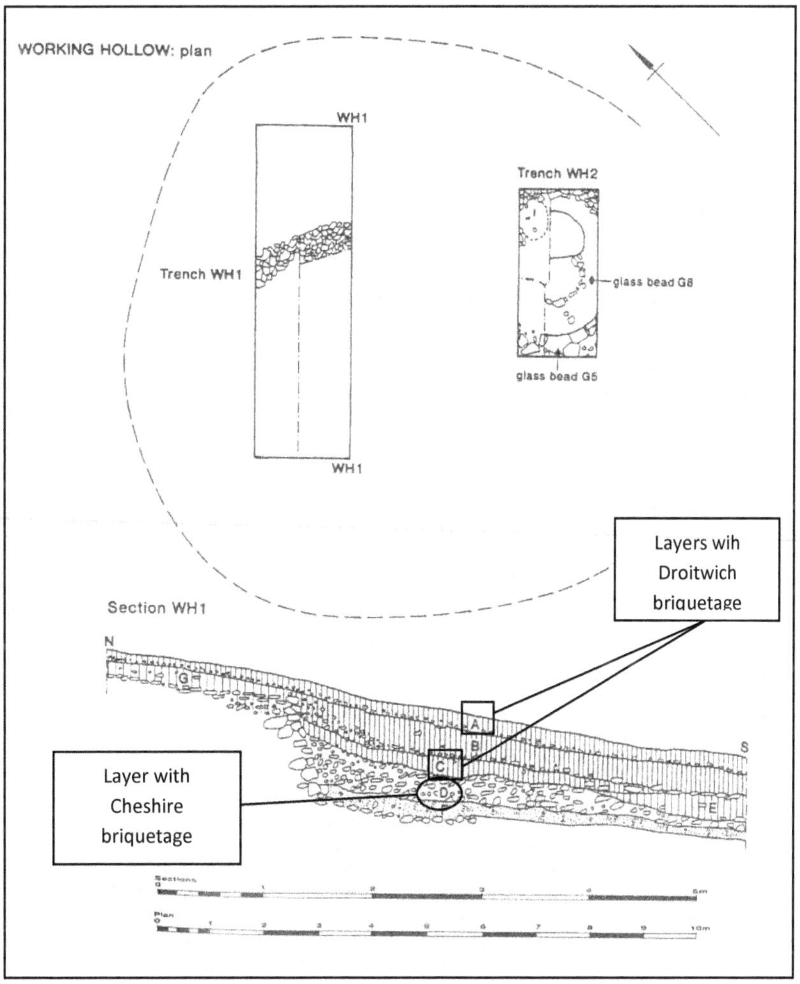

Figure 61. Working Hollow with Droitwich and Cheshire briquetage finds.

After Thomas, 2005: 111, Figure 34. Reproduced by permission of Nicholas Thomas and the Council for British Archaeology. Scale shown is 5 metres for the]#secton and 10 metres for plans.

Trench Location	Number of Sherds	Total Weight	MSW
Western Rampart Trenches	15	157	10.5
Central Area, North Entrance	2	28	14.0
Ramparts at North Entrance	1	45	45.0
South Entrance	6	15	2.5
Central Entrance Cross Rampart	3	19	6.3
Central Entrance	4	134	33.5
Totals/Average MSW	31	398	12.8

Table 12. Briquetage finds in Conderton Camp rampart and entrance way trenches.
After Morris et al. , 2005: 293, 296, 297, 298 and 299.

5.3 Oxfordshire

In comparison with the Worcestershire study area the Thames Valley study area, as shown in Figure 62, is congested with excavated Iron Age sites without briquetage, a reflection of excavation related to both modern development and gravel extraction, as well as the presence of Oxford University archaeologists and active amateur archaeological societies. The ratio of briquetage finds to sites without briquetage, 1:16.7, is a stark contrast to that seen in the Worcestershire case but not unanticipated so far from salt production sources. The six briquetage find sites in this study area are listed in Table 13 and the 100 negative sites are detailed in Appendix 7.

As shown in Figure 63 two of the finds sites are in close proximity in central Abingdon although it has been difficult to establish any relationship between them. They are surrounded by multiple negative sites. While this example illustrates the multiplication effect on site numbers when many small parcels are redeveloped in an urbanised area, it also demonstrates the problem of attempting to see large scale activity in urban areas.

Although few in number, the briquetage finds span the entire Iron Age, from the Early Iron Age at Gravelly Guy (Lambrick and Allen, 2004) to the early Roman period at both Gravelly Guy and Watkins Farm (Allen, 1990), suggesting that there may have been continuous, ongoing salt exchange in the area throughout Iron Age. With the exception of the single find of Hampshire coast briquetage at Abingdon Town Centre associated with a Late Iron Age/early Romano-British *oppidum*, the other finds are from Droitwich. Little can be said of the yet-to-be published find at the Castle Hill hillfort (Lambrick and Robinson, 2009: 211-212), however both the *oppidum* and the hillfort would normally be considered to be sites that differ from ordinary settlements. The other sites, though, are considered to be unexceptional settlements typical of the Thames Valley area. This highly ambiguous pattern suggests the possibility of widespread salt availability, but as was the case with the previous example, six sites is far too small a sample to draw statistically significant conclusions.

Of the locations identified in Table 13, detailed information about the distribution of briquetage at the site has been fully published only for Gravelly Guy and Watkins Farm, and, as will be discussed below, the find at Mount Farm.

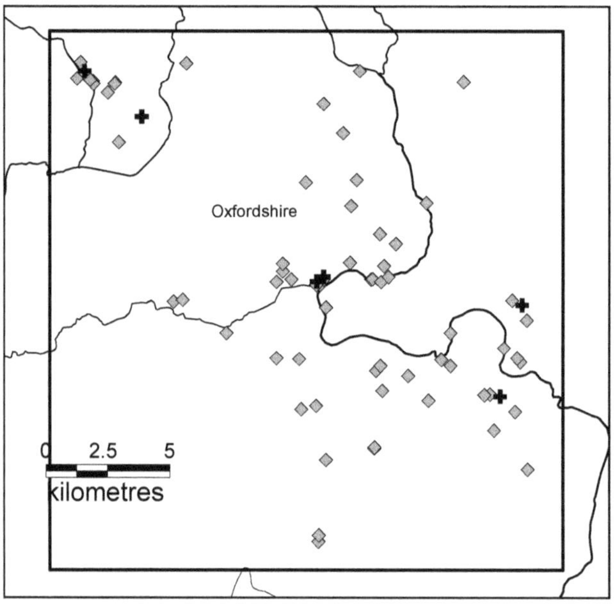

Figure 62. Oxfordshire Thames Valley study area.
Key: Briquetage finds (crosses) and excavated Iron Age sites without briquetage (diamonds). Underlying map © Crown Copyright Ordinance Survey.

Mount Farm

Excavated in 1933, the Mount Farm find may be an example of the accidental publication of briquetage, having been discussed in the excavation report as pottery (Myres, 1937). As has been noted elsewhere, the misidentification of briquetage by archaeologists was a typical occurrence until the late twentieth century. At the time of this excavation, Oxfordshire prehistoric pottery was being closely studied as new sites on the Thames gravels were regularly yielding sherds which clarified the typology of the region, a fortuitous circumstance which led to the publication of an illustration and detailed fabric description of what seems to be the base of a Droitwich salt container on the assumption that it was a previously unknown type of coarse pottery, unusually "decorated"

Find Site Name	Description	Location	Active Period
Gravelly Guy, Stanton Harcourt	settlement	SP 403 053	EIA to R-B
Watkins Farm, Northmoor	enclosed settlement	SP 426 035	MIA, R-B
Mount Farm, Dorchester	enclosure and pits	SU 579 961	MIA
The Vineyard, Abingdon	settlement	SU 499 973	MIA
Castle Hill, Little Wittenham	hillfort	SU 570 925	MIA
Abingdon Town Centre	*oppidum*	SU 4962 9702	LIA-ERB

Table 13. Oxfordshire study area briquetage find sites.

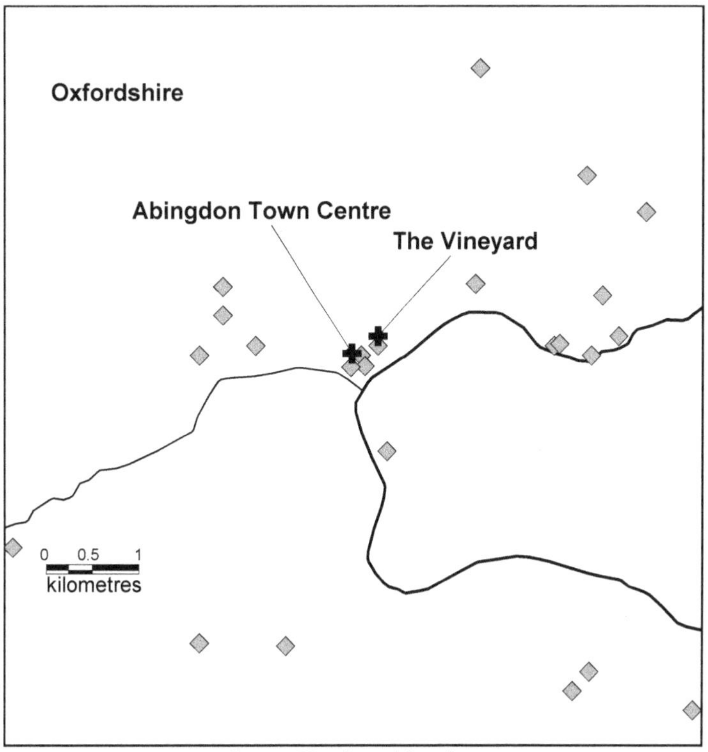

Figure 63. The Abingdon area.

Key: Briquetage finds (crosses) and excavated Iron Age sites without briquetage (diamonds). Underlying map © Crown Copyright Ordinance Survey.

on the exterior with finger marks. If this base sherd had been attached to a larger section of container wall, perhaps the comment might have been quite different. Although not a quantitatively large find, it is worthy of note due both to its early date, albeit unrecognised at the time, and its occurrence 97km from Droitwich. A more recent briquetage find from the nearby Castle Hill hillfort at Little Wittenham (Lambrick and Robinson, 2009: 211-212) dates to the same period and helps to support the case for Droitwich briquetage reaching this area.

A significant amount of information has been provided by Myres about the published sherd. It came from the fill of Pit α, shown in Figures 64 and 66, a pit within a pit cluster east-north east of a circular enclosure where it was found with more than 100 other sherds of pottery, mainly from the late Iron Age A and Iron Age B phases, although Romano-British sherds had been recovered from the topsoil. A fabric description suggests the sherd was made of sandy Droitwich briquetage (Myres, 1937: 34). The flat bottomed vessel base illustrated as Figure 8, α 16, by Myres (ibid.: 33) is shown below as Figure 65. As was the custom at the time, non-diagnostic sherds were not retained or commented upon, so it is impossible today to know whether other examples of the material were found during the excavation.

Gravelly Guy

As shown in Table 10, the briquetage assemblage recovered from this site, 80km from Droitwich, was the third largest by weight which has been found in Britain away from production sources. Context level reporting was not provided in the site monograph (Lambrick and Allen, 2004), but it was reported that briquetage was found in 166 different contexts in all areas of the site other than Block 2 (Morris, 2004), a region of scattered pits, as illustrated below in Figure 67. The amount of briquetage being found was similar throughout the Early Iron Age and Middle Iron Age phases there, while at many other sites there are signs of an increasing presence from the Early Iron Age to the Middle Iron Age,

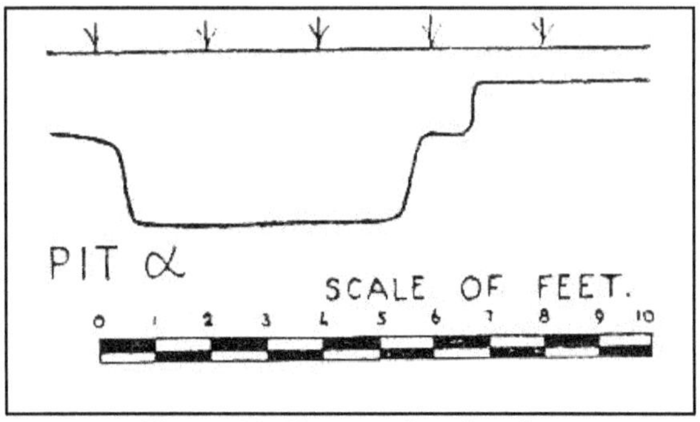

Figure 64. Pit α section drawing
After Myres, J. N. L. (1937), A Prehistoric and Roman Site on Mount Farm, Dorchester. *Oxoniensia,* II, 17, Figure 4.

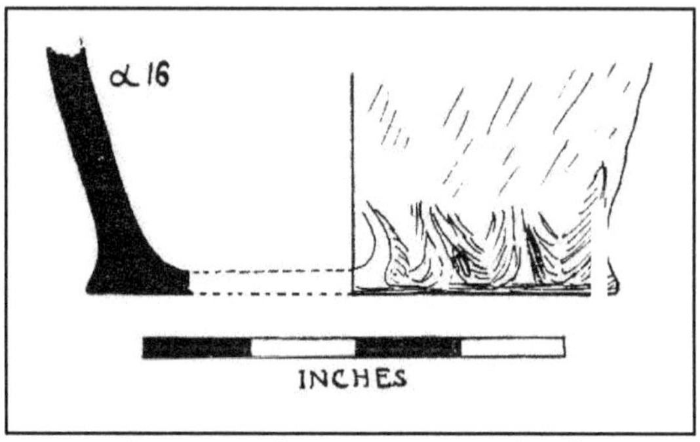

Figure 65. The Mount Farm sherd.
After Myres, J. N. L. (1937), A Prehistoric and Roman Site on Mount Farm, Dorchester. *Oxoniensia,* II**,** 33, Figure 8.

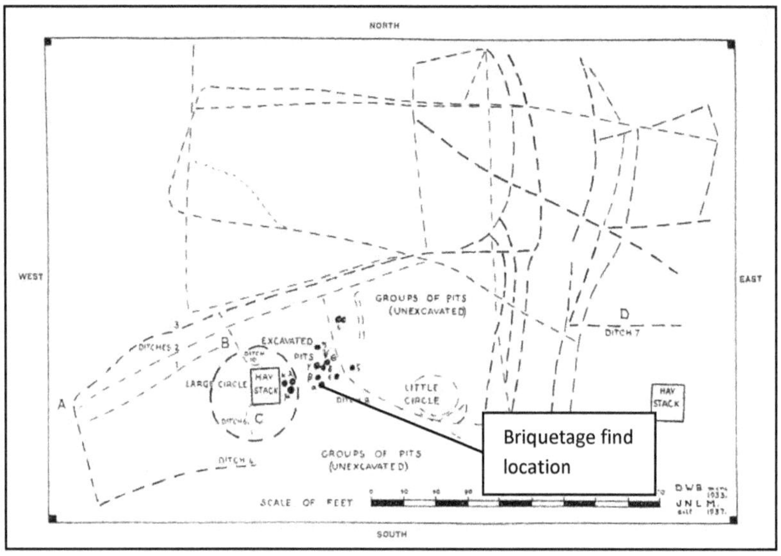

Figure 66. Plan of Mount Farm, Dorchester.
After Myres, J. N. L. (1937), A Prehistoric and Roman Site on Mount Farm, Dorchester. *Oxoniensia,* II**,** 14, Figure 3. Scale shows 270 feet.

highlighting the intensity and continuity of the salt exchange at this site noted by Morris. Despite the total weight of the assemblage, using Poole's container weight estimated of 500gm, the excavators pointed out that the assemblage need not exemplify a large quantity of salt reaching the site (Lambrick and Allen, 2004: 340). There is no reason to see this location as a central place for salt exchange such as Conderton Camp despite the total weight of the assemblage.

There was no evidence that briquetage was involved in ritual deposition activities at this site, with briquetage disposal seeming to occur with routine removal of other occupation debris. Malvernian pottery, often found at sites where Droitwich briquetage was present (Morris, 1983), was absent in the earliest phases here, with only Malvernian B1 sherds found in the Late Iron Age/Early Roman period (Lambrich and Allen, 2004: 341), leading Morris to suggest that the salt trade area was larger than the pottery trade area in this region during the earlier period (Morris, 2004: 360).

The material culture assemblage at Gravelly Guy suggested that this was a low status site (Lambrick and Allen, 2004), an assessment which would fit well with the interpretation of the presence of limited amounts of salt, presumably a rare commodity in this region due to the distance to all large scale potential salt sources. At the same time, though, there was evidence of a grain surplus being produced at the site and indications that horses were being raised here in sufficient numbers to have been traded. Indeed, one ramped hollow at the site was interpreted as a possible stock showing area (ibid.*:* 487). The artefactual evidence from the site would imply that any "wealth" being generated from possible exchanges of these commodities may have gone to some other location, or perhaps archaeologists are simply unable to identify the material or social benefit that accrued from these potential exchanges.

Watkins Farm

The assemblage of Droitwich briquetage from this site was much smaller than at Gravelly Guy, only 47 sherds with a total weight of 986gm, yet the monograph provided context level detail for the finds (Allen, 1990). Briquetage was recovered from 27 contexts, all but 2 dated to the Iron Age. The majority of contexts which contained briquetage are illustrated in the plans shown as

Figures 68, 69 and 70. From Site A, only briquetage containing contexts 170/1 and 417 are not shown, nor are the two contexts, 206/1 and 235/2, which contained a total of 50gm of briquetage (roughly 5% of the site total) from the northern portion of the site, referred to as Site B, which was of Roman date.

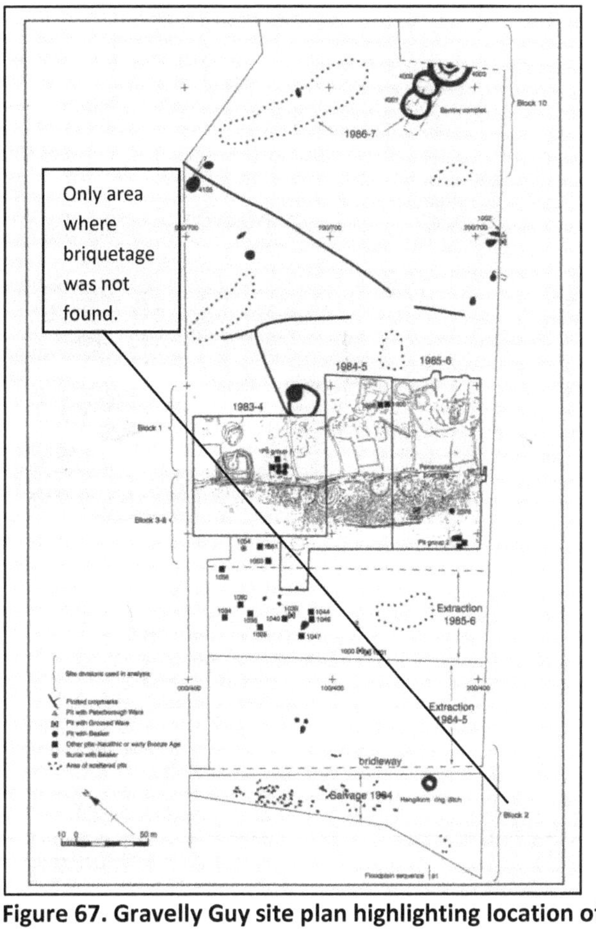

Figure 67. Gravelly Guy site plan highlighting location of Block 2.

After Lambrick and Allen, 2004: 4, Figure 1.2. © Copyright Oxford Archaeology. Scale equals 50 metres.

The sherds were reported to have come from at least five different containers. However, the phased plan of the central enclosure in Figure 70 shows that briquetage was found in all of the activity phases identified for the central roundhouse enclosure, suggesting that perhaps far more than five containers might have been exchanged during the estimated 200 year life span of the site before its reoccupation in the Roman period. As at Gravelly Guy, bone evidence suggests that horse breeding had been a major economic activity at this site and the possibility that salt in briquetage containers acted as a high status exchange good for horses, also generally regarded as high status animals in Iron Age Britain, has been discussed in Chapter 4. No briquetage at this site was utilised as an obvious special deposit, although the possible use of salt which had been removed from the container being used for this purpose remains a possibility here, as at other sites, that cannot be demonstrated archaeologically.

More about the Oxfordshire study area

Clearly, finding briquetage in this part of the Thames Valley is an unusual occurrence. Site dating for the briquetage negative sites, as shown in Table 14, is frequently imprecise, with 22% of the sites dated broadly to the Iron Age. A further 17% appear to have experienced a variety of occupation and abandonment phases between the end of the Bronze Age and the Romano-British period. As discussed in Chapter 3, although it is unlikely that any briquetage would be encountered on a Bronze Age site, it was not totally impossible as briquetage use extends back to the Middle Bronze Age. Similarly, as briquetage use seems to extend into the early Roman period in some areas, these later dates are not themselves exclusionary of the material. It is also worth recalling that the briquetage finds, as shown in Table 13, span the entire Iron Age and extend into the Romano-British period. Thus temporality is not a sufficient explanation for the negative sites, as none can be considered as being either too early or too late to have yielded a briquetage find.

Table 15 summarises the terms used to describe the excavated Iron Age sites in the Oxfordshire study area, a list which collectively helps to characterise the sites in this group. The numbers in the table exceeds the number of negative sites, as several terms may have been used to describe a single location. Perhaps, then, as was suggested in the discussion of Worcestershire above, a consideration of the nature of the negative sites might yield an explanation of why there was no briquetage. Here, however, the lack of a standard language to draw upon in judging functions of reported features makes the task difficult. For example, is a "field boundary" (e.g. Case and Sturdy, 1961/2 , Hamlin, 1963) really just another name for a ditch? Are "enclosures" in one text sometimes the same as "enclosed settlements" in another, and how do they differ from "hillforts" which are, after all, very large enclosures? Despite the ambiguity of the archaeological language, it is still possible to draw some conclusions about the probability of the presence of salt from the terms used to interpret the negative sites as shown in Table 15.

One might suggest, based upon the earlier Worcestershire case study, that there is no reasonable expectation of a find of briquetage in a linear pit group, particularly when considering the date of the excavation (Hamlin, 1963, Case, 1963) which precedes widespread recognition of briquetage as a material associated with salt distribution. The excavation that yielded evidence of quarrying (Tannahill, 2006) might also be thought an improbable location to encounter briquetage based upon the nature of the site, with little reason for salt to be used there. With no known examples of briquetage recovered from inland waterlogged sites, its absence from the waterholes (Mudd, 1993, Roberts, 1995) and wells (Everett and Eeles, 1999, OCC, 2009) might also not be considered

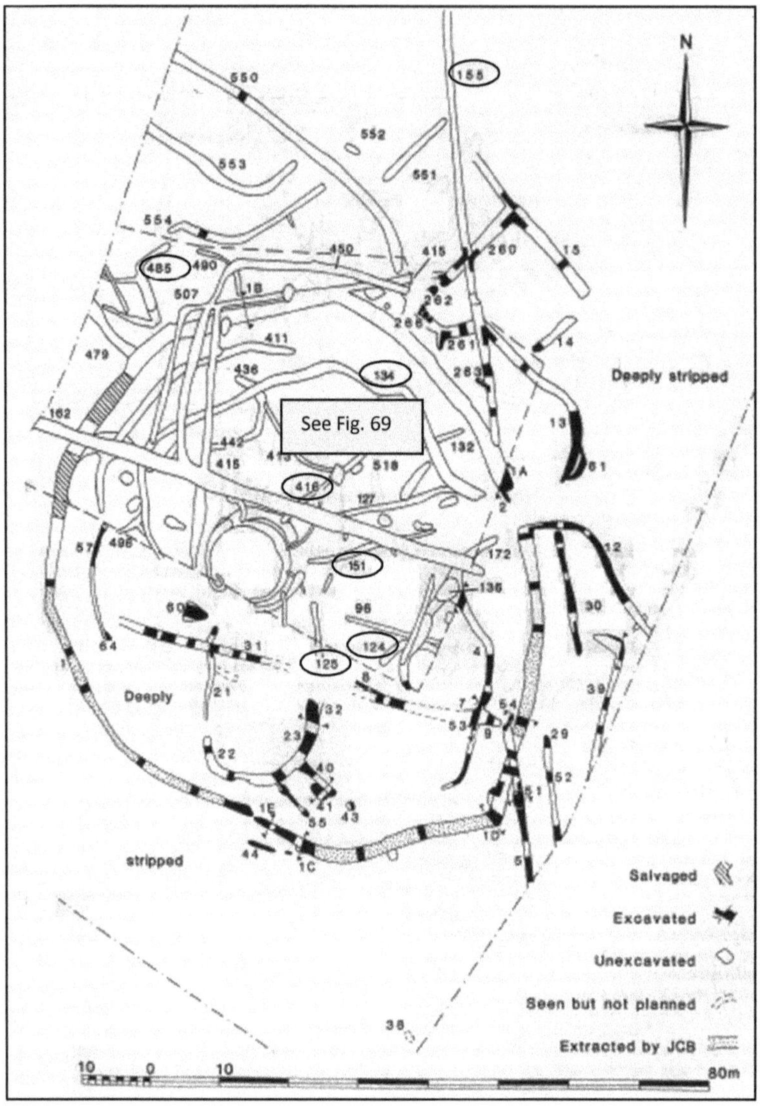

Figure 68. Watkins Farm Site A, the Iron Age enclosure, with circles indicate contexts with briquetage.
After Allen, 1990: 5, Figure 3.

Time Period	Number
Either Bronze or Iron Age	1
Late Bronze Age – Early Iron Age	3
Early Iron Age	9
Early to Middle Iron Age	11
Middle Iron Age	7
Middle to Late Iron Age	3
Middle to Late Pre-Roman Iron Age (Belgic)	1
Middle Iron Age to Early Romano-British	4
Middle Iron Age to Romano-British	1
Late Iron Age	5
Late Iron Age to Early Romano-British	6
Late Pre-Roman Iron Age (Belgic)	4
Iron Age to Romano-British	6
Generic Iron Age	22
Discontinuous multi-period sites	17
Total	100

Table 14. Briquetage-negative site dating in Oxfordshire study area.

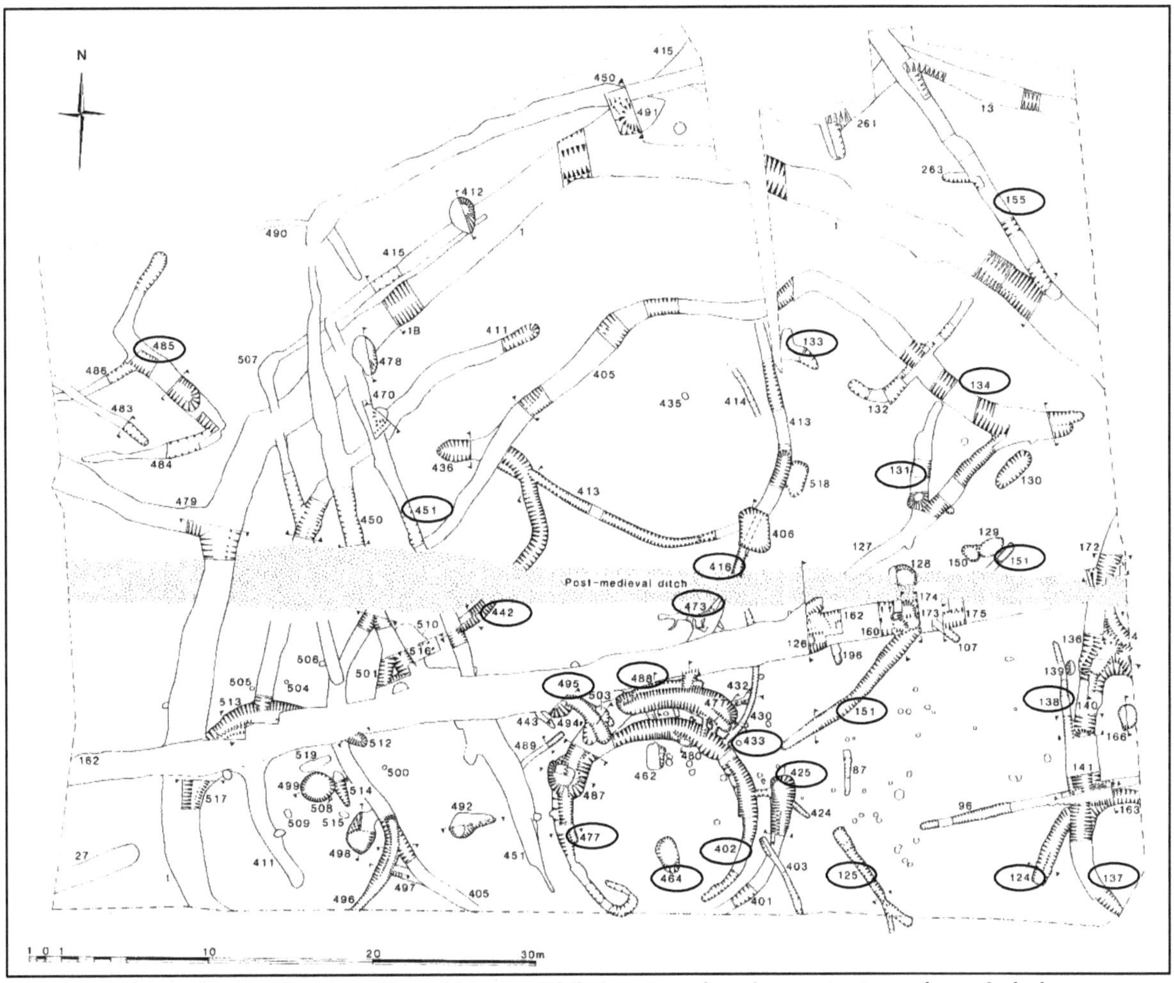

Figure 69. Northern portion of Site A with briquetage-bearing context numbers circled.
After Allen, 1990: 6, Figure 4. © Copyright Oxford Archaeology. Scale shows 30 metres.

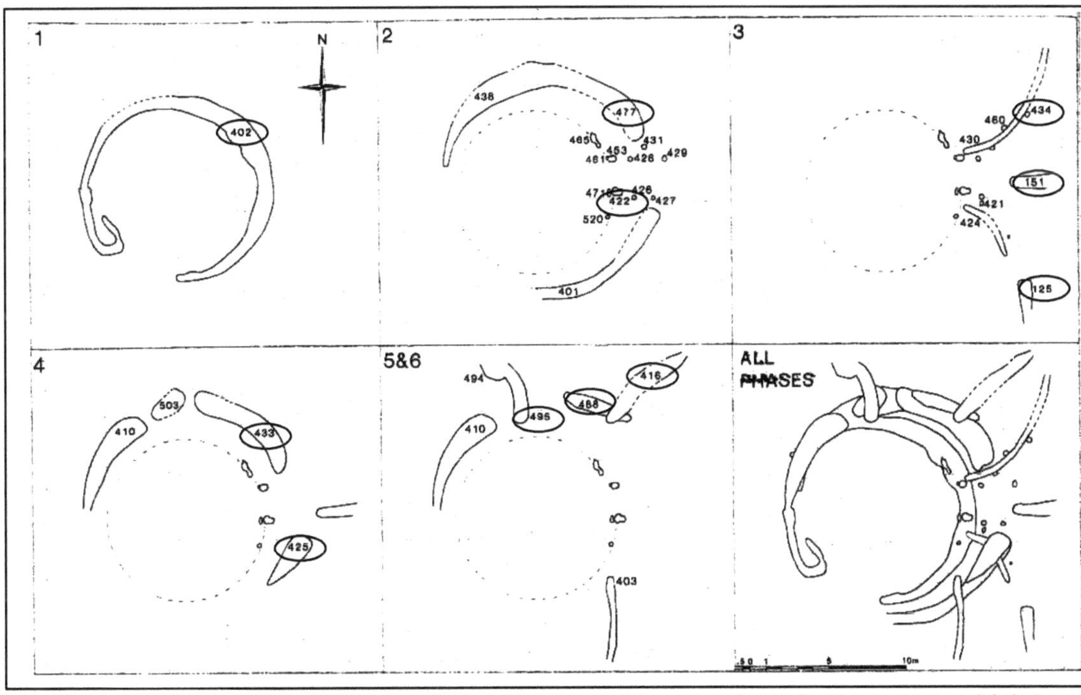

Figure 70. Phased illustration of the central roundhouse enclosure; circles indicate contexts with briquetage.
After Allen, 1990: 12, Figure 9. © Copyright Oxford Archaeology. Scale shows 10 metres.

Feature/ Site/Structure Description	Find Sites	Negative Sites
Total number of sites	6	100
Boundary	2	2
Building foundation	0	1
Cobbled surface	1	1
Cremation(s)	1	3
Ditch(es)	6	32
Enclosure(s)	4	23
Field systems	1	0
"Four poster"	0	2
Gully(ies)	5	19
Hearth or oven	0	1
Hillfort	1	1
Inhumation(s)	2	12
Linear pit group	0	1
Midden	1	0
Oppidum	1	0
Pit(s)	6	47
Pit cluster(s)	3	6
Post hole(s)	3	14
Pottery (as an isolated find)	0	6
Quarry	0	1
Ramped hollows	1	0
Ritual deposit (as an isolated find)	0	1
Ritual pit	1	0
Roundhouse or hut	4	11
Settlement	3	32
Trackway	0	3
Waterhole	1	2
Well	1	1

Table 15. Descriptive terms applied to the sites in the Oxfordshire study area.

unusual; while these features do occur at Mount Farm (Myres, 1937), they are not the source of the known briquetage sherd. The absence of briquetage on a cobbled surface that may have been used for industrial purposes (Hindmarch, 2002) or on a trackway (Hinchliffe and Thomas, 1980, Everett and Eeles, 1999, Coleman, 2001) might seem reasonable as well, although it was found in a trackway at Aston Somerville, Worcestershire (Coleman et al., 2003) and there was a cobbled surface found at the Vineyard (Lambrick and Robinson, 2009), although as an unpublished site it is impossible to say whether or not it held briquetage.

As noted in the introduction to this chapter, briquetage might well have been ignored if present by those who discovered the Sutton Courtenay warrior burial with sword and shield in 1826 (Case, 1963), though its absence within the other inhumation and cremation sites in the dataset discovered later (e.g. Moore, 2003) does pose an unanswerable question as to whether or not salt, outside of the briquetage container, was part of death rituals in this region, a use which ethnography suggests is common (e.g. Jones, 1923, Kurlansky, 2002). Yet there is a general overlap of features such as ditches and pits and structures such as four posters between the negative sites and the study area briquetage find sites listed in Table 13. For example, even though most enclosed settlements in the study area have not yielded briquetage, Watkins Farm did (Allen, 1990). Thus, while some specialised sites and features are more likely to not have briquetage in the study area, its presence in the more common Iron Age sites such as enclosures, and features, like ditches and pits, while statistically unlikely (6 sites with versus 100 without), is still a possibility in this study area.

A further theory to explore relative to the pattern of negative sites and find sites is site "status" as interpreted by archaeologists. It might be possible to suggest that the Abingdon Town Centre *oppidum* (Allen, 2000) and the Castle Hill hillfort (Lambrick and Robinson, 2009) had briquetage present because they were high status sites which may have functioned as "central places" for salt distribution as discussed in Chapter 4. Blewburton Hill, a hillfort at the southern edge of the study area, however, has been excavated and failed to yield briquetage (Collins, 1947), though its absence there could be attributed to a failure to recognise it due to the date of its main excavation phase, a time when it could easily have been missed or ignored, as was mentioned above relative to the Sutton Courtenay warrior burial; the distance from both Droitwich and the Hampshire coast or a combination of these factors. Despite this example, however, the "high status" link may not be without merit as a partial explanation of briquetage distribution locally, as 4 of the 16 briquetage finds (25%) in Oxfordshire came from either hillforts (Morris, 1983, Poole, 2005) or the Abingdon *oppidum*. Any of these sites may well have functioned as "central places" for salt distribution as discussed in Chapter 4.

At the other end of the status scale, however, the situation is more ambiguous. The finds from Gravelly Guy (Lambrick and Allen, 2004) and Watkins Farm (Allen, 1990) were discussed above, noting that the overall material culture assemblages suggest they were low status farming sites despite the evidence for horse breeding. The ceramic assemblages found at these locations are closely comparable to that from the Farmoor site (Lambrick and Robinson, 1979) where no briquetage was found. As other evidence such as finds of finely worked copper alloy harness fittings dating from the Bronze Age onwards and horse images on Late Iron Age Celtic coins suggest, horses may have been treated as high status animals. It is possible that a special good, such as salt in a briquetage container, was sometimes required in payment for another type of high status good, in this case horses. Alternatively, the briquetage may be no more than evidence of the use of salt as part of the horses' diet, with horses needing five times as much salt as humans for good health (Kurlansky, 2002).

As the discussion above has illustrated, there is little to differentiate the sites in this study area that have yielded briquetage from those where none has been found. This would suggest that briquetage salt containers were rare objects in this region in the Iron Age, though possibly associated at times with high status activity, and unlikely to be encountered under most circumstances by most individuals. This need not reflect the actual availability of salt in the study area. In addition to the proven salt sources from Droitwich and the Hampshire coast evidenced by briquetage finds in the study area, the possibility exists that the Thames allowed for transport of salt from Kent and/or Essex, both areas where briquetage was not normally used for distribution. These transport distances would have been comparable to the overland movements in evidence (Lambrick and Robinson, 2009), and may have actually been an easier mode of transport for a bulk commodity as was discussed in Chapter 4.

5.4 Essex

Essex is the second study area that is a salt production region yet, as can be seen in the map in Figure 71, the distribution of Iron Age briquetage finds is quite unlike that seen in the Worcestershire study area, where finds clustered at the Droitwich salt spring. The Essex study area briquetage find sites are far more diffuse, with finds occurring at 41 Iron Age sites, 30 of which were Red Hills or other sites used for salt production, a pattern which might be anticipated from the earlier discussion in Chapter 3 of the large number of production sites known in Essex. Of the non-production briquetage finds listed in Table 16, only that made at Layer-de-la-Haye (ECC, 2003) is unaccompanied by at least one site in very close proximity without briquetage. Yet even in Essex the number of Iron Age sites without briquetage exceeds the number where it has been found. Appendix 5 contains a list of 70 excavated Iron Age sites in the study area where it has not been recorded, clear evidence that even close to the production sources there briquetage was not generally used for salt transport in this region, a practice that was discussed in Chapter 4.

In a region where so much briquetage has been discovered, it may be unsurprising that few of the publications of the non-production sites where it does occur provide explicit context-level descriptions of the finds, although interpretive discussions of the sites in Table 16 do tend to note the relative rarity of such finds away from production sites (e.g. Crummy, 2007). An exception to this general rule, however, is the assemblage from Elm Park House, Ardleigh. As was shown in Table 10 above, this site yielded the sixth largest collection by weight of non-production briquetage that has been found, particularly intriguing in light of the general absence of briquetage use for salt distribution in this area.

Elm Park House, Ardleigh

The landscape adjacent to the village of Ardleigh, Essex, contains a cropmark complex, shown in Figure 72, well known since the 1950s. The history of the exploration of these cropmarks, dating from the Neolithic to the Roman period, is discussed by Brown (1999). It was not until 1976 that briquetage of Late Iron Age or early Roman date was located in the Ardleigh cropmarks. 281 sherds with a total weight of 5,891gm were recovered from 129 contexts (Major, 1999) during that excavation. Subsequent to these finds, a small trench excavated in 1981 to allow for the installation of a lift in Elm Park House revealed a shallow gully that yielded nine further sherds of briquetage dated to the same time period (Thompson and Barford, 1986).

The largest group of sherds, 47 with a total weight of 754gm, came from a small trench, Area 23, which had been opened to verify the location of ditches visible in the much larger adjacent trench, Area 20 (Brown, 1999). Area 23, though small, also had significant concentrations of baked clay and Roman tile (Major 1999: 158). Otherwise, the briquetage was fairly evenly distributed across the site according to Major (1999: 157). Although no detailed plan of Area 23 was published, the detailed plan of Area 20 that appears below as Figure 73 shows a portion of ditch 7857. This ditch defined the southern boundary of a trackway, C49, and dated to the mid-first century AD, the end of the Late Iron Age or the early Roman period. Also visible in this plan is ditch 7424, of similar date, which defined the north side of trackway C49, the only other feature specifically identified by Major because of the multiple briquetage rim sherds recovered within it. Section drawing S.774 of the south western edge of Area 20, shown as Figure 74 here, shows the profile of ditch 7857, although not fill 7592 which actually contained the concentration of briquetage. Similarly, drawings S.736 and S.786 in Figure 75, show

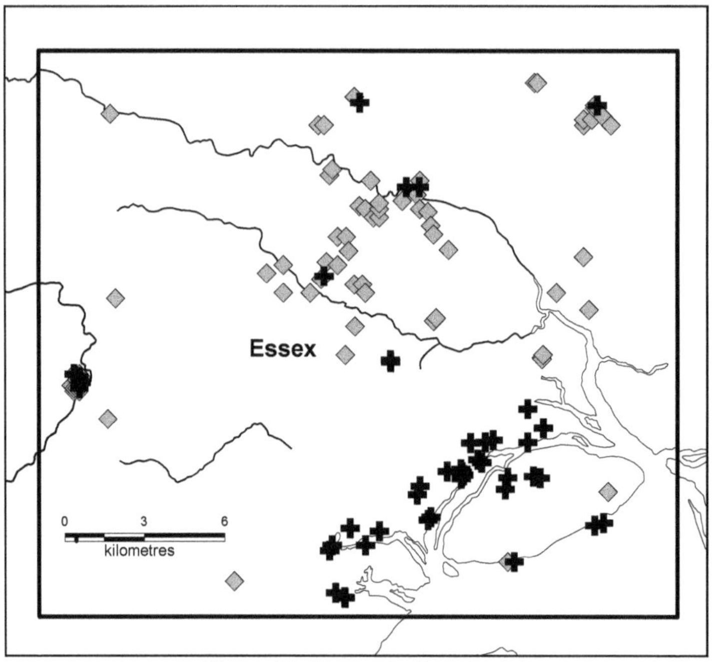

Figure 71. Essex study area.

Key: Iron Age salt production sites and briquetage finds shown as crosses and excavated Iron Age sites without salt evidence as diamonds. Underlying map © Crown Copyright Ordinance Survey.

Name	Description	Location	Active Period
Gosbecks Farm Site B	ditches & pits	TL 968 288	LIA
Kelvedon	settlement	TL 864 190	LIA
Kelvedon Eddy Trench A	ditch & pits	TL 864 186	LIA
Kelvedon Eddy Trench B	ditch	TL 864 187	LIA
Kelvedon Eddy Trench C	ditch	TL 865 188	LIA
Elm Park House, Ardleigh	settlement	TM 055 287	LIA-ERB
Kelvedon Doucecroft (KL4)	enclosures & roundhouse	TL 862 191	LIA-ERB
Camulodunum (Sheepen)	*oppidum*	TL 9850 2578	LIA, RB
Camulodunum (Stanway)	mortuary enclosure ditches	TL 955 226	LIA, RB
St Helena School, Sheepen Road, Colchester	find site	TL 9898 2578	LIA, RB
Hill Farm Malting Barn, Layer-de-la-Haye	settlement	TL 979 196	IA

Table 16. Non-production briquetage find sites in the Essex study area.

ditch 7424 but not context 7639 that contained the group of container rim sherds.

No particular significance is attributed to the disposal of the briquetage in ditches that defined a stock trackway in Areas 20 and 23 by either Brown or Major, which suggests, in the absence of such a discussion, that the authors may have assumed that it was disposed of as routine occupation debris. This conjunction may, however, indicate that in the salt-rich Essex area, the mineral was regularly provided for the animals using the trackway. The other possibility raised by the concentration of briquetage at Ardleigh is that the occupants there were also salt makers (Rodwell, 1979), producing briquetage near the Red Hills, but sometimes bringing briquetage containers back to Ardleigh, though whether filled with salt or empty, for human use or as animal salt licks (Barford, 2000), is unknowable. Rodwell (1979: 160) also suggested that salt makers may have occupied Kelvedon and Camulodunum, which might make this explanation relevant to the briquetage finds at those locations.

The Essex study area sites

In a pattern reminiscent of the clustering seen in the Worcestershire map, Figure 71 shows that the negative find sites are not distributed evenly across the landscape in Essex. Of the 70 negative sites, 32 (45.7%) are found in Colchester with another 6 (8.6%) in nearby Stanway; 7 negative sites (10%) are in Kelvedon on the west central

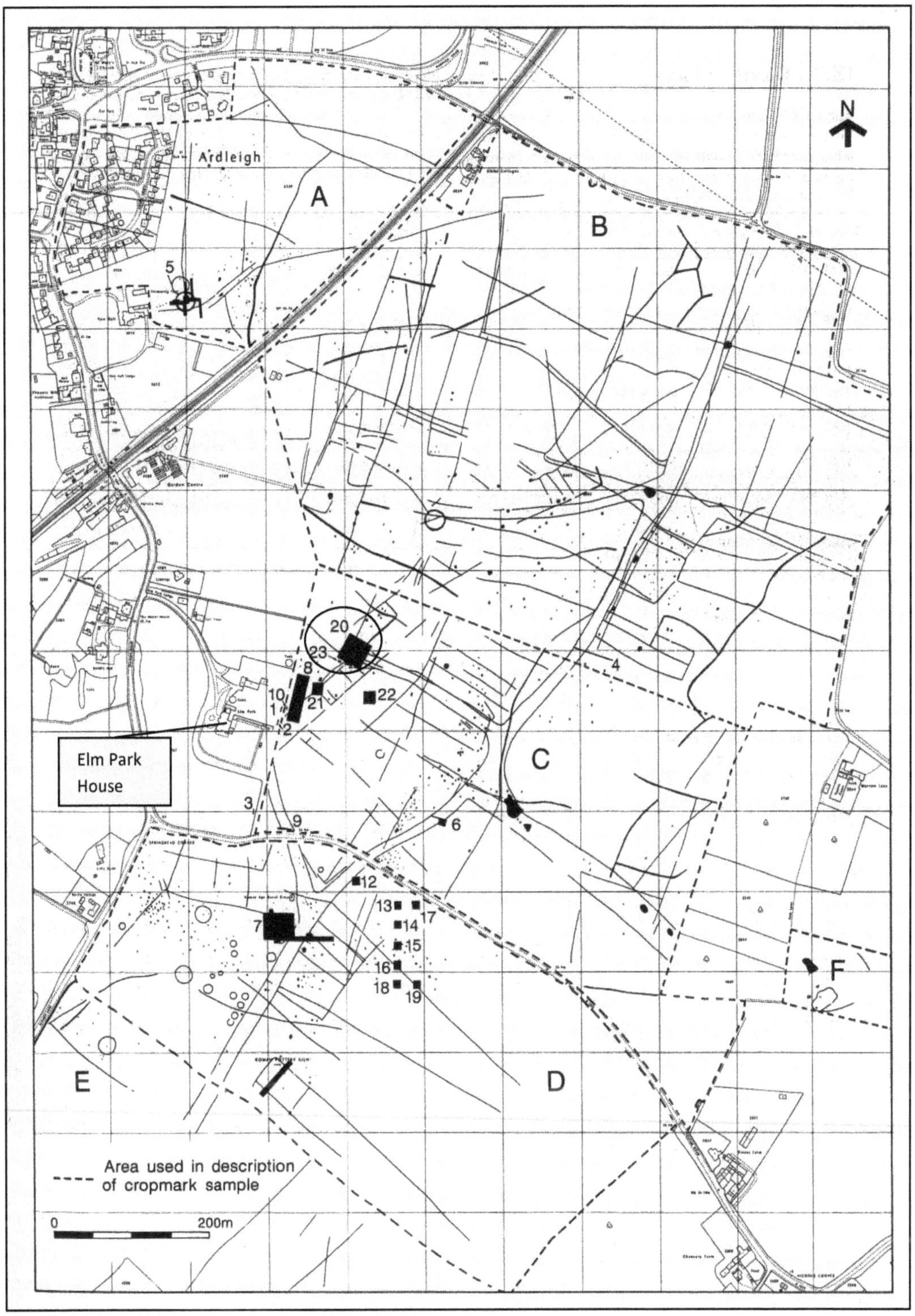

Figure 72. The Ardleigh landscape showing the location of the Central Excavation Unit trenches.
After Brown, 1999: 15, Figure 7, (c) Essex County Council. Circle marks trenches known to have briquetage finds.

boundary of the study area, and 9 negative sites (12.8%) are near Ardleigh in the northeast of the study area. Thus 54 of the negative reports (77.1%) are concentrated in these small sections of the study area. The concentration of sites in the Colchester area results from nineteenth and twentieth century exploration of the Late Iron Age *oppidum*, its environs and the later Romano-British city occurring as the town was redeveloped and housing areas expanded, when its identification as the capital of the Catuvellauni and/or Trinovantes at Camulodunum (e. g. Hawkes and Hull, 1947, Cunliffe, 2005) in the Late Iron Age was confirmed and the destruction of the early Roman town there during the Boudiccan revolt of AD 60 or 61 was archaeologically verified. Modern Kelvedon had a much earlier phase as a Late Iron Age settlement before becoming the site of the Romano-British town Canonium on the Colchester-Chelmsford road which was excavated during the second half of the twentieth century

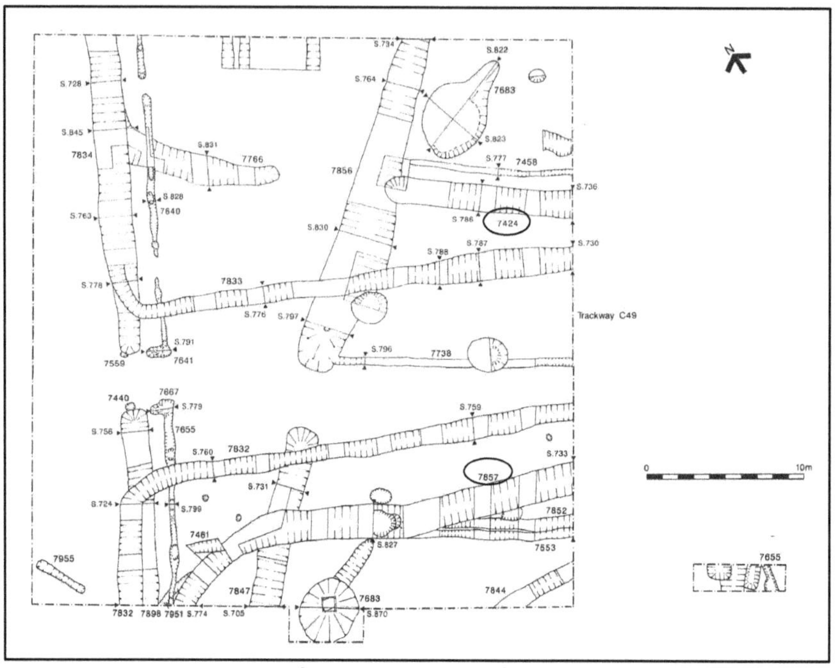

Figure 73. Area 20 plan.
Circles show briquetage-containing ditch 7424 and ditch 7857, whose continuation in Area 23 held briquetage.
After Brown, 1999: 46, Figure 31, (c) Essex County Council. Scale shows 10 metres.

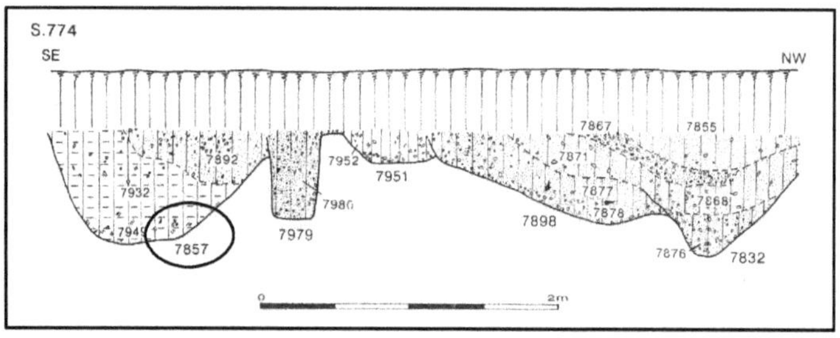

Figure 74. Section drawing showing ditch 7857 in Area 20.
After Brown, 1999: 48, Figure 33, (c) Essex County Council. Scale shows 2 metres.

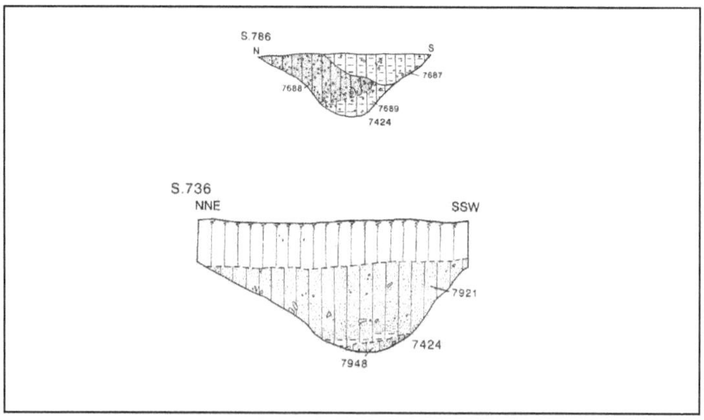

Figure 75. Two section drawings of ditch 7424 in Area 20.
Afters Brown, 1999: 47 & 48, Figures 32 & 33, (c) Essex County Council. Same scale as Figure 74.

Time Period	Number
Either Bronze or Iron Age	1
Early Iron Age	3
Early to Middle Iron Age	2
Middle Iron Age	2
Middle to Late Iron Age	1
Middle Iron Age to Early Romano-British	1
"Pre-Belgic" Iron Age (Early or Middle Iron Age)	1
Late Iron Age	25
Late Iron Age to Early Romano-British	9
Late Iron Age to Romano-British	5
Late Iron Age or Romano-British	1
Iron Age to Romano-British	5
Either Iron Age or Romano-British	1
Generic Iron Age	10
Discontinuous multi-period sites	3
Total	**70**

Table 17. Negative site dating in Essex study area.

(Rodwell, 1988). The Ardleigh area, as was noted above, has had significant archaeological attention, much of it involving the Colchester Archaeological Group, since the late 1950s when aerial photography began to reveal a high density of cropmarks to the south and east of the modern village (Erith, 1960). However, while the wider landscape shows a continuity over this long period near Ardleigh, few, if any, sites in this part of Essex show direct continuity of use between the Middle and Late Iron Ages (Hawkes and Crummy, 1995). The transition period is also characterised by the appearance of the so-called "Belgic" or Aylesford-Swarling pottery tradition (Sealey, 2007) and new burial customs, as demonstrated at the Lexden Belgic Cemetery (Hawkes and Hull, 1947). In Essex the archaeology strongly suggests a period of major social discontinuity as new behaviours and forms of material culture supplanted the old.

Table 17 considers the reported dating of the briquetage-negative sites. For the purposes of compiling this table in the most straight-forward manner possible and to allow for comparisons with the other study areas, the terms "La Tène III", "Iron Age C", "Belgic", "Aylesford-Swarling", "Late Pre-Roman Iron Age" and "Late Iron Age", that have each had their period of currency in archaeological literature about Essex during the past 130 years or so, have all been subsumed here into the term "Late Iron Age". As used here, the term refers to the period beginning *c.*50 BC and extending until the Roman conquest *c.*AD 43 characterised by the emergence of distinctive burial patterns that have parallels on the continent (e.g. Laver, 1927) and the production of wheel made, grog-tempered pottery that accompanies them (Gibson and Woods, 1997).

As the Oxfordshire and Essex study areas both have statistically significant numbers of negative sites, they can usefully be compared; although coastal Essex was a salt production area, many negative sites in the study area would have been importing salt, though not over the same distances, as negative sites in Oxfordshire. Yet there are distinct differences between the Oxfordshire and Essex study areas with respect to site dating. Dating of Essex sites is somewhat more precise with only 14.3% classified very broadly as "Iron Age" compared to 22% of the Oxfordshire sites. A full 57% of the Essex sites are of Late Iron Age date or later, while only 15% of the Oxfordshire sites fall into that category. Also in sharp contrast to Oxfordshire, the Early Iron Age is not well represented in this study area, with only 7% of the negative sites definitely related to this period compared to 23% of those in Oxfordshire. Intriguingly, the expansion of negative site numbers in the study area in the Late Iron Age coincides with the general expansion of salt production site numbers in Essex discussed in Chapter 3, which allows for speculation that there may have been a possible link between greater salt production and the other changes noted at this time. Alternatively, it is possible that the population density in the Essex study region was far lower in the Early and Middle Iron Age than in the later period and thus there are fewer sites to find, that Early and Middle Iron Age sites in the study area are of types that are difficult to locate archaeologically with aerial photographs so have not yet been located and excavated, or that because Late Iron Age sites are particularly likely to be near the Romano-British sites of interest to many archaeologists they are more likely to be discovered when these sites were excavated.

Table 18 is a summary of the site types, features and structures at the sites in this study area noted in the original source material. In the case of the Red Hill briquetage finds, however, they have been lumped as a single category with no further enumeration of any features or structures that might have been excavated there. All of the other figures in the "find sites" column in this table refer to the 11 non-production sites listed in Table 16. As was the case in Table 15 one might easily question the typological categories and argue for different separations or combinations than used here, yet the excavators tried to convey their impressions of the sites with these terms, so the continued separation of generic ditches from field ditches and linear features has been carried forward here. Similarly, as one might anticipate in comparing areas of Iron Age England, there are many categories which overlap. Enclosures, pits, ditches and settlements occur in both the Oxfordshire and Essex study areas, though not always at the same frequencies, and these variations help to characterise what seem to have been distinct cultural differences between the inhabitants of these areas. In both areas, the *oppida* have yielded briquetage, suggesting a possible role in the salt trade for both sites. Continuing the comparison of the negative sites, while the number of enclosures found were similar, appearing at 23% of the Oxfordshire sites and 21% of the Essex ones, pits were noted at 47% of the Oxfordshire study area sites, but at only 23% of those in Essex, the difference possibly due to waterlogged conditions in some of the Essex study area sites making pits less useful than in Oxfordshire. A surprising similarity in the datasets, as hillforts are not as common in Essex as in Oxfordshire (e.g. Cunliffe, 2005), both study areas contain a single excavated hillfort - Pitchbury Ramparts in the case of Essex (Hawkes and Hull, 1947, Hawkes and Crummy, 1995) – that did not contain briquetage, albeit from a rather limited ceramic assemblage in this instance.

A major variation between the two areas, though, can be noted in the treatment of the dead. In the Oxfordshire study area, negative sites with inhumations outnumbered cremations by a 4 to 1 ratio. Neither method of body disposal, however, would be thought of as particularly common features in the archaeological record there, with only 12% of sites having an inhumation and 3% of sites reported as having a cremation. In Essex, the ratio is reversed at 2.4 to 1 in favour of cremations, with 8.6% of negative sites having inhumations and 21.4% of sites having cremations reported. Further, five negative sites or 7.1% of the Essex total were considered to represent cemeteries while no similar sites are reported within the Oxfordshire study area. These archaeologically visible practices in Essex date mainly to the Late Iron Age and are part of the package of behavioural changes in that period discussed earlier. Despite the seeming availability of salt in Essex, there is little evidence to suggest that salt had a role in the rituals associated with the disposal of bodies there. Only the singular find of briquetage in a mortuary enclosure ditch at Stanway (Crummy, 2007) suggests that this appearance might be erroneous and reflects the general movement of salt in non-briquetage containers in this locale.

Looking for a clear association of status with the presence or absence of briquetage is difficult in Essex as the ceramic is only rarely found away from production sites, as has been discussed above and in earlier chapters. With 30 find sites being Red Hill salt production sites where its presence would be ubiquitous (Fawn et al., 1990), its strongest association seems to be with low status industrial locations. At the same time, though, two of the 11 inland finds, or 18% of this small non-production subset, are at the opposite end of the status scale: the Camulodunum *oppidum* with its rich collection of ceramics and early Roman imports (Hawkes and Hull, 1947) and the Stanway cemetery known for its rich burials (Crummy et al., 2007). Both can be considered to be high status sites. There is a potential contradiction with this concept of status, however, and Rodwell's previously noted idea that salt makers lived at Camulodunum (1979: 160), as actual saltmaking is unlikely to have been a high status activity. One might reconcile these opposites with the proposition that low status individuals, perhaps slaves, living at Camulodunum made the salt which was traded by, and for the benefit of, the higher status individuals living at the site. Alternatively, it is also possible that the presence of briquetage at these locations was symbolic rather than accidental (Chapman, 2000), perhaps showing control of the salt production process by the residents of Camulodunum in the Late Iron Age.

Of the briquetage negative sites 11, or 15.7% of the total, can be considered to represent high status activity locations. This group includes the previously mentioned hillfort, the 6 dykes listed in Table 18 and 4 rich burials. The dykes fall into this grouping due to the ability of some authority or elite class, either political, economic and/or religious, to muster the labour required for their construction. This group are the excavated subset of the even larger number of Late Iron Age dykes within the study area. The specific dyke excavation sites included here lie on Grymes Dyke (Crummy, 1978), two separate segments of the Berechurch Dyke (Hawkes and Hull, 1947, EHNMR, 2007a), Shrub End Dyke (EHNMR, 2009a), Lexden Middle Dyke (EHNMR, 2007b) and Lexden Tapswoods Dyke (EHNMR, 2009b). The burials include the insertion of an Early Iron Age cremation into a Bronze Age barrow at Ardleigh (Erith, 1975), and Late Iron Age sites including the Lexden Tumulus, thought by the excavator to be the burial place of a noble or chieftain who died during the reign of Cunobelin (Laver, 1927), known from pre-Conquest coinage and from Roman histories, the Kelvedon Warrior Burial (Sealey, 2007) and Stanway Hall Farm where two cremations in elaborate wood-lined chambers with burial goods, reminiscent of the Lexden ritual, were located (Gilman, 1992). While briquetage was certainly a low status industrial material in Essex, the pattern of inland finds and its absence makes it difficult, if not impossible, to understand the relationship of salt to the elite levels of society. For example, while there is now sufficient evidence to suggest that it is likely that salt in briquetage had no role in the construction of dykes, it is impossible to know that

Feature/ Site/Structure Description	Find Sites	Negative Sites
Number of sites	41	70
Bank	0	1
Building (not a roundhouse)	2	0
Burial (inhumation)	1	6
Cemetery	1	5
Cremation(s)	1	15
Ditch(es) - Generic	11	18
Dyke	0	6
Enclosure(s)	3	15
Farmstead	0	6
Field ditch(es)	0	4
Gully(ies)	2	5
Hearth	0	1
Hillfort	0	1
House	0	1
Kiln	0	1
Linear boundary earthwork	0	2
Linear features	0	1
Mortuary enclosure ditches	1	0
Oppidum	1	0
Oven	2	1
Pit(s)	7	16
Post hole(s)	1	2
Pottery (as an isolated find)	0	5
Quernstone (as an isolated find)	0	1
Red Hills	30	0
Ring ditch	1	1
Roundhouse	1	1
Settlement	4	6
Trackway	1	5
Well	1	1

Table 18. Descriptive terms applied to the Essex study area sites.

salt was not used in magical rituals to add to the defensive strength of these barriers.

Despite having a statistically significant number of both find sites and negative sites to consider, the Essex study area does not provide a clearly discernable pattern about the usage of salt or its distribution during the Iron Age in this region. Information on the overall presence, distribution process and uses of salt in the area are ambiguous at best and no clear view of the role of salt in the everyday life of the area is generated by the data which can be presented here. However, the occurrence of briquetage finds away from production areas in the study area seems to be a phenomenon of the Late Iron Age, an intriguing finding in itself as the use of briquetage in the region dates to the Middle Iron Age (Fawn et al., 1990). As was pointed out above, there may have been a symbolic meaning attached to the presence of these briquetage sherds related to social relationships between the salt producers and those receiving the salt (Chapman, 2000). If that was the case, then perhaps the absence of briquetage at the many other sites could be an indicator of alternative types of social relationships between the producers and users of salt. Unfortunately, it is equally possible that the briquetage only demonstrates that relationships were the same in both proposed cases, but the archaeological evidence is insufficient to illuminate this point.

5.5 The Danebury region of Hampshire and Wiltshire

The final study area is the region surrounding Danebury hillfort in what once was the most densely populated area of prehistoric Wessex (Sharples, 2010). Unlike the other study areas, this 400km^2 area straddles two modern counties, lying primarily in Hampshire but extending into western Wiltshire. In Chapter 4 it was noted that although salt was produced on a wide swath of the south coast between Dorset and East Sussex, most inland finds of briquetage cluster in a relatively narrow band, running between the Southampton area inland to the west and through central Hampshire northwards towards what is now southern Oxfordshire, a route which passes through the study area. As shown in Figure 76 the study area contains 10 briquetage find sites which are listed in Table 19, and 60 excavated Iron Age sites found in Appendix 5 where no briquetage has been identified. The area is largely rural, with the town of Andover, lying in the north eastern section of the study area, and the village of Longstock in the south east, both near the A34 dual carriageway, being focuses of modern expansion and road building schemes. Unlike the other study areas there has been long term, recent work on sites here by a single investigator, Sir Barry Cunliffe, which has produced large quantities of data at a fairly consistent level of detail, not only on Danebury hillfort excavated for 20 seasons (Cunliffe, 1984a, Cunliffe, 1984b, Cunliffe and Poole, 1991a, Cunliffe and Poole, 1991b, Cunliffe, 1995), but on 9 further sites, published in 7 volumes, as part of the Danebury Environs Programme (Cunliffe and Poole, 2000a, Cunliffe and Poole, 2000b, Cunliffe and Poole, 2000c, Cunliffe and Poole, 2000d, Cunliffe and Poole, 2000e, Cunliffe and Poole, 2000f, Cunliffe and Poole, 2000g).

In Table 19 Balksbury Camp is described as a "hillfort" but only using the broadest possible sense of the definition of this term, as it is neither on a hill nor likely to have been a good defensive fortress. It should be noted that the excavators Wainwright and Davies (1995) suggested the term "plateau enclosure" as a description for the site, but for the sake of simplicity the better known description of the site has been used here. Although all of the sites in Table 19 have yielded briquetage, the quantities have not always been particularly large and might well have been missed without a rigorous review of the ceramic assemblage. New Buildings, for example, had 1 piece of briquetage; sufficient to put it into the finds table but not in the same class as the nearly 7.6 kilos of briquetage found at Danebury (Poole, 1984b, Poole, 1991). Among the Table 19 find sites far more material has been found within pits than in other features, a common pattern at Iron Age sites where pits are often the most prolific sources of artefact finds. Pits were the most common locations for briquetage finds made at Woolbury Ring (Poole, 2000e), Bossington (HCC, 2006c), where the site consisted entirely of a pit, Bury Hill (Poole, 2000f), Suddern Farm (Poole, 2000a) and Houghton Down except in the Late Iron Age-Romano British phase there (Poole, 2000b). At Danebury, although briquetage occurred in all areas of the hillfort, it was also most commonly found in pits (Poole, 1984b, Poole, 1991).

The use and calculation of the Specific Salt Container Index (SSCI) (Morris, 1983) was presented in Chapter 4

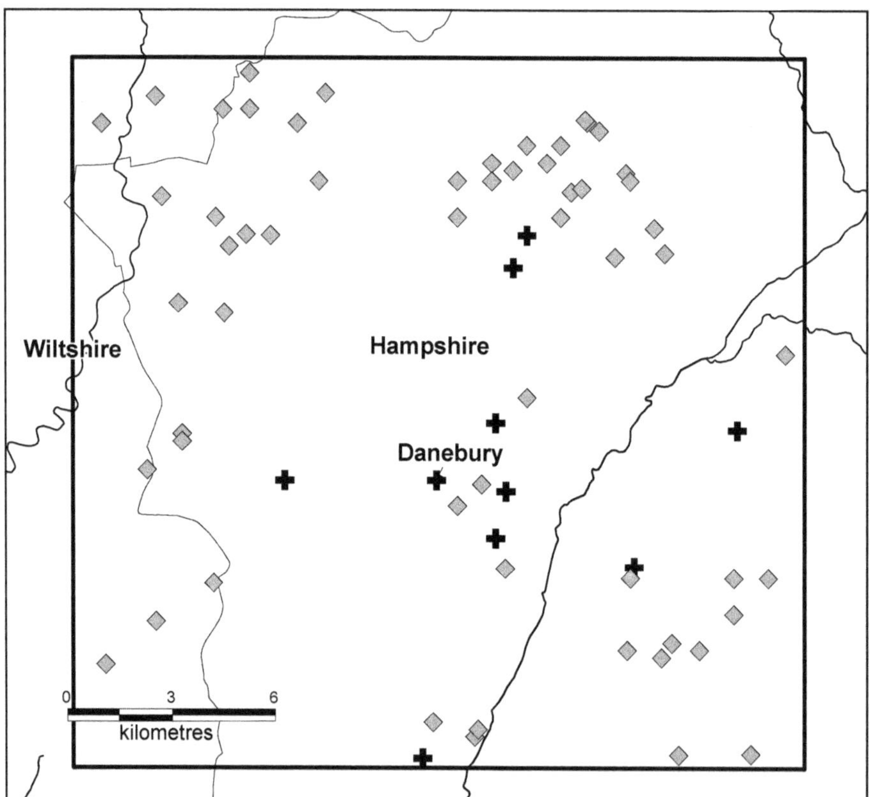

Figure 76. Briquetage find sites and excavated Iron Age sites without briquetage.
Key: Briquetage finds (crosses) and excavated Iron Age sites without briquetage (diamonds). Underlying map © Crown Copyright Ordnance Survey.

Name	Description	Location	Active Period
Balksbury Camp	hillfort	SU 350 445	LBA, EIA
Houghton Down	settlement	SU 341 361	eEIA-MIA,LIA- RB
New Buildings	enclosures and pits	SU 344 374	EIA
Danebury	hillfort	SU 324 377	EIA-LIA
Chilbolton Down	enclosure	SU 411 391	E-MIA, MIA
Woolbury Ring	hillfort	SU 381 353	MIA
Bossington	pit	SU 32 30	M-LIA
Bury Hill	hillfort	SU 346 436	M-LIA
Suddern Farm	double/triple ditched enclosure	SU 280 377	M-LIA, RB
Nettlebank Copse	banjo enclosure	SU 341 393	LIA, RB

Table 19. Briquetage find sites in the Danebury region.

NB: Active period dating relates only to finds of briquetage, not complete life cycle of find site.

as a means of comparing sites which have briquetage present. Morris has recently (2010) emphasised the need for valid SSCI comparisons to be based upon information from sites sharing the same time period, and this approach is used in Table 20. This table utilises the data from the first ten years of excavation at Danebury hillfort and five of the find sites reviewed during the Danebury Environs Programme which also are in the study area. This particular Danebury data subset was selected as the information from the 735kg of pottery accumulated during the second ten years of excavation was reported with an emphasis on what was totally new information (Brown, 1991b) compared to the more detailed information provided for the 675kg of pottery reported in Danebury Volume 2 (Cunliffe, 1984b). Calculation of the SSCI for the full assemblage would have necessitated a detailed reconsideration of all pottery sherds collected in the second ten years to identify weight in each ceramic phase, which is beyond the scope of this research about Iron Age salt and its material surrogate, briquetage. Cunliffe theorised that Danebury acted as a "special place" for trading activity for salt as well as other commodities, such as quernstones and Kimmeridge shale, being brought into the area (1995: 91-94). He also stated that briquetage seemed to be uncommon at non-hillfort sites (ibid.: 94). Although Cunliffe eschewed quantitative comparisons between Danebury and other sites due to the disparity in sample sizes (ibid.: 93), a review of the SSCI values for study area sites for which they could be calculated, though using very small briquetage samples, generally supports his conclusion. Only Woolbury, another hillfort, has an SSCI approximating that of Danebury during the Middle Iron Age. For the other non-hillfort sites in the table, regardless of date, briquetage was roughly 10 times as common in the ceramic assemblage at Danebury as any other nearby site where it was reported.

As did Tables 15, and 18, Table 22 attempts to capture the site description, features and structures noted in the study area. Clearly ditches, enclosures, pits, post holes, roundhouses and settlements are descriptions of features that could be found in any of the study areas and reflect areas of commonality. However, pits are found at 100% of the find sites in Oxfordshire and the Danebury study areas, but only 63.6% of the 11 non-production briquetage find sites in Essex. Similar statistics can be calculated for the negative sites as well, with 47% of the Oxfordshire sites and 33% of the Hampshire sites having pits, but only 22.9% of the Essex sites reporting these features. Although present in Oxfordshire, Essex and the Danebury area, neither inhumations nor trackways, can be considered common in any of the excavations in these areas. Other regional distinctions can also be noted. Though all three areas have sites described as hillforts, only the present study area has multiple hillforts in both the briquetage finds and without briquetage categories, reflective of the greater overall density of hillforts that characterises much of Wessex (Cunliffe, 2005: 348).

While ritual deposits were noted in Oxfordshire and special deposits in this region, they were not highlighted as major features in the Essex study area, perhaps indicative, along with the distinctive Late Iron Age methods for body disposal there, of alternate ritual practices unknown in these central areas. It is also possible that some of the apparent distinctiveness in defined features results from the unique historic archaeological precedents in each of the areas which have carried localised terms and sensibilities into the present period.

While the SSCI data suggests a hierarchy existed among the sites where briquetage was found, consideration of status at the briquetage negative sites in this region is not straightforward particularly in the Early and Middle Iron Ages in this region. The ambiguity of the potential status indicators found within enclosures and hillforts, such as house size, has recently been discussed by Sharples (2010: 76-77). Yet some sites and structures, as was argued above, can be seen as being of higher status based on labour content for their construction, as was the case in Essex. Perhaps the three hillforts would routinely be considered high status sites given the effort required for their construction and maintenance along with the relative material wealth found during excavations there. The linear ditches and the bank and ditch features might also be considered to represent higher status sites for this same reason, although there is little material culture evidence

Site Name	Activity Period(s)	SSCI
Houghton Down	eEIA – MIA	0.0004
New Buildings	EIA	0.0006
Danebury	EIA	0.005
Woolbury	MIA	0.002
Danebury	MIA	0.005
Suddern Farm	M-LIA, RB	0.0003
Nettlebank Copse	LIA-RB	0.0003
Danebury	LIA	0.009

Table 20. Comparison of Specific Salt Container Indices for selected phases of Hampshire-Wiltshire study area briquetage find sites as calculated by the author.

Time Period	Number
Bronze Age and Early Iron Age	1
Bronze Age to Middle Iron Age	1
Late Bronze Age to Early Iron Age	1
Earliest Early Iron Age to Romano-British	1
Early Iron Age	4
Early to Middle Iron Age	1
Early to Late Iron Age	1
Early Iron Age to Early Romano-British	1
Middle Iron Age	4
Middle to Late Iron Age and Romano-British	2
Middle Iron Age to Early Romano-British	1
Late Iron Age	2
Late Iron Age and Romano British	1
Late Iron Age or Romano-British	1
Generic Iron Age	30
Iron Age to Romano-British	4
Iron Age and Romano-British	2
Discontinuous multi-period sites	2
Total	60

Table 21. Briquetage-negative site phasing in the Hampshire-Wiltshire study area.

found in these features to support such a designation. In sharp contrast to Essex, or even the Sutton Courtenay warrior burial (Whimster, 1977), there are no rich warrior burials here to identify personal status for archaeologists, suggesting that in this area greater emphasis was placed upon status at a group level, demonstrated by the construction of boundaries of many types, than on personal status at the time of death.

In common with the Oxfordshire and Essex study areas, the Danebury-centred study region illustrates that evidence for the presence of salt appears at only a minority of excavated Iron Age sites which are not directly involved in salt production. With respect to south coast salt distribution, as was discussed in Chapter 4, Danebury hillfort lies within a rather atypical group of inland sites which happen to have briquetage included in their ceramic assemblages. The presence of this material allowed for the construction of a "central place" theory by Cunliffe (1978) for the distribution of salt which worked with the evidence in this region as illustrated in Table 20; but in light of further information about briquetage distributions, even in this locale, not all hillforts appear to be part of this exchange network, supporting an idea put forward more recently by Sharples that there were varying uses for Wessex hillforts (2010: 3), some of which may have never served as the location for commodity exchanges.

An effort was made to develop briquetage distribution maps for the sites in this study region based upon available published data. It was hoped that all ten sites could be mapped. However, insufficient information was available for three of the sites for this purpose. No context level distribution detail was available for the finds from Bossington (HCC, 2006c), unsurprising for an Archaeology and Historic Buildings Record (AHBR) entry. The find at Chilbolton Down (Champion, 1984) is only described within the pottery fabric listings without further detail being provided. The unquantified small

Feature/Site/Structure Description	Find Sites	Negative Sites
Number of sites	10	60
Bank	0	1
Building	1	1
Burial (inhumation)	4	6
Ditch(es) - Generic	2	18
Ditch and Bank	5	3
Enclosure(s)	7	10
Enclosed settlement	2	4
Exclusively small finds: ceramics, quernstone, bones, charred grain	1	1
Field ditch(es)	0	3
"Four poster"	5	3
Gully	5	0
Hearth or Oven	2	0
Hillfort	4	3
Hollow(s)	1	2
Linear ditches	4	2
Linear pit alignment	1	0
Occupation area	0	2
Pit(s)	10	20
Post hole (s)	6	6
Pottery (as an isolated find)	0	5
Quarry	5	1
Roundhouse	3	3
Settlement	2	9
Six post structure	1	0
Special deposits	7	4
Trackway	2	1
"Two poster"	3	2

Table 22. Descriptive terms applied to sites in the Hampshire-Wiltshire study area.

amount of briquetage found at Balksbury Camp (Wainwright and Davies, 1995) figures in discussions of trade, but is not otherwise detailed within the published information about the site. As a result of these information deficiencies, the development of maps relied on the consistency of information provided by the reports on Danebury hillfort and the Danebury Environs Programme monographs.

The other seven find sites listed in Table 19 were all excavated by Cunliffe and his team as noted above. The detailed information provided in the relevant monographs was utilised in producing the briquetage find distribution maps. Most, though not all, briquetage finds from sites dug as part of the Danebury Environs Programme were defined to context level, however as is typical for ceramic finds, the location was not defined in three dimensions. As a result, although distribution maps can identify particular contexts or pit layers where finds occurred, the information which follows should be construed as indicating the general area, rather than the exact spot, where the briquetage find was made. In the case of some ditch section finds, this can mean that the level of precision achieved can be no better than within several cubic metres of the actual find.

New Buildings

As noted above, the New Buildings Enclosure B site yielded a single 10gm sherd of briquetage which the text informs came from enclosure ditch F122/2, context 1 (Poole, 2000c), the uppermost fill of the ditch. The location of this feature is illustrated in Figures 77, 78 and 79. The briquetage in the upper fill layer of the ditch could have entered the archaeological record for the first time following the final recut of the ditch during activity phase 3c, which roughly dates to the sixth century BC (Cunliffe and Poole, 2000d: 46, 59). It is also possible that the sherd was part of the spoil removed during the final recutting of the ditch and was later redeposited in ditch F122, and thus may be even earlier in date. There was no evidence that the briquetage, or the other occupation debris which accompanied it, had entered the ditch through any process other than casual deposition.

Nettlebank Copse

The next largest find by sherd count was made at Nettlebank Copse. The four briquetage sherds weighing a total of 38gm found there, all of which were located in either the main enclosure or north antenna ditches, can also be identified to specific context by use of data included on the microfiche set included with the text, as shown in Figures 80 through 84 (Cunliffe and Poole, 2000e, Poole, 2000d). A maximum of three recuts of these ditches were identified in the excavated sections,

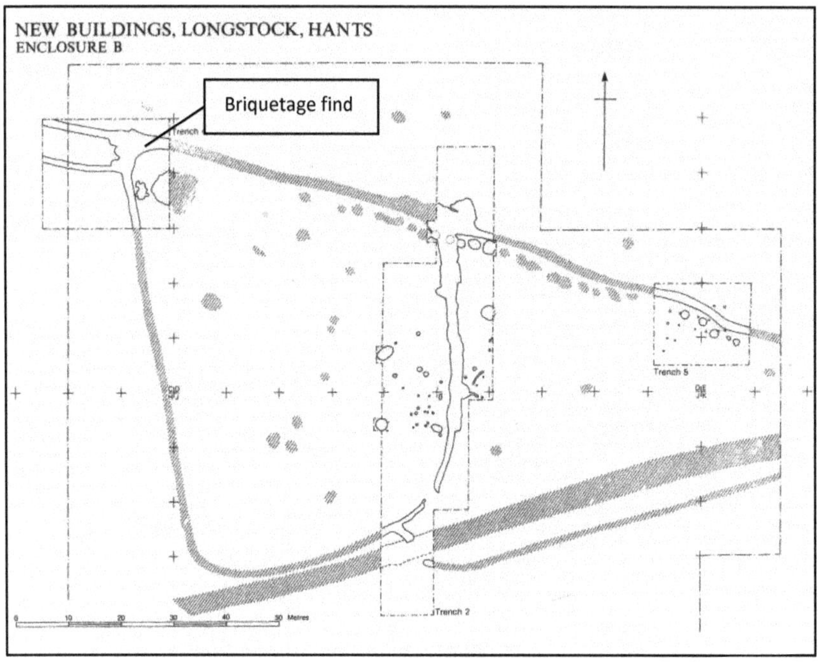

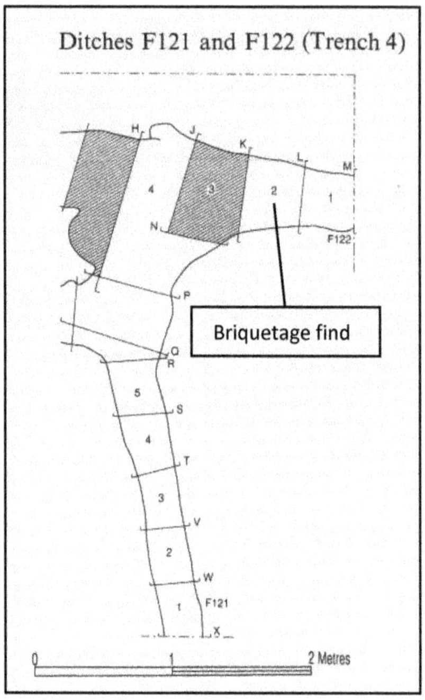

Figure 77

Figure 78

Figure 77. New Buildings site plan showing briquetage find location.
After Cunliffe and Poole, 2000d: 25, Fig. 4.8b. Scale shows 50 metres.

Figure 78. Detail view of Ditch F122 showing segment where briquetage was found.
After Cunliffe and Poole, 2000d: 43, Fig. 4.24. Scale shows 2 metres.

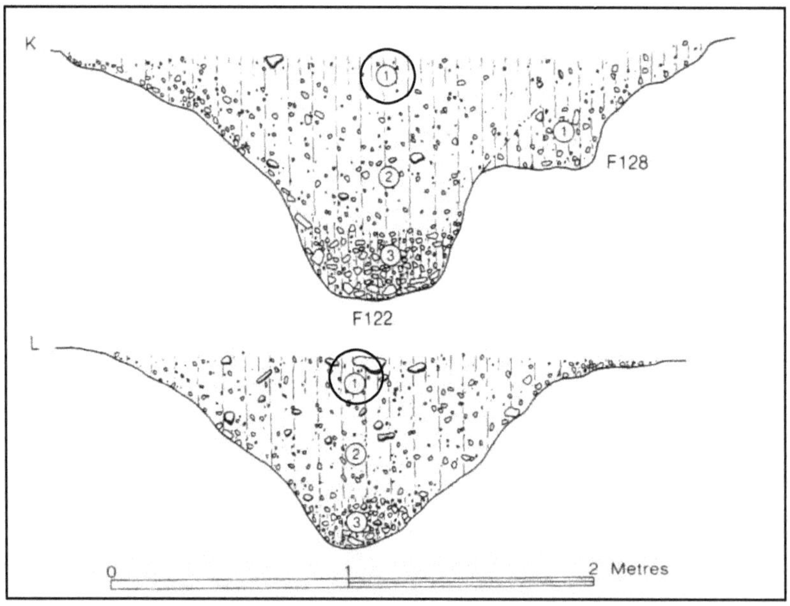

Figure 79. Section drawings of New Buildings ditch F122 segment 2; circles indicate fill layer with briquetage.
After Cunliffe and Poole, 2000d: 44, Fig. 4.25.

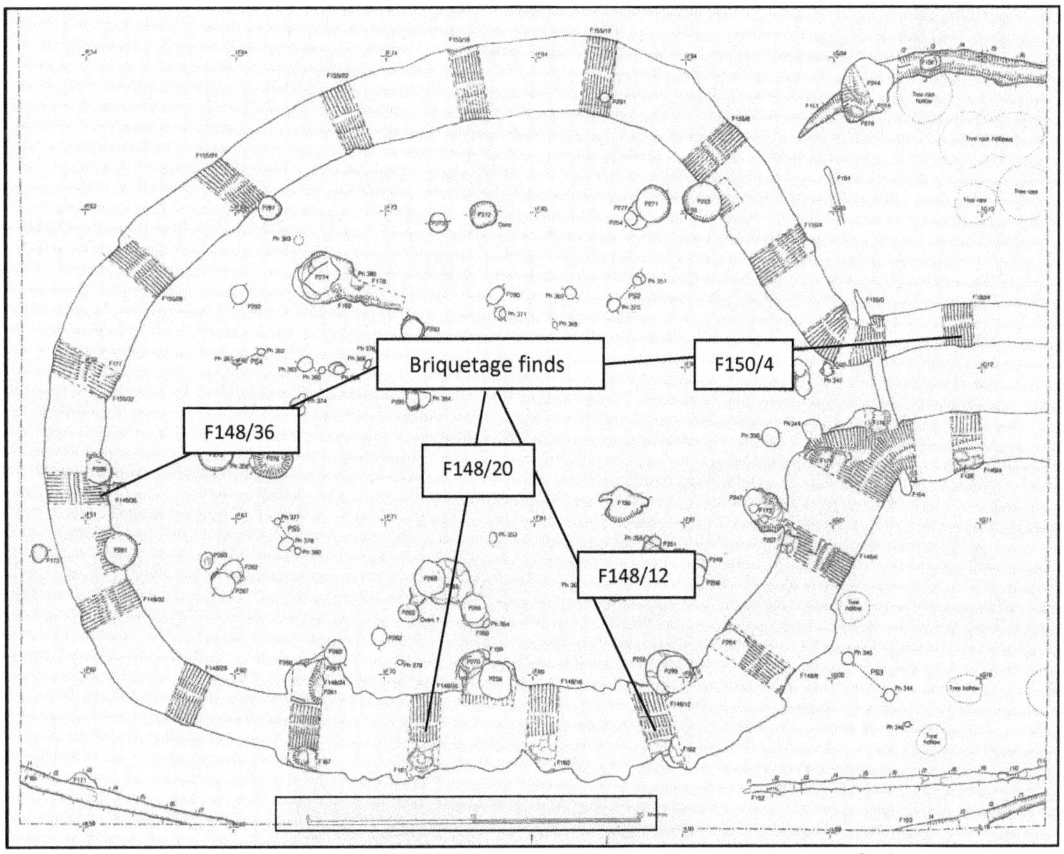

Figure 80. General Nettlebank Copse site plan showing briquetage finds.
After Cunliffe and Poole, 2000e: 14, Fig. 5.4. Scale bar shows 20 metres.

with periods of sedimentation occurring between the recuttings, although not all activity phases were identified in all sections of the ditches. Judging from pottery found within the ditches, the first ditch cutting at Nettlebank Copse probably occurred near the start of the Middle Iron Age at the beginning of the third century BC, while the top fills included Roman pottery of the early first century AD, so the ditches were thought to be in use for 300 to 350 years (Cunliffe and Poole, 2000e: 29-42). As briquetage would have been in use for salt production on the south coast during this entire period, there is no necessary requirement to assume residuality for the briquetage in any particular fill to explain its presence there. The discussion about the potential for redeposition of earlier material, while equally relevant to dating of these finds in these recut ditches, will not be repeated here.

As shown in Figure 80, the finds generally occurred in the southern portion of the enclosure ditch, an area rich in occupation debris disposal during the first century BC to mid first century AD during the Late Iron Age (Cunliffe and Poole, 2000e: 134). Unfortunately there is no varying stylistic temporal sequence to apply to briquetage finds in this region, so dating must be based upon any other datable artefacts with which it is found, and most often this is pottery. Dating based on the accompanying pottery with which it is found clearly need only act as a *terminus ante quem* for the briquetage. Use of this dating method does also raise the issue of the possible curation of briquetage and its delayed disposal, although this seems unlikely for what is essentially an industrial by-product, and it seems particularly unlikely at this site as the briquetage finds are considered to be part of the occupation debris disposed of routinely rather than being accumulated in a midden, a process similar to that just described with respect to New Buildings. With that caution in mind, based on the associated pottery, three of the finds were dated to Danebury ceramic phase (cp) 8, the Late Iron Age, *c.*100 BC to AD 50, and the find in ditch segment 148/12 was considered to be of early Roman date (Poole, 2000d). Although special deposits were located elsewhere in the ditches, they included human remains, animal bones, pottery, burnt chalk, oven daub, ceramic oven bricks, and charcoal, but never briquetage (Cunliffe and Poole, 2000e: 93).

Suddern Farm

The next largest assemblage by sherd count was recovered from the Suddern Farm site and consists of 8 sherds with a total weight of 58gm. The sherd find locations are identified in the main text of the monograph down to context level within the pits or scoop where they were discovered (Poole, 2000a), as illustrated in Figures 85 to 90. It appears that not all of the briquetage entered the archaeological record at the same time at this site. Although the site seems likely to have been continuously occupied from about 750 to 300 BC (Cunliffe and Poole, 2000c: 201), none of the briquetage finds are earlier than the late fourth century - early third century BC. Pit P132 was thought to be the earliest of these four features, being

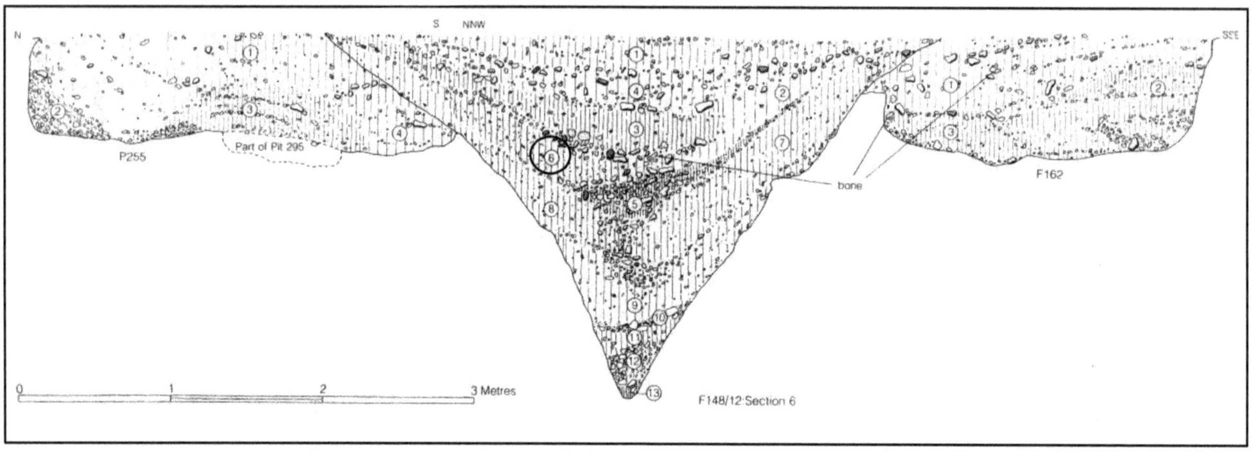

Figure 81. Nettlebank Copse enclosure ditch segment F148/12; circle shows fill layer of find.
After Cunliffe and Poole, 2000e: 31, Fig. 5.13. Scale shows 3 metres.

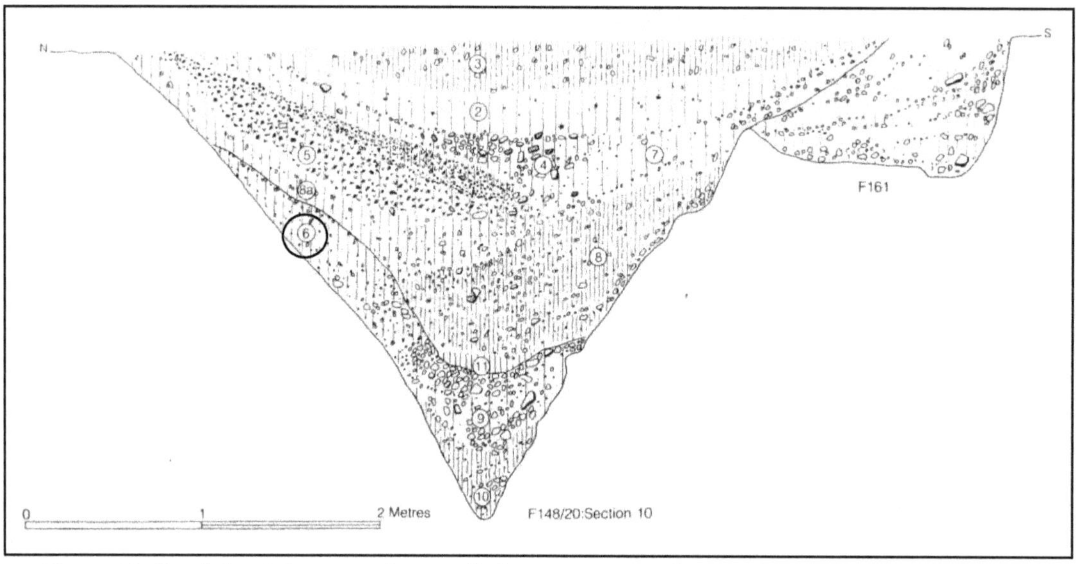

Figure 82. Nettlebank Copse enclosure ditch segment F148/20; circle shows fill layer of find.
After Cunliffe and Poole, 2000e: 32, Fig. 5.14. Scale shows 2 metres.

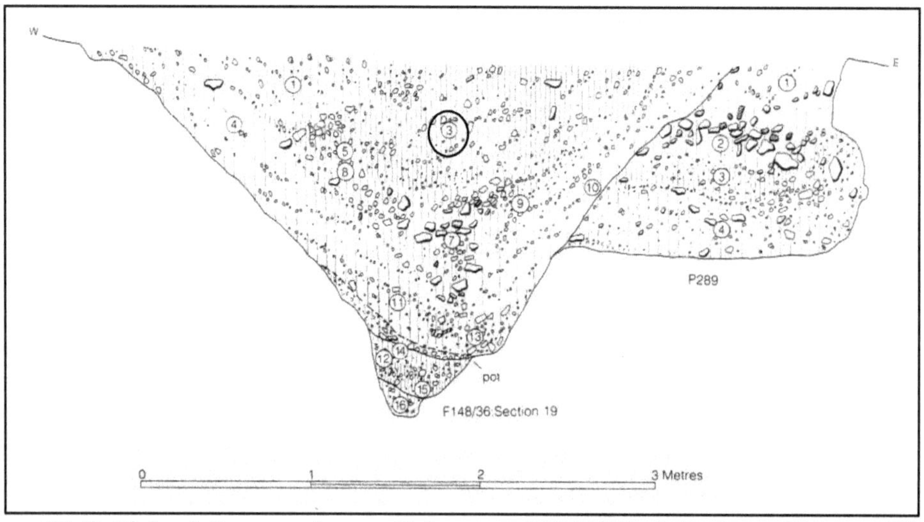

Figure 83. Nettlebank Copse enclosure ditch segment F148/36; circle shows fill layer of find.
After Cunliffe and Poole, 2000e: 34, Fig. 5.17. Scale shows 3 metres.

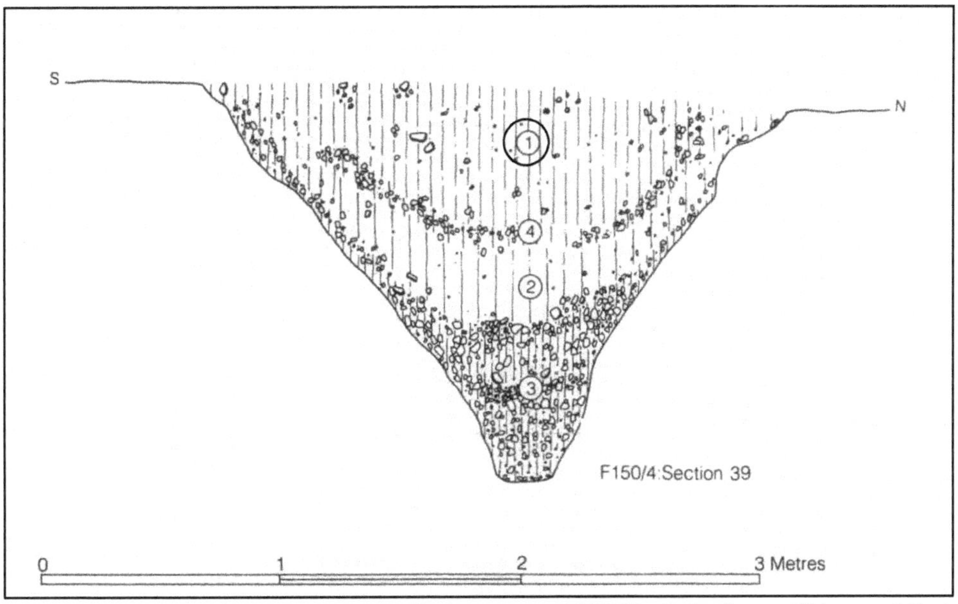

Figure 84. Nettlebank Copse antenna ditch segment F150/4; circle shows fill layer of find.
After Cunliffe and Poole, 2000e: 38, Fig. 5.20.

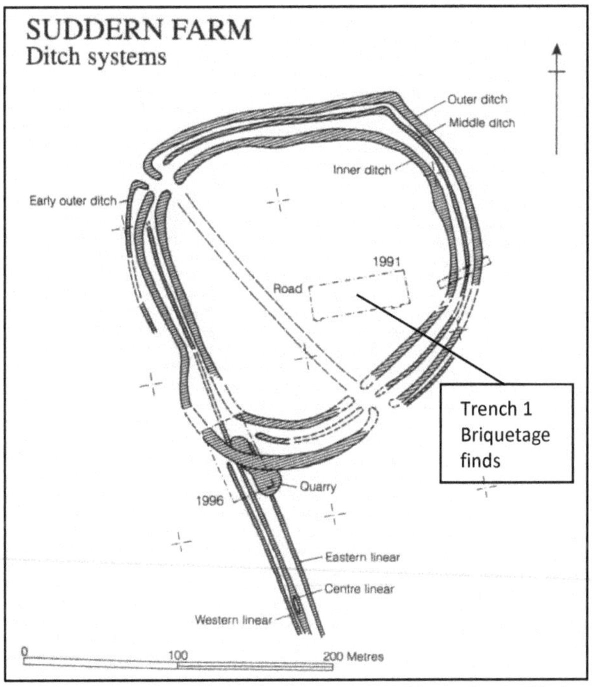

Figure 85. Suddern Farm site plan showing general location of briquetage finds.
After Cunliffe and Poole, 2000c: 14, Fig. 3.5.

dated to Danebury cp 6, or roughly 310-270 BC in the Middle Iron Age, followed by Pit P92, less tightly dated to cp 6-7, or 310 to 50 BC (Cunliffe, 1995: 17-18), from the Middle to Late Iron Ages. However, this broad dating for the P92 sherd is problematic, as elsewhere in the monograph we are told that ceramic dating for the site "...indicates that there was a cessation of occupation in cp 7 (*c.*270 - 50 BC)" (Cunliffe and Poole, 2000c: 201), which would suggest that the briquetage might not, in fact, be later than the Middle Iron Age, perhaps no later than the early third century BC. Scoop P136 was a Late Iron Age-Early Romano-British feature of uncertain purpose, dated to cp 8-9, equivalent to 50 BC to AD 50 (Cunliffe, 1995: 18) which would still allow for the briquetage to be representative of then-contemporary salt trade from the south coast. However, the majority of briquetage found at the site, 5 sherds weighing 24gm, was found within the upper context in P101 accompanied by Roman pottery of the mid-third and fourth centuries AD, a point in time when briquetage had been out of use for south coast salt production for at least two centuries. In this instance, the briquetage must be considered as residual material in the upper fill of pit P101. As in the prior examples, none of the briquetage was thought to have entered the archaeological record as an act of deliberate deposition.

Woolbury

Although a similar level of detail to that shown above has been provided for briquetage finds made at Woolbury hillfort, shown in Figure 91, the published information contains contradictions between the specialist finds report in the printed text and the detail on the supplemental microfiche (Poole, 2000e: 60 & fiche 1:F6). The finds report in the monograph states that 11 sherds in three fabrics, without a specified total weight, were recovered at this site from five contexts, all of Middle Iron Age date. The fiche detail agrees that three fabrics were identified and gives a total weight of 28gm for the finds, but then enumerates 12 sherds in total. Even more problematic for this mapping exercise, while the text identifies Layer 26 as being the source of one of the finds, the fiche specifies that Layer 56 had two briquetage sherds of early Roman date. On the assumption that the printed monograph has been more thoroughly scrutinised than the information included in the fiche, which may not

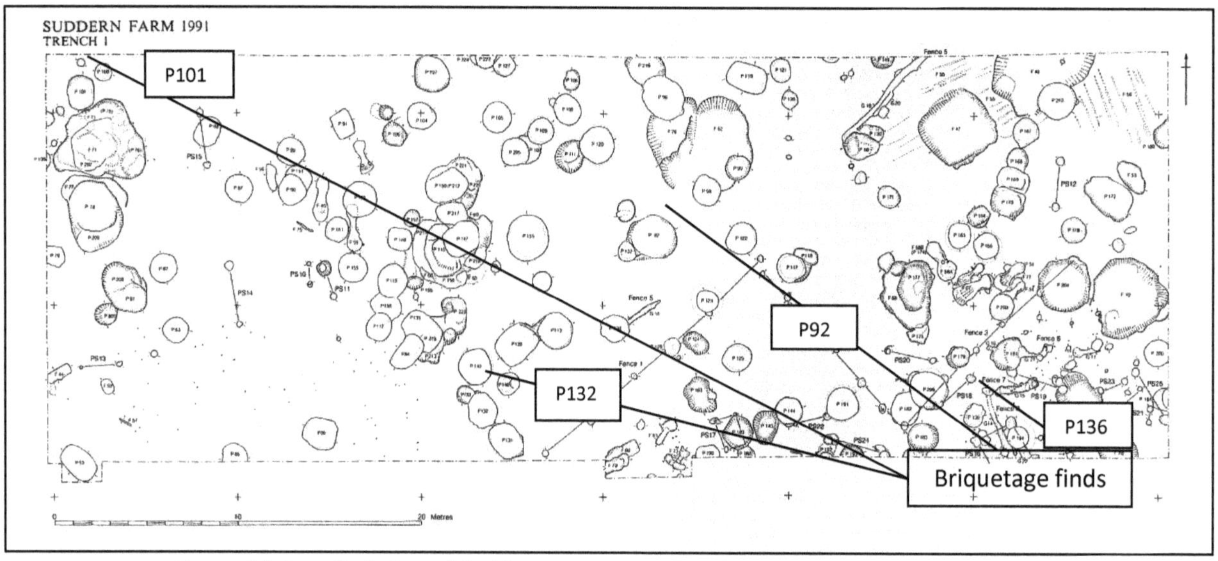

Figure 86. Detailed plan of Suddern Farm Trench 1 showing briquetage-containing pits.
After Cunliffe and Poole, 2000c: 25, Fig. 3.9. Scale shows 20 metres.

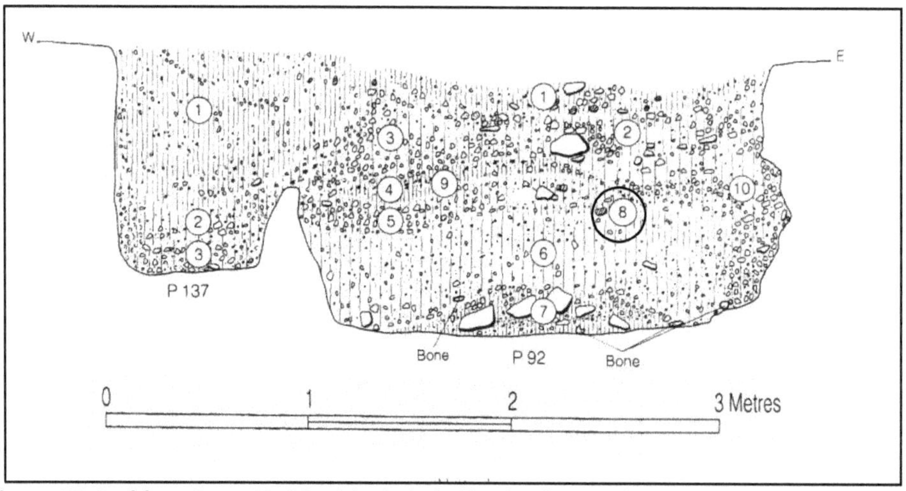

Figure 87. Suddern Farm Pit 92 with circle indicating layer where briquetage was found.
After Cunliffe and Poole, 2000c: 32, Fig. 3.16. Scale applies to Figures 87 through 89.

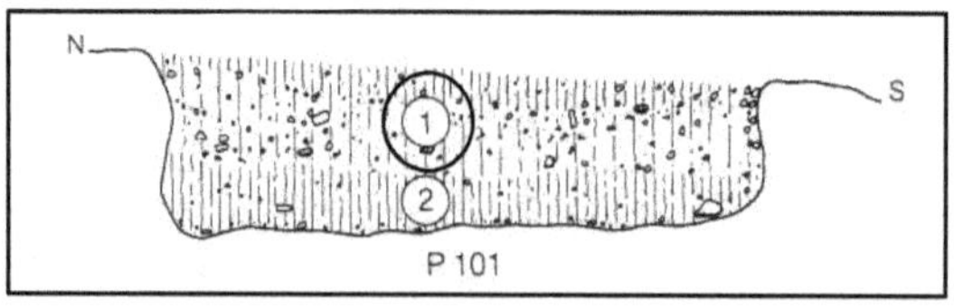

Figure 88. Suddern Farm Pit 101 with circle indicating layer where briquetage was found.
After Cunliffe and Poole, 2000c: 32, Fig. 3.16. Same scale as Figure 87.

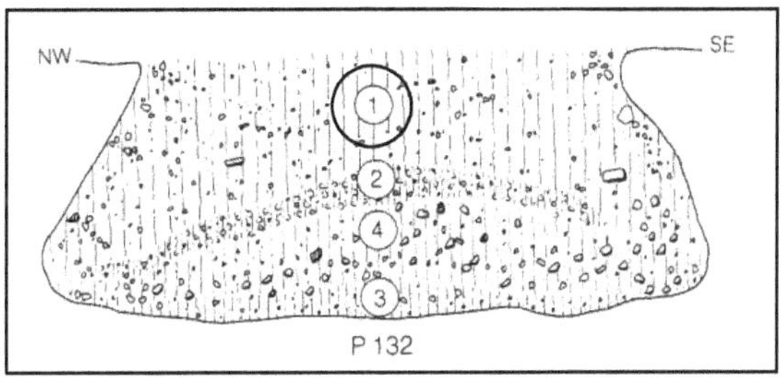

Figure 89. Suddern Farm Pit 132 with circle indicating layer where briquetage was found.
After Cunliffe and Poole, 2000c: 33, Fig. 3.17. Same scale as Figure 87.

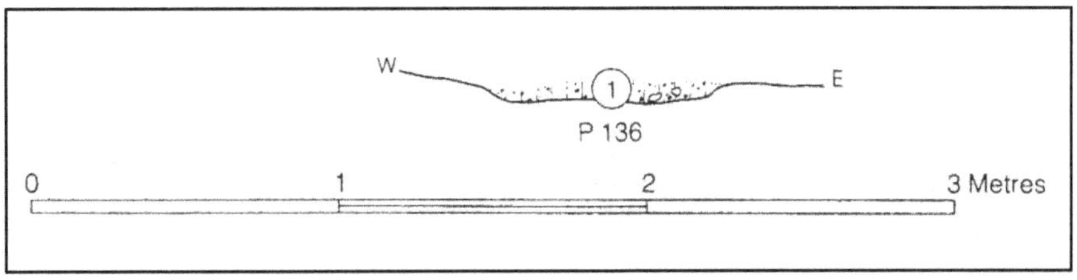

Figure 90. Suddern Farm shallow scoop P136 with a single fill that contained briquetage.
After Cunliffe and Poole, 2000c: 40, Fig. 3.22.

in fact be true, the comments and illustrations below will assume that Layer 26, in the bottom of a lynchet (Cunliffe and Poole, 2000a: 39), rather than Layer 56, a later spread of cobbles, was the source of 1 sherd of briquetage with a total weight of 2gm, making the total sherd count for this site 11, rather than 12. Unfortunately, both layers are of Roman date, contradicting the comment that all briquetage is of Middle Iron Age date. It would appear that 10 sherds are of Middle Iron Age date and 1, or possibly 2, is/are from the early Roman period.

Figures 92 through 94 show the details of the Woolbury finds as discussed above. Pits F5, F6 and F15 are all classified as being of Middle Iron Age date. Per the fiche, F6 is the earliest of these features, considered to be from cp 6, or 310 - 270 BC, while F5 and F15 are of cp 7 date, covering the period 270 - 50 BC, which some might argue actually extends into the early part of the Late Iron Age. No particular function was associated with the shallow single-fill pits F5 or F6. Of the three pits, F5 had the largest amount of briquetage from a single feature: 7 sherds with a total weight of 14gm, or 50% of the briquetage found at Woolbury during this excavation. F15, a pit of beehive form which may have initially been created to store grain, had 8 fill layers which the excavators interpreted as resulting from a slow fill cycle evidenced by layers that were composed of eroded material alternating with layers of soil and occupation debris (ibid.: 27). While F15 was one of only two pits at this site with a special deposit, fill layers 2 and 4 which contained briquetage did not contain any special deposits, and the inclusion of briquetage in these layers can be considered to represent another example of casual deposition. In contrast to these pits, Layer 26 is part of a negative lynchet resulting from the build up downhill of ploughsoil from a Roman field. As shown in Figure 94, Layer 26 was sealed below at least 3 other layers so it was securely dated by the Roman pottery sherds it contained. It is this early layer which the writer believes may have contained the 1 or 2 sherds of briquetage. It is possible that the briquetage reached the adjacent field initially as an inclusion in household waste being used as fertiliser on the field, but this is speculation.

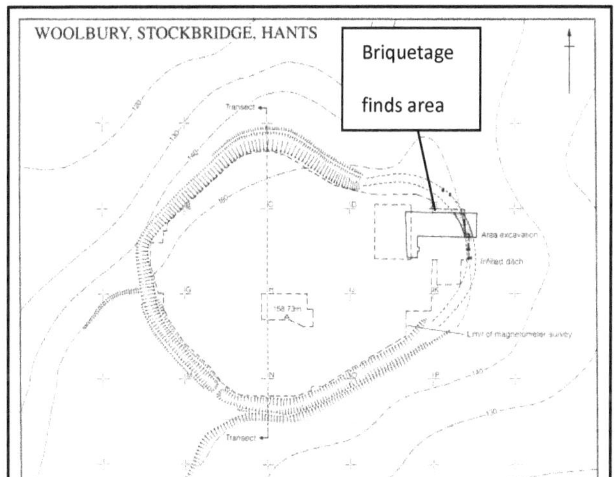

Figure 91. Plan of Woolbury showing trench where briquetage was found.
After Cunliffe and Poole, 2000a: 9, Fig. 1.2. Scale shows 400 metres.

Houghton Down

The next site in the study area's hierarchical progression of briquetage assemblage size as measured by sherd count is Houghton Down, which yielded 12 sherds found in 6 different features covering the entire time span of site occupation. The assemblage weighed a total of 56gm (Poole, 2000b: 119 & fiche 14:A12). Briquetage-bearing feature locations are shown in Figure 95, while the phasing of these features and the metrics of the finds are shown in Table 23. Given that the briquetage entered the archaeological record over a period of more 800 years, it is unsurprising that the finds "... formed a random scatter across the site tending to lie towards the central area of the enclosure..." (Poole, 2000b: 119), with no particular pattern of deposition being noted. The details of the individual features where finds were made are illustrated in Figures 96 through 103. No section drawing of Pit P312 was included in the site monograph or on the associated microfiches (Cunliffe and Poole, 2000f). All of the briquetage samples at this site are interpreted as having entered the archaeological record as a result of routine disposals of occupation debris, even when occurring in the same fill layer as the special deposit of a flint nodule in pit P342 as shown in Figure 103.

Although the assemblage from Houghton Down is not particularly large in either sherd count or weight, it did contain some interesting pieces. Two of the sherds had flattened rims that exhibited finger-tip depressions, a decorative feature known to occur on briquetage of this region, though more commonly associated with pottery. One of these sherds found in pit P342 and dated to the Early Iron Age had been modelled sufficiently to create "...an oval scalloped pattern along the rim" which, even more unusually, was accompanied by "...burnished lines running parallel to the rim around the interior surface" (Poole, 2000b: 120 & fiche 14:A12). Burnishing rarely occurs on briquetage from any region; Poole notes that the combination of these possibly decorative features had caused debate as to whether or not this sherd

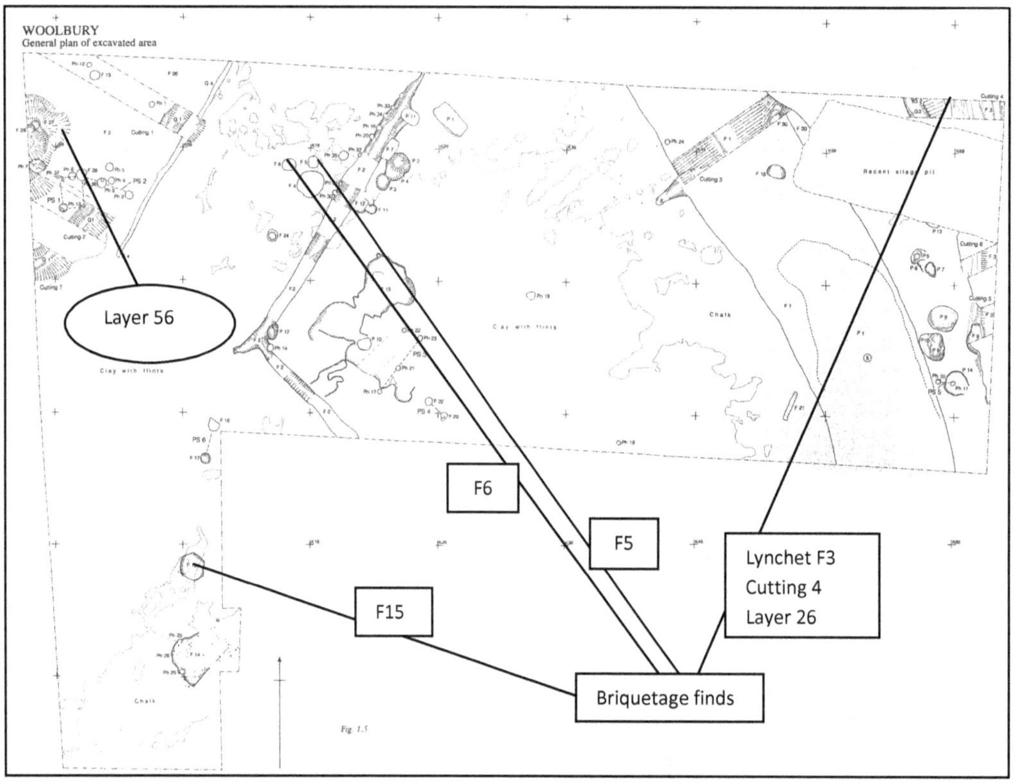

Figure 92. Trench plan of Woolbury showing features where briquetage was found.
After Cunliffe and Poole, 2000a: 17, Fig. 1.5. Distance between small crosses on plan is 10 metres.

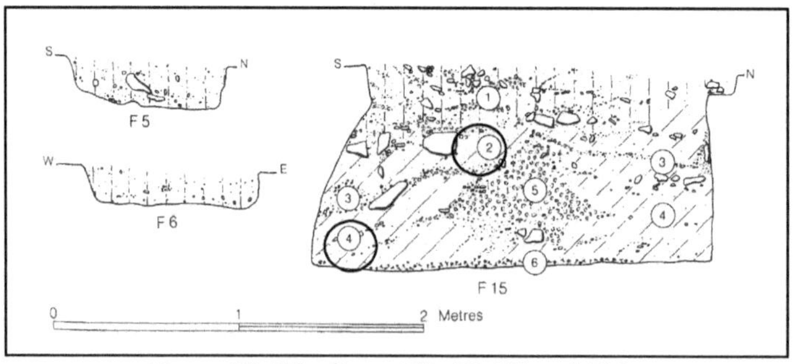

Figure 93. Woolbury Pits F5, F6 and F15.
After Cunliffe and Poole, 2000a: 27, Fig. 1.10. The single fills of Pits F5 and F6 contained briquetage. Finds in Pit F15 came from circled fills.

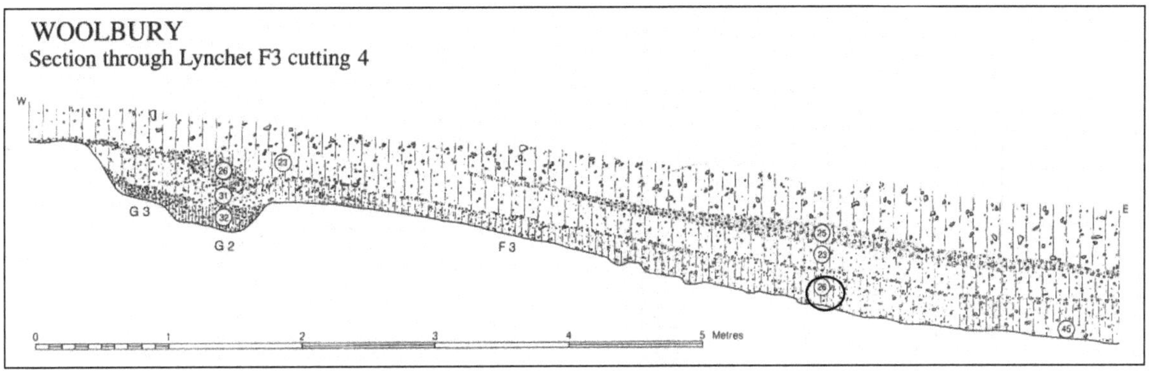

Figure 94. Section through Woolbury Lynchet F3 cutting 4 with circle indicating briquetage-bearing Layer 26. After Cunliffe and Poole, 2000a: 39, Fig. 1.18. Scale shows 5 metres.

was pottery rather than briquetage. Although not stated by Poole, it may be that this is a rare example of a special container designed for presentation of salt as a gift. However, an equally viable explanation is that the briquetage had been created by someone more familiar with the production of pottery than industrial ceramics that had simply utilised the production techniques with which they were familiar in creating this particular vessel.

Unlike the other Houghton Down features that contained briquetage, the early Middle Iron Age pit F241, shown in Figure 99, was located within what had been a circular structure, CS3, which was dated to the early Early Iron Age (Cunliffe and Poole, 2000f: 24), though there is no suggestion that the structure was still standing when the pit was dug. F241 was one of the largest pits on the site, with the base 1.8m deep and the total volume of the pit estimated at 7.52m^3 (ibid.: 33 & fiche 11:E11). In addition to being the source of the only early Middle Iron Age briquetage find, this pit also held the only possible example at Houghton Down of complete pots being thrown into a pit to break them during deposition. Four possible instances of this practice were identified in F241 (ibid.: 125). The two sherds of briquetage found in the top context of F241 were made of an atypical fine, sandy clay fabric and were fired black with a thin brown surface veneer (Poole, 2000b: fiche 14:A12), adding to the overall uniqueness of this pit's contents. Although this was not identified as a special deposit in the site monograph (Cunliffe and Poole, 2000f: 123-24), the unique nature of this briquetage and the other atypical practices associated with ceramics found in this pit seem to leave open the possibility that this may be an example of something other than casual disposal of briquetage.

Bury Hill

To this point, all of the discussions of briquetage finds made during the Danebury Environs Programme Iron Age phase have consisted of assemblages that were too small to allow for statistically meaningful analysis of the distribution patterns. It was hoped, therefore, that the 33 sherd assemblage from Bury Hill hillfort (Cunliffe and Poole, 2000b) would allow for the development of a more sophisticated model using statistical techniques to discuss the deposition of this material into the archaeological record. Unfortunately, the published information for this assemblage is inadequate for this task (Poole, 2000f). The difficulty may result from the authors' decision as to how significant the assemblage was relative to the other sites they had excavated. Although this site had the largest sherd count of the sites excavated in this project, most of the assemblage consisted of undiagnostic body sherds with a total weight of 116gm and a mean sherd weight of only 4.5gm. The generally uninformative nature of the assemblage may have been the reason that few details were included in the site discussion. The briquetage was not been treated in great detail in the text and there is no sherd-context level report included on the fiche for this site. Table 24 summarises the information provided by sherd fabric, which is the only data presentation which allows calculation of sherd count or assemblage weight, as neither were stated directly in the specialist finds report in this volume.

Unlike the other Danebury Environs Programme sites published by Cunliffe and Poole, only two features at Bury Hill, shown in Figure 105, can be identified as having been the source of briquetage, Pit P53 and Layer 164, though the text specifies that 11 samples of briquetage were located, "...nine from seven pits, one from a post-hole and one from a layer" (Poole, 2000f: 62). This specialist finds report identifies Layer 164 as the source of sherds that might have been from a bowl shaped vessel, rather than the more typical cylinder or half cylinder used for salt transport in this region. Integrating this information with a subsequent comment that 15 sherds of fabric 3 briquetage may have come from a single fragmented bowl, it is possible that Layer 164 was the context from which this was retrieved, as only this context was discussed explicitly as having yielded briquetage bowl form fragments in this text. Layer 164 was described as

> *"... a churned area.... roughly 4 x 2 m. ...of yellow sandy clay, slightly soily and mottled, with flecks and fragments of charcoal, pottery and an above average concentration of struck flints... essentially the natural clay disturbed or trampled with anthropogenic material incorporated"* (Cunliffe and Poole, 2000b: fiche 2:A4).

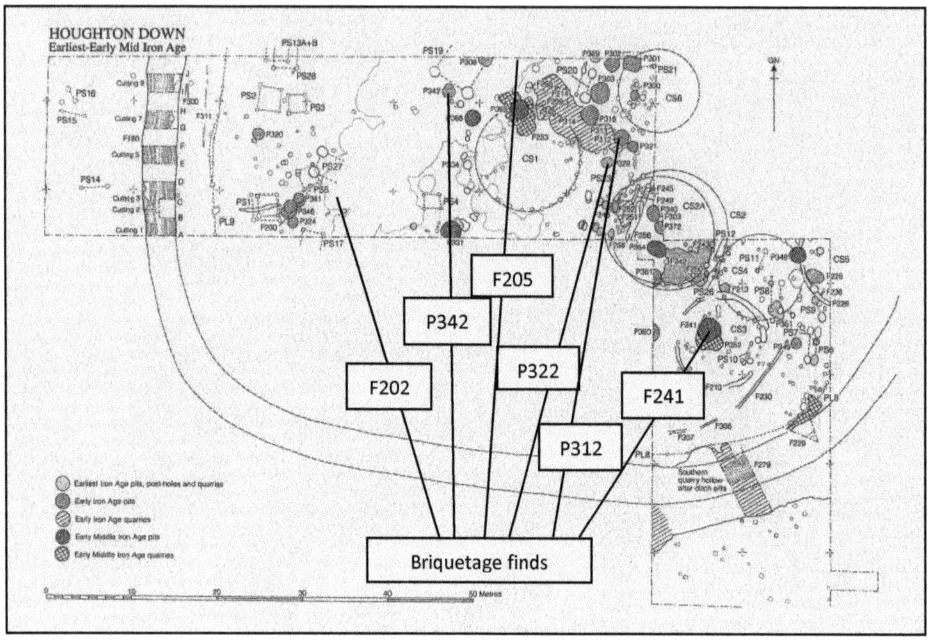

Figure 95. Plan of Houghton Down showing location of features with briquetage finds.
After Cunliffe and Poole, 2000f: 21, Figure 6.5. Scale shows 50 metres.

eature & Layer	Period	Number of Sherds	Total Weight (gm)
Pit P322 (1)	Earliest Early Iron Age	1	4
Pit P312 (9)	Early Iron Age	2	14
Pit P342 (7)	Early Iron Age	1	14
Pit F241 (1)	Early Middle Iron Age	2	6
Quarry Hollow F205 (1)	Late Iron Age	5	14
Quarry Hollow F202 (3)	Late Iron Age/Roman	1	4
Totals		**12**	**56**

Table 23. Houghton Down briquetage finds by context.
Based on Poole 2000d: fiche 14:A12.

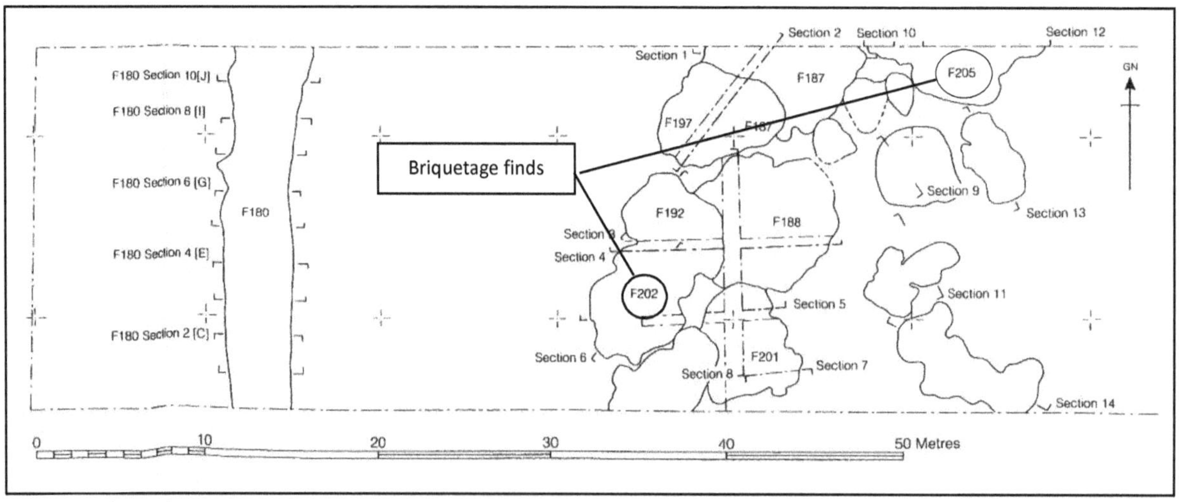

Figure 96. Plan of Late Iron Age western quarry group at Houghton Down showing features with briquetage finds.
After Cunliffe and Poole, 2000f: 43, Figure 6.21.

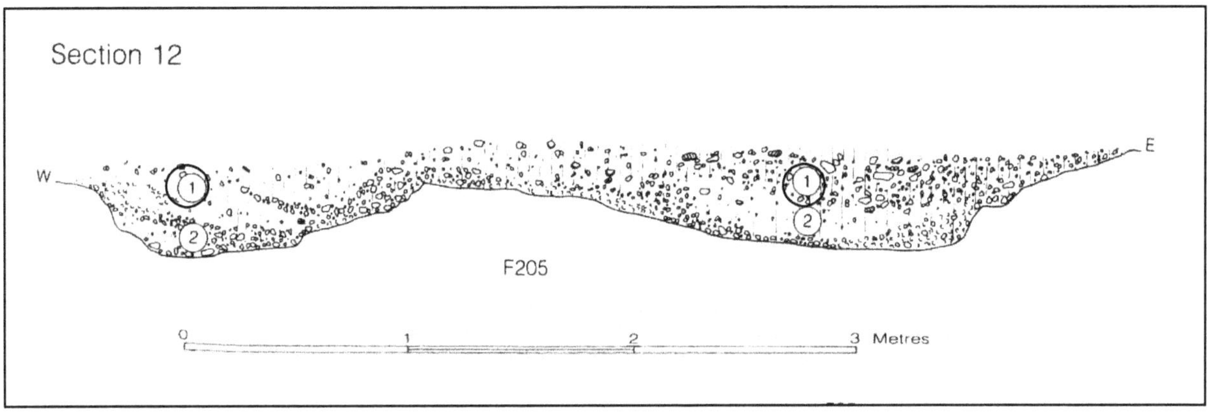

Figure 97. Section drawing of Houghton Down quarry scoop F205.
After Cunliffe and Poole, 2000f: 45, Fig. 6.22. Circle indicates fill with briquetage.

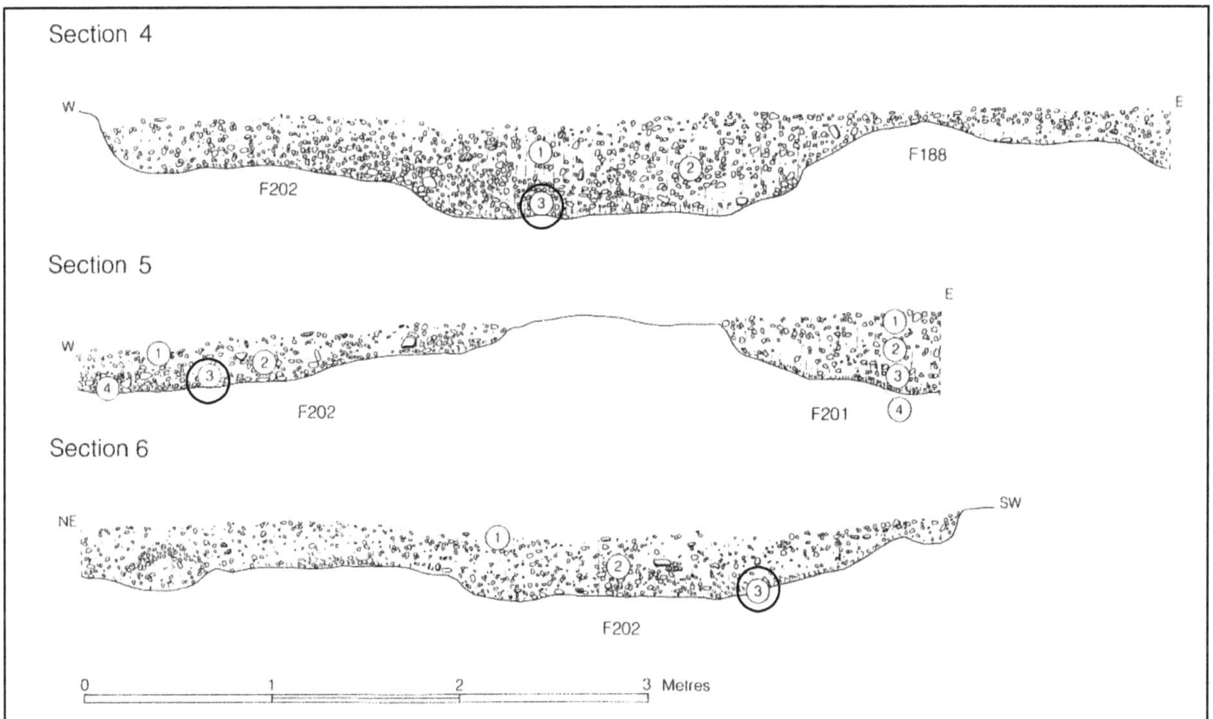

Figure 98. Section drawings through F202 at Houghton Down.
After Cunliffe and Poole, 2000f: 43, Figure 6.21. Circles indicate fill with briquetage.

Although the bowl form is relatively unusual away from the salt production areas of the south coast, there is nothing in this description to suggest that the sherds were treated as anything other than occupation debris when they entered the archaeological record. Pottery found in this layer was identified as Danebury cp 8/9, or approximately 50 BC to AD 50 in the Late Iron Age (Poole, 2000f).

The identification of Pit P53 as a source of briquetage resulted from a review of the microfiche discussion of pit fills (Cunliffe and Poole, 2000b: fiche 2:C3). Within the context of the full set of Iron Age Danebury Environs Programme microfiches, it is unusual that the description of this pit's fills contained a direct statement of the presence of briquetage. In fact, this appears to have been a unique occurrence among the layer by layer descriptions of pit fills in this set. In all of the other instances, briquetage finds documented elsewhere in the text or on other fiches were usually lost within the fill descriptions, subsumed into a general comment about the presence of occupation debris within a layer. This is not a critical comment as such, merely a statement of fact and representative of the challenge in identifying the briquetage find locations at this site despite the relatively large number of sherds in the ceramic assemblage.

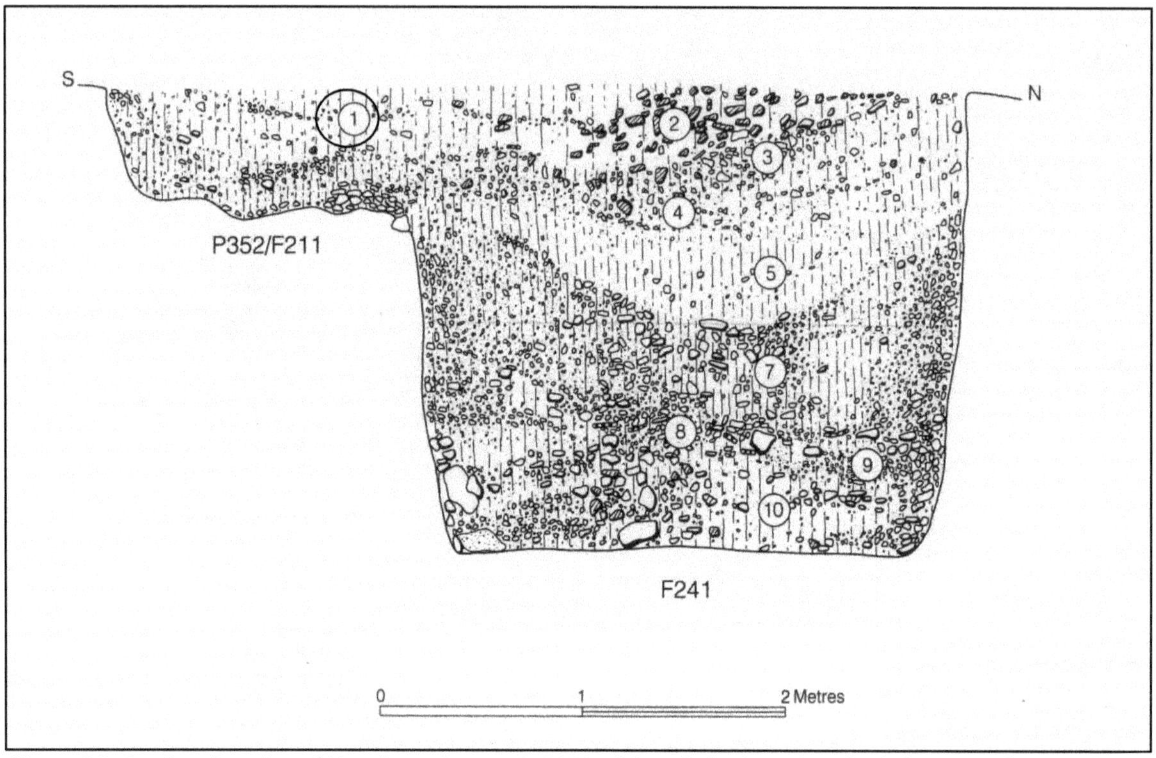

Figure 99. Section drawing of Houghton Down pit F241 with a circle denoting the briquetage-bearing layer.
After Cunliffe and Poole, 2000f: 35, Fig. 6.16.

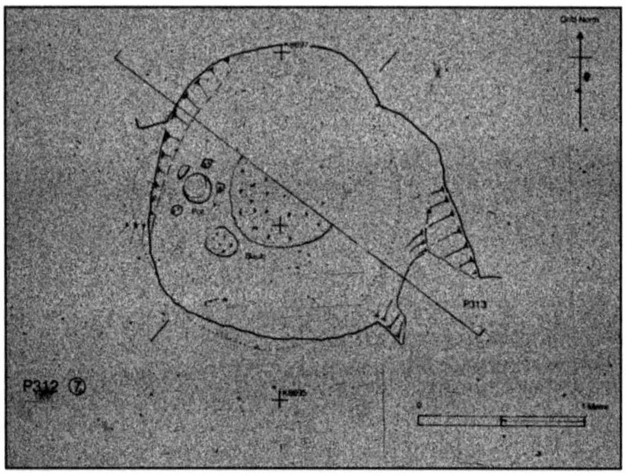

Figure 100

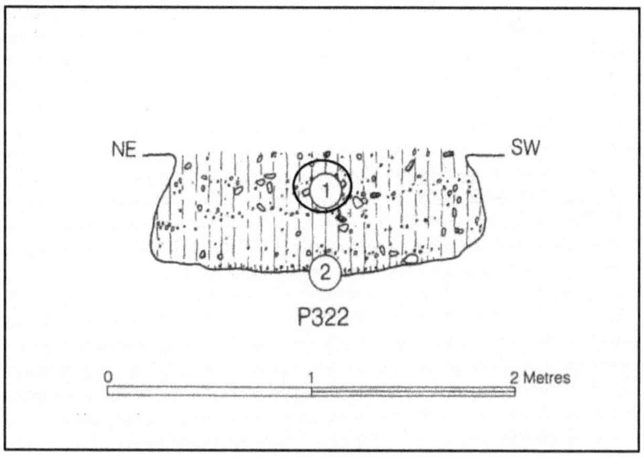

Figure 101

Figure 100. Detailed plan of Houghton Down pit P312 showing the special deposit in layer 7.
NB: Briquetage was found in Layer 9, below the illustrated layer. The poor quality of Figure 57 is a result of its reproduction here from microfiche as it does not appear in the published monograph. From Cunliffe and Poole, 2000f: Fiche 11:G8. Scale shows 1 metre.

Figure 101. Section of Hougton Down pit P322 with circle showing layer where briquetage was found.
After Cunliffe and Poole, 2000f: 34, Fig. 6.15.

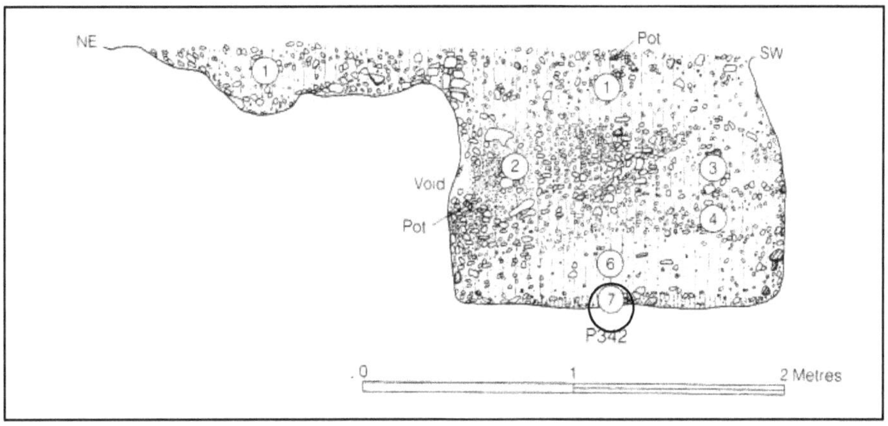

Figure 102. Section of Houghton Down pit P342 with circle showing fill layer of briquetage find.
After Cunliffe and Poole, 2000f: 36, Fig. 6.17.

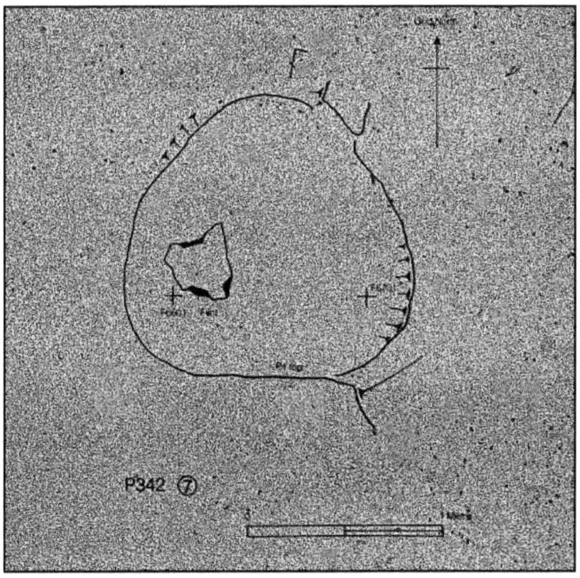

Figure 103. Plan view of Houghton Down pit P342 showing special deposit of flint in Layer 7.
NB: Briquetage was also found in Layer 7, P342, though it was not considered to be part of the special deposit. The poor quality of the image in this figure is a result of its reproduction here from microfiche as it does not appear in the published monograph. From Cunliffe and Poole, 2000f: Fiche 12:A7. Scale shows 1 metre.

Fabric Description	Sherd Count	Total Weight (gm)
1 – high chaff inclusion count	11	70
2 – silty/fine sandy clay with fewer chaff inclusions	3	14
3 – coarse sandy clay, few or no inclusions	19	32
Totals	**33**	**116**

Table 24. Details of the Bury Hill briquetage assemblage.
After Poole 2000b: 62.

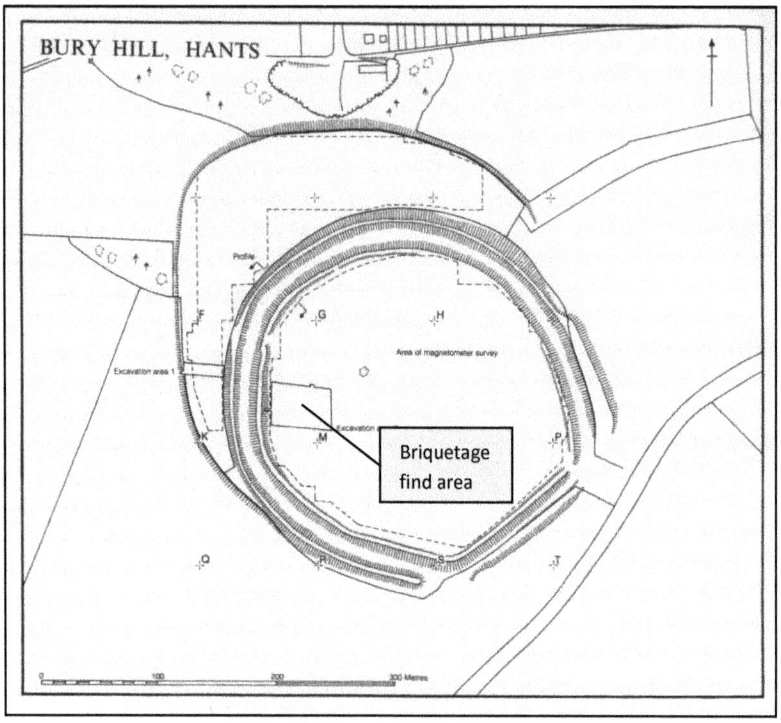

Figure 104. General plan of Bury Hill with interior trench noted.
After Cunliffe and Poole, 2000b: 10, Fig. 2.2. Scale shows 300 metres.

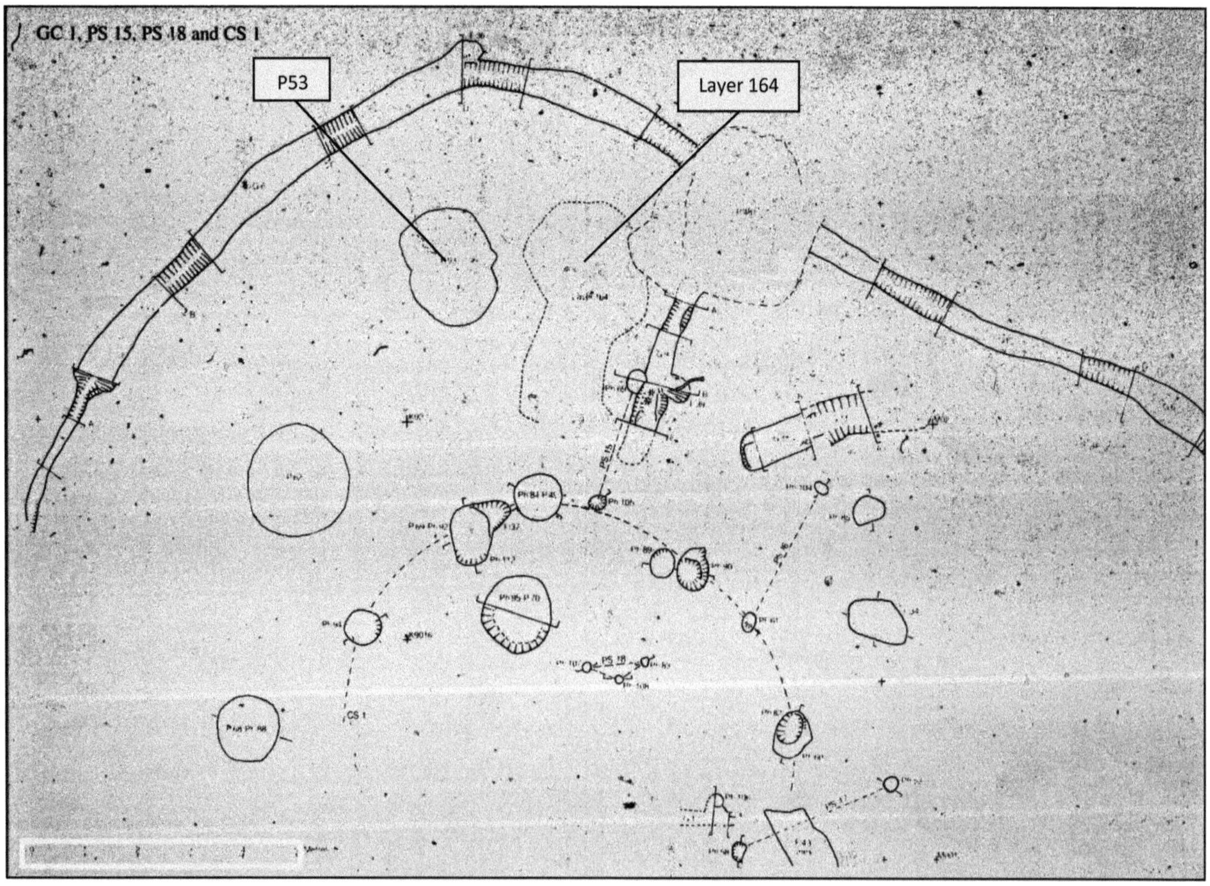

Figure 105. Plan showing the location of Layer 164 and pit P53 at Bury Hill.
After Cunliffe and Poole, 2000b: fiche 2:A5-6. The poor quality of this figure is a result of its reproduction here from microfiche as it does not appear in the published monograph. Scale bar (white rectangle) in lower left shows 5m.

Most context sheets for on-site field use only provide a series of tick boxes to note the presence of finds such as pottery, metal, bone or worked stone, and these sheets would have acted as the source document for the fill descriptions. Further, it is unlikely that briquetage would have been recognised in the field during these excavations, occurring in the small quantities which have been noted here, and thus would have escaped special mention on the context sheets by the excavators.

Unfortunately little can definitively be said about Pit P53, as it was only excavated to a depth of 0.1 m to show the shape of the pit in plan, and that a find of briquetage occurred in the course of the excavation of the top context. It was thought possible that this layer represented the slump of a midden of occupation debris into the pit. This small sample of pit fill was dated rather broadly to Danebury cp 3-6, which would identify it as one of the five pits at Bury Hill which contained both briquetage and pottery of cp 6 date (Poole, 2000f), approximately 310 - 270 BC, on the boundary between the Early and Middle Iron Ages. The other two unspecified pits with briquetage were dated to cp 7, or 270 - 50 BC in the Middle Iron Age and the unidentified single post hole with briquetage could not be assigned to a ceramic phase. Although Bury Hill seems to have been constructed in the Early Iron Age, there is little evidence for occupation associated with this portion of the site's history (Cunliffe and Poole, 2000b) and no briquetage was found in features of this early date. While it may yet be possible to reconstruct the location to context level of all briquetage found at Bury Hill by reviewing the original documentation and finds from the excavation, it is not possible to do so through use of the published monograph.

Danebury

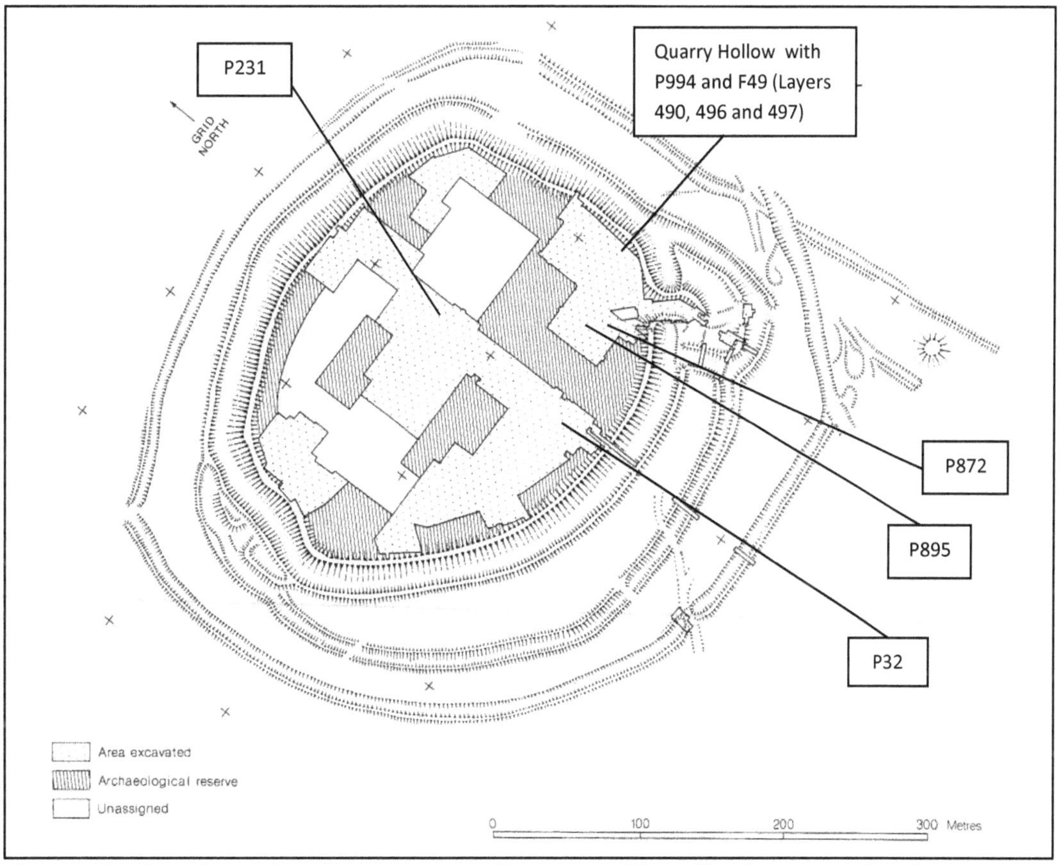

Figure 106. Plan of Danebury showing general locations of identified briquetage bearing features.
After Cunliffe and Poole, 1991a: 4, Fig. 1.3.

The final briquetage assemblage recovered within the Hampshire-Wiltshire study area to be considered here comes from Danebury hillfort, which had the luxury of 20 seasons of excavation under the direction of a single site director and has been fully published in 6 volumes, complete with additional information provided on microfiches accompanying each volume. The importance of the site rests, in part, upon the rigorous, and relatively early, analysis of the site features and artefacts using computers to handle the voluminous quantity of data available (Cunliffe, 1984a). Highly sophisticated statistical techniques were utilised to maximize the reliability of ideas put forward to interpret the site, particularly with respect to the taphonomy of artefacts, dating and identification of centres of particular types of activity.

Over a period of twenty years, 1600 sherds of briquetage with a total weight of nearly 7.6kg were recovered at Danebury (Poole, 1984b, Poole, 1991). As was shown in Table 10 it is one of the largest assemblages from any source not associated with a salt production site in Britain, and is the largest currently known resulting from sea salt distribution. Given the level of attention to detail used in the recording of data at this site and the availability of microfiches to provide additional data beyond the texts, the writer anticipated that developing a briquetage finds distribution map for Danebury would be difficult only due to the volume of data which would need to be addressed in the exercise. This has not proven to be the case, however, as it has been impossible to reconstruct the full distribution of briquetage finds from the published resources.

It is certain that some categories of highly detailed data about briquetage were collected during at least the first 10 years of excavation, 1969-1978, as a short paper by Lock explaining how the data for these finds was converted to 320 lines of coded data for use in a COBOL programme is included in the microfiches with Volume 2 (1984: fiche 11:C4). The coded data file for briquetage he mentions, however, does not appear on any microfiche, in contrast to the presentation of coded information on pits that is provided in a subsequent volume (Cunliffe, 1995). The Lock paper never suggests that data on the context of finds was part of this coding exercise, as the apparent objective was to be able to manipulate the quantitative information along with the more qualitative traditional ceramic descriptive parameters of form and fabric types along with ceramic phase information. The lack of information on briquetage context cannot be considered unusual, as this level of detail is not typically provided for large assemblages of ceramic finds, unless it arises in the discussion of atypical find, such as a complete pot, a special deposition or the presence of an unexpected form or fabric.

There is some evidence that the nature of the distribution of the briquetage finds was given consideration, possibly with respect to finding concentrated activity areas within the site. Assessing the first ten years of work, Poole notes that briquetage had been found in all ceramic phases of the hillfort with more than two thirds of it associated with cp 7 and 8 (270 BC – AD 50) and it came primarily from stratified deposits with other sherds being found in pits or, less often, in post holes (1984b: 426). Within the discussions of forms and fabrics, finds from 5 specific pits, though not their context numbers, and 3 layers were mentioned: pits P32, P231, P872, P895 and P994, layers L410, L496 and L497 (ibid.: 428-29). Locations for the pits within the site grid were listed in the pit data table provided with Danebury Volume 6 (Cunliffe, 1995: fiche 32:A7-D7) and locations of all the mentioned layers other than Layer 410 were discovered among the section drawings in Volume 2. An additional briquetage bearing layer, L490, was noted in a discussion of F49, a gulley or ditch fragment, which is also the location of the previously mentioned Layers L496 and L497 (Cunliffe, 1984a: 149). While the volumes on Danebury are comprehensive, not all of the layers from stratified deposits or the 1,062 pits excavated during the first 10 years, let alone all of the 1,122 pits located in that period, are illustrated in the publication or on the microfiches. Illustrations were, however, located for pits P231 and P994, as well as Layers L490, L496 and L497 and appear here as Figures 107 through 111.

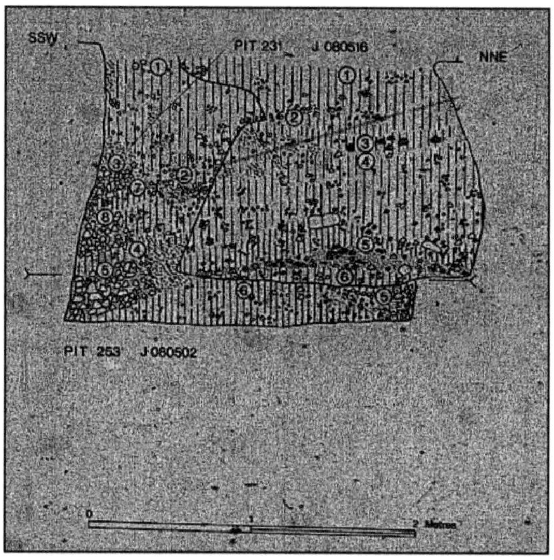

Figure 107. Cross section of Danebury intercutting pits P231 and P253.

From Cunliffe, 1984a: fiche 4:C14. The poor quality of the image is a result of its reproduction here from microfiche as it does not appear in the published monograph. Scale shows 2 metres.

Pit P32, although not illustrated here, was said to have held enough briquetage that the material may have represented two half-cylinders, while pit P994, shown in Figures 108 and 109, may have held the remains of a further half cylinder of salt (Poole, 1984b: 428). The presence of these nearly complete containers is consistent with the SSCI evidence presented earlier in Table 20 and the total sherd weight evidence in Table 10 demonstrating that salt was reaching Danebury in large quantities directly from the south coast production areas where the half cylinder forms originate. The other site features that were mentioned by Poole (ibid.: 428-29) were noted as they contained more diagnostic sherds, such as rims (L410, L496, L497, P872), bases (L410, P231, P872) or longer cylinder wall segments (L410, P872). In the case of Layers L496 and L497, some of the rim sherds from these adjacent contexts, as shown in Figure 110, could be refit to form a longer segment.

The briquetage finds for the second ten years of excavation, 1979-1988, were even less thoroughly published, as the second set of monographs highlighted new information uncovered in this period. Once again Poole (1991: 404) states that briquetage had been found in all ceramic phases, with cp 7 (270 - 50 BC) accounting for 63% of the sherds by number, with a second statistical peak in cp 4 (360 - 310 BC) caused by the presence of a single group of sherds that composed either one or two nearly complete containers. Although the detailed data is not provided, the distribution of briquetage was commented upon and appears to be similar to the earlier description:

> *"Spatially the briquetage occurred throughout the fort with no obvious concentrations. Usually only a few sherds were found in any one context, apart from the one large cp 4 sample mentioned above. The majority (65%) occurred in 91 pits, about 1% in seven post-holes, 2% in six gullies and nearly 32% in 74 layers* (ibid.: 404)."

None of the enumerated features is further identified in the text or microfiches, and briquetage is not mentioned in passing in any descriptions of other features. Like the other sites in this study area, there is no reason to believe that at Danebury, a site rich in special deposits, briquetage entered the archaeological record through any method other than casual deposition with occupation debris.

The summary interpretive volume published on Danebury, Volume 6 (Cunliffe, 1995), included a number of reports dealing with specific categories of artefact distribution, including archaeobotanical information and other evidence of crop processing, pottery distribution and the distributions of artefacts associated with textile production, woodworking and metalworking. A study reproduced in the microfiche accompanying this volume contains maps of small finds of iron, bronze, metal tools, metalworking evidence, iron spears/arrowheads, horse gear or cart fittings, glass, amber and shale, bone tools/objects and clay objects, quernstones, whetstones and worked bone, but none of briquetage distributions (Osgood, 1995). The Danebury briquetage assemblage appears to have been somewhat neglected compared to other categories of material during the publication process, taking a lesser place compared to other types of finds, including daub (Poole, 1984a) which was regularly used in structured depositions at the site (Pryor and Brown, 1984) and features frequently in detailed pit layer descriptions. Certainly many of the other artefact classes have attracted wider scholarly interest within the archaeological community than briquetage and this may have influenced the analytical depth provided in the publications.

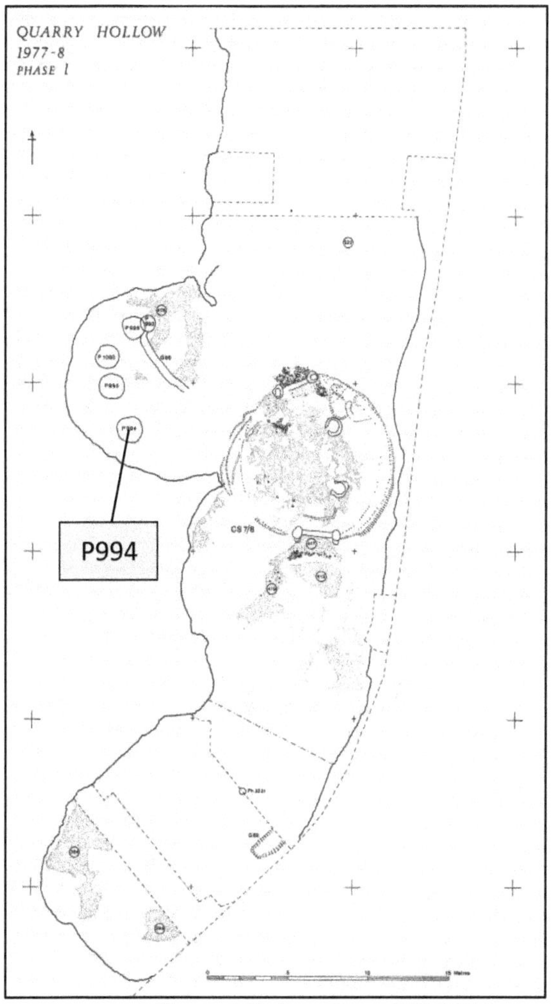

Figure 108. Location of Pit P994.

After Cunliffe, 1984a: 156, Fig. 4.106. Scale indicates 15 metres.

5.6 Conclusions

Chapters 3 and 4 discussed the regionally unique aspects of salt production and distribution during the Iron Age. In this chapter, moving beyond the study of salt to consider other aspects of life and death during the period, detailed reviews of the excavated archaeology of four areas have further demonstrated the strongly regional nature of the Iron Age material culture evidence in Britain. Although the present analysis has used the social geography of site features as discriminators, a similar pattern of distinctive regionalism in relatively small areas could have just as easily been demonstrated by focusing upon other aspects of material culture, such as pottery styles (e.g. Cunliffe, 2005). This wider perspective, which moved away from an exclusive focus on salt in the Iron Age, showed regionalism as a major theme of this period in Britain, and was needed to contextualize the localised nature of salt production and distribution systems and assist in seeing them as totally integrated within the mainstream of other contemporary social behaviours.

The effort to provide briquetage distribution maps for generally well published sites has demonstrated the significant barriers which exist to their development utilising only these published information resources. None of the maps presented here contains enough points to individually represent a statistically meaningful pattern, although the general absence of briquetage among the numerous structured depositions located in all of these sites collectively suggest that briquetage, though not necessarily the salt it had once contained, did not represent a suitable object for inclusion in such a deposit. It would appear that most briquetage, at least in the sites detailed here, entered the archaeological record as domestic rubbish, the contents being far more important than the container. While this chapter has focused on salt containers, the following chapter will discuss the many uses for salt in the Iron Age.

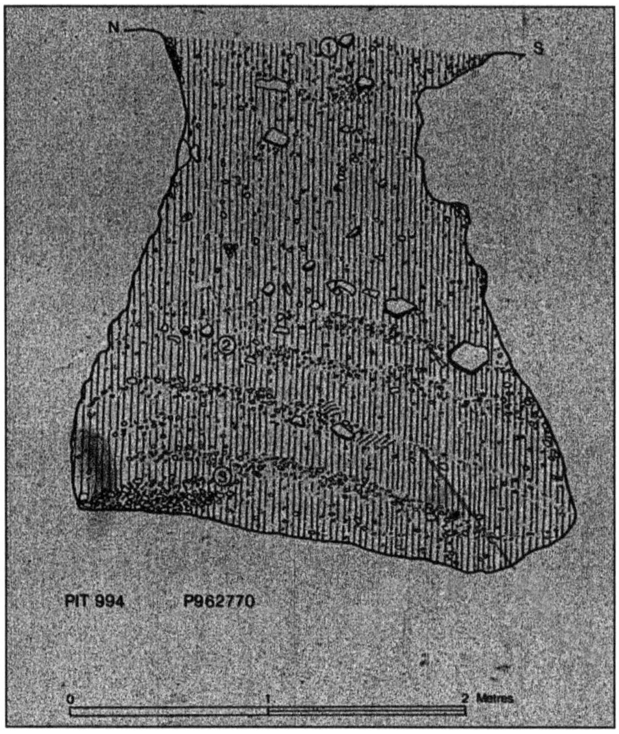

Figure 109. Cross section of Pit P994.

After Cunliffe, 1984a: fiche 5:E7. NB: Linear feature in lower right corner due to scratch on fiche, dark area at lower left corner of pit is part of a finger print, not a pit fill characteristic. Scale shows 2 metres.

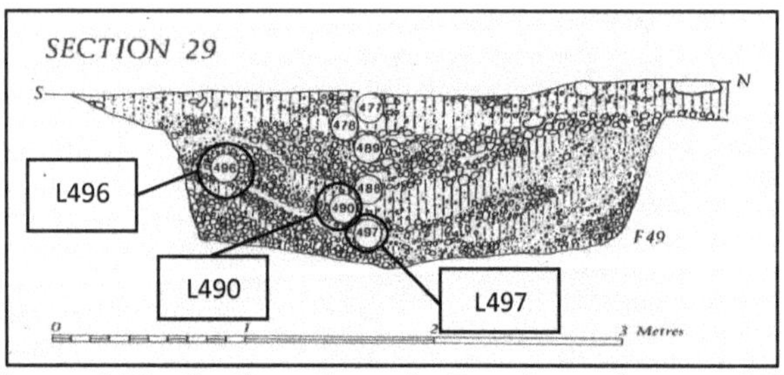

Figure 110. Section through Danebury feature F49, showing briquetage bearing fill Layers L490, L496 and L497.

From Cunliffe, 1984a: 156, Fig. 4.107. Scale shows 3 metre distance.

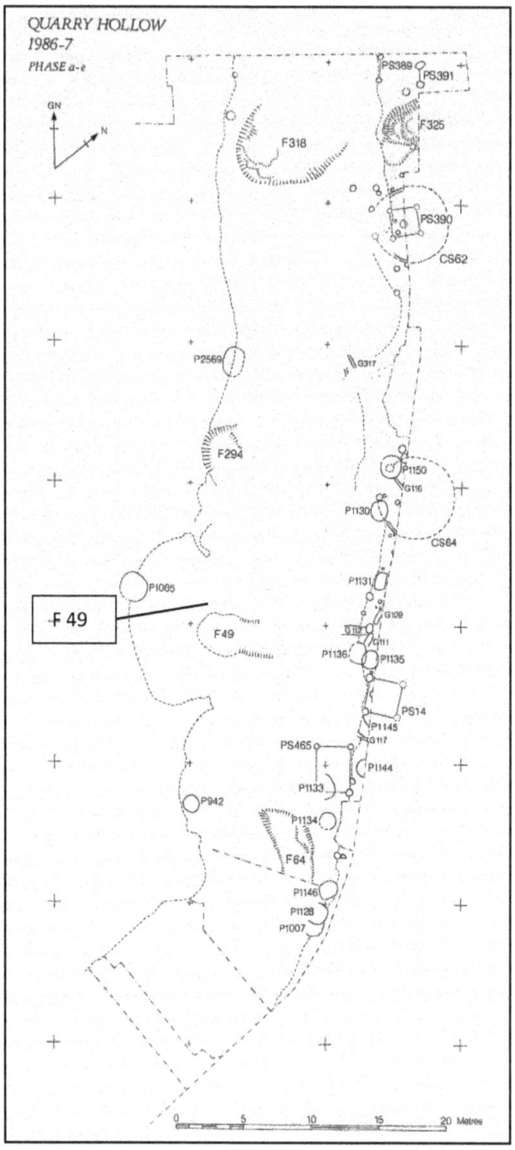

Figure 111. Danebury feature F49, location of Layers L490, L496 and L497.

From Cunliffe and Poole, 1991a: 172, Fig. 4.105. Scale shows 20 metres.

Chapter 6
Using salt: food preservation and so much more

6.1 Introduction

In both the discussion of salt production and its distribution during the Iron Age in the previous chapters, reference was made to the need to accept that, at times, it was necessary to believe in the archaeology of the invisible, i.e., behaviours or practices which cannot be directly demonstrated through artefacts or other archaeological evidence, but must have occurred none the less. Discussing Iron Age salt usage is another such instance of archaeology of the invisible, one in which most evidence is provided by analogy or through modern ethnographic parallels. This same conclusion was reached by Hopkinson almost 35 years ago, who wrote:

> *"We can expect to find no evidence on the use of salt in the preservation of food like butter, cheese, fish, meat, as well as skins and leather goods, and can only observe methods among primitive societies where ancient customs have persisted over long periods of time* (1975: 10)."

While new technology and intervening archaeological discoveries, not to mention changed sensibilities towards non-industrialised societies, may have made some of this comment obsolete, the central problem, the absence of actual salt in the archaeological record, still remains. Archaeology by analogy, to be certain, is far less satisfactory in many ways than archaeology where the discussion can focus on concrete objects such as sites, artefacts and landscapes filled with monuments, but it is sometimes necessary if one is to develop a nuanced and rounded view of life in the past. After all, considering salt usage can, as will be demonstrated below, sometimes assist one in arriving at alternative explanations for artefacts, sites and the social relationships implied by these objects.

Iron Age salt usage is likely to have involved aspects of society far afield from the technological. The term used here for these activities is "social salt", a phrase meant to capture the various possible aspects of the sociology of salt, including areas such as social status and gift exchange, ritual activity and trade networks. While only some of this activity can be directly perceived archaeologically, much can be theorised based upon various analogies that may provide models for these Iron Age behaviours.

Briquetage finds, while generating useful evidence of salt production and distribution, only infrequently provide evidence of the use of the salt. One such example, as discussed in Chapters 4 and 5, occurs at Stanway, Camulodunum, Essex, where briquetage was found in mortuary enclosure ditches (Crummy, 2007), suggesting use of salt in the burial rituals. In comparison, when briquetage is recovered from a roundhouse gulley, as at Shorncote Quarry, Gloucestershire (Brossler et al., 2002), little specific information, beyond a probable association of the salt with unspecified domestic activity, is the typical interpretation. Therefore, in this chapter most of the information on the use of salt in the Iron Age will come from sources other than briquetage finds or will rely on associations with other artefacts, such as animal bones, or examples drawn from Classical literature, anthropology, experimental archaeology or, in the case of food, even cook books.

Consistent with usage in earlier chapters, the term "briquetage" can apply to material from any salt production source or distribution–related find, while the term "VCP" will apply only to material from Cheshire.

6.2 Salt for domestic animals

Observation of salt licks in modern fields or personal experience with animals may have provided some archaeologists with the knowledge that some farm animals require salt in their diets to maintain good health. Few, however, would possess the specific information on how much salt the animals actually require for optimum conditioning. Yet successful animal domestication necessitated the development of the methods and techniques to meet the physical requirements of livestock. While such practices surely began to develop in the Neolithic period, the rise of small scale salt production in Britain, beginning by at least the Middle Bronze Age, had the potential to make meeting the salt need of livestock easier.

As will be discussed in more detail below, Iron Age domestic animals generally were smaller than their modern counterparts, but as it is impossible to directly measure the dietary salt requirements of these ancient animals, modern information will be the basis for estimating their salt requirements. Modern cows, for example, require a minimum of "...0.46% sodium chloride in the total diet dry matter..." (Anon., 2001), which computes to a minimum of about one ounce (28gm) per adult animal per day according to the Animal Science Department at Oklahoma State University (Gill, undated). Cattle, however, have a great fondness for salt and adults can ingest almost 2kg before experiencing any toxicity (ibid.). While cattle can meet their minimum dietary need for salt through a combination of natural salt contained in the plants they eat and licking urine patches in the soil, they readily take salt presented to them in more concentrated form. Where available, it is possible that damaged or discarded briquetage may even have served as animal salt licks (Barford, 2000), such use occurring as either an intentional practice of the animal keepers or the result of instinctual salt-seeking behaviour of the domestic animals with access to briquetage, though clear evidence to support this idea is lacking.

Access to a reliable salt supply would have allowed pastoralists to consider whether salt might be used to modify the animals' inherent salt-seeking behaviour to their own benefit. This behaviour was certainly known to the first century AD Roman writer Columella, whose good farming manual advised

> *"Salt sprinkled on the stones and water-courses near the enclosures contributes to the good bodily health of the cattle and they gladly have recourse to it after they have eaten their fill...,"* (*DRR:* VI, XXIII, 2).

Columella also described how salt could be used as a reward to train bullocks to a yoke in only three days (*DRR*: VI, II, 7). In another example of salt-induced animal behaviour management, though occurring many centuries later, drovers were said to use a handful of salt to attract the lead cow, with the rest of the herd following her, along the road to market (Fielding and Fielding, 2006). In citing these examples, no attempt is being made to claim that these exact practices were in use in Iron Age Britain; however, it would be odd if British herdsmen had failed to notice how readily cattle would take salt and turned this information to their own benefit in some way.

Sheep, as well as cattle, need regular amounts of salt in the diet. North Dakota State University suggested 0.5% of the daily mixed diet be salt, with ewes needing between 0.25 and 0.40 ounces (7 - 11.2gm) per day (Anon., 1996). Salt was apparently widely used by shepherds to influence the behaviour of the flock. The Greek philosopher Aristotle (*Hist. of Animals*: VIII, 10) wrote in the fourth century BC that salt was mixed with the feed of ewes to make them thirsty, with the ensuing extra water intake leading to increased milk production. A similar practice is also cited by the first centuries BC and AD Roman poet Vergil (*Georgics*: III, 394-97), who also said it would work the same way with goats. Columella and Pliny both made this observation, additionally discussing the use of salt keep sheep satisfied with a repetative diet (Columella, *DRR*: VII, III, 20) or to lure sheep into a new pasture (Pliny, *NH,* XXXI, vii, 40, 88). As was the case with cattle, undoubtedly some of these methods had become known in Britain.

Horses, though far less common then either sheep or cattle in Iron Age Britain, would have also required dietary salt. Horses normally eat the equivalent of 1.5 to 3.0% of their body weight daily and need approximately 0.25% of their daily diet as salt for good health (Veitch and Eades, 2006). Assuming a live body weight for an Iron Age horse of between 800 and 1000 pounds, or 364 to 454kg, this would translate as an optimal daily salt intake of between 0.46 and 1.2 ounces, 13 to 34gm, per horse per day. While some of this requirement could be met through the natural salt content of their fodder, supplemental salt intake would have been needed in many cases, particularly as horses lose sodium and chloride ions when they sweat (Sykes, 2009), as they would have done as the result of hard work or in hot weather.

Little work has been done on the dietary salt requirement of European wild pigs, although modern domestic pigs can be fed a minimum, by weight, of 0.2% dietary salt and stay in good health. However, at least one study has shown that diets enriched with up to 1% salt reduce the incidence of biting among closely kept pigs, a behaviour thought to be driven by a desire for salt, obtained from the blood of the injured animals (Brooks, 2009). If Iron Age pigs were allowed to roam relatively freely, there may have been little need for providing them with salt; however, if they were restrained in small sties then it is likely that they would have needed supplemental dietary salt to avoid biting within the group. This connection between salt and behaviour may have escaped detection as the need for dietary salt supplementation among pigs only became part of modern animal science literature in the 1940s (Brooks, 2010 pers. comm.); it is quite possible that this dietary requirement was unknown in the Iron Age.

In addition to the behaviour modification uses of salt noted above, Columella (*DRR*: VI, VII & VIII) provides numerous examples of salt being used either alone or in combination with various herbs or honey as veterinary medications for cattle, horses, sheep, goats, pigs and chickens, though the efficacy of some of these preparations is highly questionable. Although salt would have been a valuable commodity in Iron Age Britain, livestock would also have been valuable, and diversion of some of the available salt for veterinary medical use may have occurred, both to insure animal health and in attempts to restore it. Unfortunately it is impossible to estimate the total percentage of available salt which would be diverted for purposes associated with livestock from the artefactual remains. Certainly the health of animals would have been a major concern among farmers, reliant on oxen for traction, cows for milk, and sheep for milk and wool and on all of these, along with pigs, for meat, not to mention horses for their use as traction animals, food or as status symbols. However, there is no clear assurance from the archaeology that the care and breeding of animals in Iron Age Britain was the same as in these Roman examples. The bone evidence for both cattle and sheep shows that there was a significant size increase of the average animal between the Iron Age and the Roman period in Britain, and the same pattern can be found in central and eastern Europe (Maltby, 1981). This may be due to the introduction of new breeds as well as a general improvement in animal care and breeding under the Roman pastoral regime. This improvement may well have been related to the existence of texts on the best farming practices such as that written by Columella cited above, allowing recommended best practices to be widely copied.

6.3 Salt for food preservation

This is by far the most commonly mentioned usage for salt that appears in archaeological texts. In the following discussion, the absence of salt as an artefact will require the use of alternate resources for information about how it might have been used in the past as a food preservative.

Sources that might be considered to be historical, anthropological or technical in nature will figure prominently in the discussion that follows.

Meat

As with domestic animals, humans require some salt in their diet to maintain good health. However, in northern Europe, unlike the tropics, this is a minimal amount and as little as a single gram per day can be sufficient (Carter, 1975). With respect to the human diet, the main usage of salt would not have been as a seasoning for foods, but as a preservative. Since the 1970's, virtually every detailed synthesis of Iron Age Britain's archaeology has acknowledged that salt was available in that period and being used for food preservation (e.g. Harding, 1972, Darvill, 1987, Cunliffe, 1991, Haselgrove and Pope, 2007). In fact, the phrase "...and salt was used for food preservation..." has now become rather like an archaeological mantra, repeated *ad infinitum* down the generations of archaeologists with little examination of its meaning. As late as the 1930s, individuals in Britain could easily purchase 30 pound (13.6kg) blocks of salt, the necessary amount required for preservation of the typical butchered pig (Kirkham, 2001), suggesting that until at least that period, archaeologists writing about the use of salt for food preservation may have had personal knowledge of the techniques required. However, with the rise of supermarkets after World War II, this link between practice and knowledge would have been largely broken. With the exception of those engaged in archaeological experiments in food preservation, the details of the technology behind the phrase would be unknown to most archaeologists.

In a brief digression before detailing the use of salt for meat preservation, it should be noted that this is not the only method of meat preservation that may have occurred in the past. For example, in modern Turkey a traditional dish called *kavurma*, made from cooked lamb or mutton preserved in fat, can stay edible through the winter with proper storage (Ryder, 1983b, Kut, 1996). While there is no direct evidence for a similar dish being part of the diet of Iron Age Britain, there is also no reason to argue that it could not have been. This example is provided to highlight the fact that salt would have been one of many preservation strategies utilised in the Iron Age.

Although modern archaeologists usually have little firsthand knowledge of salt-based food preservation technology, they have been able to develop a detailed understanding of certain aspects of the Iron Age diet. For example, detailed studies of prehistoric animal bone assemblages (e.g. Legge, 1981, Maltby, 1985, Serjeantson, 2007) have led to conclusions about the nature of both animal management in the period and the dietary implications of such practices. However, other than in Serjeantson, the link between salt and food has not figured significantly in these studies. Yet archaeozoological reports such as these, however detailed they may be, are highly sanitised information which separates the messy, physical nature of animal slaughter from the consciousness of most modern archaeologists. Individuals who could not imagine slitting the throat of a live animal to turn it into meat, this author included, have no problem excavating animal bone assemblages and equally little difficulty discussing feasting and food preservation. To the extent this gap exists, archaeologists have an unrealistic view of the past. Animal slaughter was a very physically demanding process, potentially involving more than one person in the act of capturing the animal, killing it and then, sometimes, hanging its body to drain it of fluids, removal offal and butcher the carcass.

The process leading to salt preserved meat begins with dissecting the carcass into smaller parts. It is not strictly necessary to hang an animal to butcher it, and the study of Iron Age butchery patterns at Mingies Ditch in Oxfordshire found that animals appeared to have been butchered on the ground (Wilson, 1993). However, hanging the carcass would have reduced the opportunity for soil to contaminate the fat and flesh and minimizes the risk of predation of the carcass by either wild animals or domestic ones, such as dogs. Figure 112 shows a modern example of a simple wooden frame used for hanging an animal for butchery. Careful removal of the hide, for later use in leather production, would likely have also been part of the butchery process, along with retention of those non-food parts of the animal, such a hooves, horn and tendons, which could be put to other uses.

The bone evidence suggests that Iron Age domestic animals had similarities with some modern species so they will serve as models for their Iron Age counterparts in this discussion. For example, Dexter cattle are about the same size as those of the Iron Age (Reynolds, 1995). At 12 months of age, a Dexter can yield up to 250 pounds, approximately 114kg, of meat and between 475 and 500 pounds, 216-227kg, of meat at 24 months of age, although these yields are attained with the inclusion of some grain as well as grass in the diet (Anon., 2000). As the use of grain for animal feed may not have been an Iron Age practice, cattle of that period of the same age probably weighed less and yielded less meat thant the modern quanities shown in Table 25. Even so, the meat yield was probably 45-55% of the live weight of the animal (Anon., 2004), suggesting 100 to 200 pounds, or 40 to 91kg, would be a reasonable meat yield estimate for a slaughtered Iron Age yearling and 300 to 350 pounds, 136 to 160kg, for a two year old animal. While a major communal feast or large ritual offerings might utilise meat in this quantity, in the absence of these activities, only preservation would allow for the full utilisation of the meat before it spoiled.

Similar issues would arise from the slaughter of sheep, goats or pigs, all species for which evidence of autumn culls exists. The Soay sheep is considered to be one of the best living models of a British Iron Age sheep (Reynolds, 1995) and one recent estimate of the meat yield of this species was 22 to 23 pounds, roughly 10 to 11kg, per adult animal slaughtered between one and three years of age (Weaver, 2008). Albarella (2007), however, notes that most Iron Age assemblages reveal that the sheep

Figure 112. A simple wooden structure to hoist a slaughtered animal.

From http://www.fao.org/docrep/003/x6932e/X6932E01.htm#fig1 [accessed 21 October 2008]. Used with the permission of Food and Agriculture Organisation of the United Nations.

Animal	Estimated Meat Yield in Pounds	Estimated Meat Yield in Kilograms
Modern Dexter Cattle (12 months of age)	250	114
Modern Dexter Cattle (24 months of age)	475-500	216-227
Iron Age Cattle (12 months of age)	100-200	40-91
Iron Age Cattle (24 months of age)	300-350	136-160
Modern Adult Soay Sheep	22-23	10-11
Modern Male Wild Pig (18 months of age)	185	85
Modern Female Wild Pig (18 months of age)	94	43

Table 25. Estimated meat yields for domestic animals.

were being culled well before they are at the optimal age for meat production, so the meat yield per animal was probably significantly lower than this estimate. This estimate would be applicable to goats as well, to the extent that they were kept, as they are so similar physically to sheep in this period that their skeletal remains cannot always be distinguished (e.g. Cunliffe, 2005) and such remains are often classed, therefore, as "ova-caprid" or "sheep-goat" in bone reports. Even if this lower meat yield per animal is true of most autumn culls, the killing of a number of small animals at the same time would still generate a large quantity of meat and a large number of sheep or goat skins.

Iron Age pigs were more like modern wild boars than the modern domestic pig (Reynolds, 1995). Suckling pig bones are quite common on Iron Age sites (Maltby, 1981), and a moderately large group of people could certainly devour a suckling pig before it spoiled, but again, the problem of an excess of meat would arise when a large number of piglets were killed at a single time or if adult animals were butchered. Adult wild pigs exhibit significant sexual dimorphism, and with a meat to live weight ratio of approximately 75% (Anon., 2004), the sows would yield about 94 pounds, approximately 43kg, of meat and the boars roughly 185 pounds, or 85kg, when they reach full growth at 18 months of age (Pimentel, 1998). Though there is some evidence that horses were sometimes eaten in Iron Age Britain (Maltby, 1981: 184), horse numbers were so low overall as to reasonably allow their exclusion from this discussion. Having said that, the issues related to meat preservation from horses would be the same as those with any other large domesticated animal noted above.

From the moment at which an animal is killed, various bodily processes continue for up to several hours after death which can affect the quality of the meat (Rixson,

2000). After the animal was drained of blood, which might have been retained to be eaten, initially there would be no need to worry about spoilage of the meat while it hung:

> *Certain natural enzymes of the muscles (cathepsins) are released post mortem and during the aging process these enzymes break down some of the muscle structures resulting in improved tenderness of the meat; there is also an improvement in the flavour. Aging is an autolytic process whereby tissues are broken down by the body's own enzymes and in the same way autolysis can result in spoilage of meat but from a practical point of view this takes so long that before such spoilage becomes significant meat would have undergone major spoilage due to other factors* (Rixson, 2000: 11-12).

Although removal of offal would be required shortly after death to avoid spoilage, the hanging period for the carcass was a necessary step for salt preservation to be effective and lengthier than might be expected in the days before refrigeration. For example, Kurlansky (2002: 57-58), citing a 1930 Scottish highland cookbook by Fraser, indicated that a haunch of venison needed to hang in the larder for two days prior to salting. Structures suitable for protection of the carcass during the aging process would have undoubtedly been a normal part of the environment when animals were culled with the intent to preserve the meat by salting and may account for finds of post holes that do not seem to relate to roundhouses, palisades or enclosures.

However, the post-slaughter processes are complex and multi-dimensional. The unavoidable exposure to oxygen in the atmosphere can create problems:

> *Oxidation is a spoilage problem principally in two respects: oxidation of fat and oxidation of myoglobin. The unpleasant odour and taste of rancid fat is the result of the oxidation of unsaturated fat molecules. Fats with higher levels of unsaturated fatty acids (e.g. pork fat) are more vulnerable to oxidation or rancidity than fats with lower levels or unsaturated fatty acids (e.g. mutton fat). Mutton will therefore remain palatable for longer after slaughter than pork, especially in warmer weather.*
>
> *If the meat is left it will progressively undergo oxidation of the myoglobin, changing from bright red through shades of reddish brown to brown, until finally it will become dark brown or black. Although the colour changes do not in themselves affect the eating quality they are invariably accompanied by bacterial spoilage* (Rixson, 2000: 12).

While cutting the carcass into smaller segments might seem like the logical first step after slaughter and skinning, doing so actually speeds the spoilage of the meat. First, the cutting implements will introduce bacterial contamination and, second, the smaller segments increase the surface area subject to oxidation (Rixson, 2000). Thus, further butchery must occur with a plan in mind for additional treatment of the meat which attempts to minimise further spoilage. One such approach would be to cut the meat into very thin strips, as shown in Figure 113, and then dry them quickly, producing a product like South African sun-dried biltong, although in Britain's climate, following an autumn cull, this chewy meat product would have to be made over a fire, possibly within a structure, to successfully dry the meat quickly enough to avoid spoilage. Such dried meat slices can be stored for months without spoiling if kept in lidded containers to keep flies away. Although few Iron Age ceramic storage jars seem to have had ceramic lids, a skin or cloth tied across the top would have been sufficient to protect dried meat from insect damage. Though optional, salt may have had a role in this meat preservation method: the likelihood of success of the drying process can be greatly enhanced by soaking the meat strips in a 14% salt solution for five minutes, then draining them, before hanging them to dry (FAOUN, 1990).

The above information leads to the central question about the use of salt for food preservation: what is the actual mechanism by which it works? According to Rixson (2000: 12) one of the primary conditions for the growth of bacteria on meat, thus leading to its spoilage, is the presence of surface water on the flesh. Immersion of the meat in salt water will leave a salt residue behind which will "...enhance the osmotic pressure of the meat which inhibits the growth of micro-organisms and therefore the putrefactive spoilage of uncured meat" (ibid.: 12-13). While halophilic, or salt-tolerant, bacteria do exist, they cause a much slower rate of spoilage (ibid.: 13). Thus the essence of salt preservation is the creation of a micro-environment hostile to the bacteria that cause spoilage.

There are two approaches to salt preservation, brining and dry salting (Erlandson, 2003: 10). As will be discussed below, only where salt was abundant would the use of the latter method be practical. Brining, or wet curing, where the meat was immersed in a salt solution, would have been preferable where salt was in short supply, as with this method any residual solution could be boiled to reclaim the salt (Agricola, 1556) or fresh salt could be added to the solution to bring it back to full strength (Erlandson, 2003: 18). Therefore, it is reasonable to assume that this method was more widely used in the Iron Age, though curing times could take 2-4 days per pound (0.454kg) of meat depending on the shape and thickness of the piece (Erlandson, 2003: 73-4). Presumably, brine would only be discarded when it became "ropy", or thick through bacterial action, a condition which would result in meat spoilage rather than preservation (Foxwell, 1988: 69). During wet curing, controlling the concentration of salt in the preserving solution would be crucial. The addition of too much salt would not only be wasteful, but would also toughen the meat (Erlandson, 2003); a solution of too little salt would fail in its role as a

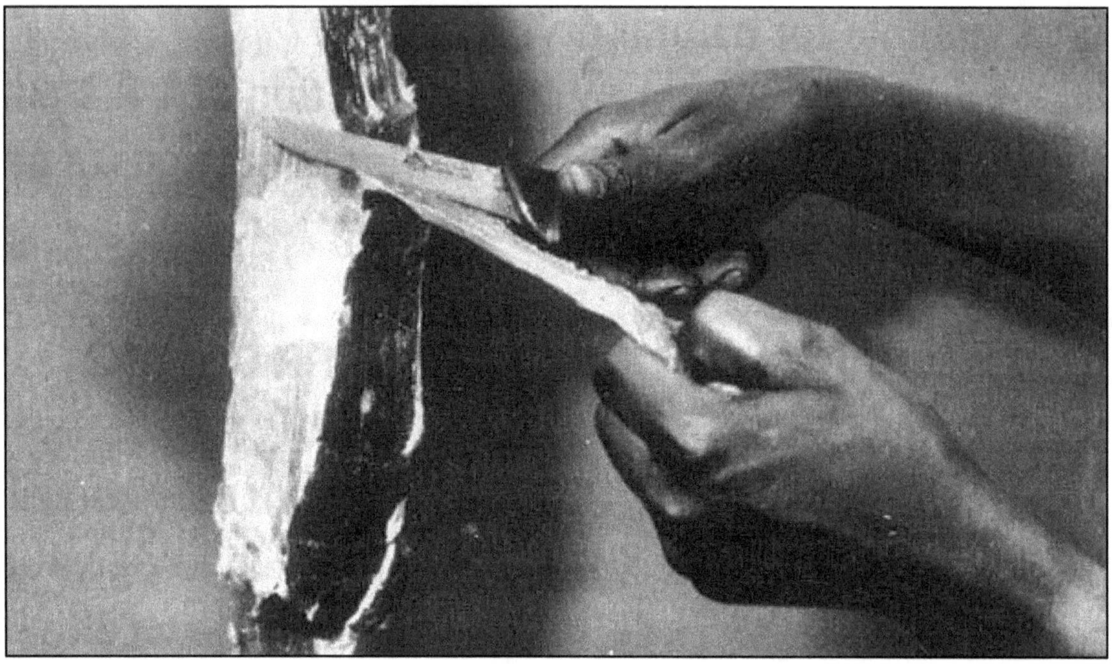

Figure 113. Cutting meat strips suitable for air drying.

From http://www.fao.org/docrep/003/x6932e/X6932E02.htm#ch2, [accessed 21 Oct 2008]. Used with the permission of Food and Agriculture Organisation of the United Nations.

preservative. Continuous monitoring of the salt concentration during the curing process would be needed, as once salt was absorbed into the meat the solution would weaken. The brine solution was thought to be strong enough when it would float an egg (Foxwell, 1988: 68), a method known to the Roman writer Cato (*DA*, LXXXVIII), and a similar process could easily have been used in Iron Age Britain. Unhygienic Iron Age water supplies used to make up the brine solutions may well have caused problems with the preservation process, though these issues would be less severe with stronger brine solutions.

Using only salt as a meat preservative would have resulted in the production of a grey-coloured end product (Rixson, 2000: 13). The pink colour of modern salt beef, hams and bacon comes from the presence of nitrates or nitrites in the preservation solution, as these interact with the myoglobin protein and prevent it from darkening while also suppressing growth of several varieties of harmful bacteria (ibid.*)*. It is possible that some pink preserved meats were available in the Iron Age, created where the local water had naturally occurring nitrite or if the salt used for preserving was tainted with nitrate, which would become nitrites in solution as a result of bacterial action (Foxwell, 1988). However, routine addition of nitrites to brine solutions at that early date is thought unlikely, as Roman writers fail to mention its use until the late Roman period (Rixson, 2000: 13). For example, a second century BC recipe for making skin-on ham, as still produced in modern Spain, reported by Cato the Elder in *De Agricultura* (CLXII, 122), requires the use of large quantities of specifically Roman salt, but there is no mention of any other additives. Alternatively, as it is also possible to minimise the greying of meat without nitrites through the addition of ascorbic acid to the preserving solution (Foxwell, 1988), the use of willow bark or rose hips as an additive could have yielded similar results in Iron Age Britain.

As hinted above, the brining method has a variety of process factors which can be manipulated. For example, in order to allow for sufficient penetration of the salt solution to forestall spoilage, the meat would have been cut into smaller segments before immersion in the solution, and the shape and size of the pieces could be varied, although bone evidence suggests that there were cultural parameters involved in selecting the "correct" way to butcher animals (e.g. Halstead et al., 1978). A strong brine solution, 80-90% concentration, can be used with little concern about control of temperature of the solution, but would yield a tough preserved meat (Erlandson, 2003: 20). Addition of a sweetener, such as honey, to a weaker brine solution, will lead to bacterial action that will tenderize the meat, but such sweet brines ideally require chilling to about 1.5°C (approximately 35°F), and the container must be held at this temperature for the entire curing cycle, a period of several days or longer, to avoid spoilage (ibid.: 19-21). Structures such as the fogous of Cornwall (Henderson, 2007) or the pits found widely in southern Britain (e.g. Harding, 1972) would have been well suited to storage of meats during wet curing when maintenance of a relatively stable temperature was important, while the inherent dampness of such structures would not be problematic for this process.

The second method, salt packing or dry curing, is straight forward, though salt intensive, and as one might imagine from the name this is a process of placing cuts of meat into a deep coating of salt, as shown in Figure 114. This process works more quickly than the wet curing process

(Foxwell, 1988), but a large quantity of salt is required, with salt used in a 1: 10 ratio, or 1kg of salt for each 10kg of meat to be preserved (e.g. Erlandson, 2003). There must be sufficient salt for it to be in contact with all the meat surfaces and to provide a barrier to separate adjacent pieces of meat. In Erlandson's dried cured bacon recipe (ibid.: 73), half of the salt is rubbed directly into the meat after its hanging period, then it is left in a cool place, 2°C (35°F) for 3 days, after which the remaining salt is applied and the meat left for 2 days per pound (0.454kg) of weight. Foxwell (1988) is less restrictive in his suggestion on the holding temperature, allowing a range from 2°-4°C (36°–39°F) and says that thinner cuts of meat may only need 1.5 days per pound (ibid.: 63) for this process step. Certainly the given optimal temperature ranges would have been most easily found in the autumn and winter during the Iron Age, coinciding with the timing of herd culls. Either ceramic pots or wooden containers would have been suitable for use with the brining or the salt packing method of preservation. Metal cauldrons would have been avoided for meat preservation, due both to the caustic nature of the brine solution which would have damaged them, and the fact that the metal would leave an unpleasant flavour behind in the meat (ibid.: 18). Both wet and dry curing methods are still widely used today. One might speculate about long term continuity of preferred methods; Lincolnshire, for example, an area of significant salt production in the Iron Age and later periods, traditionally produces dry cured bacon (ibid.: 73).

After either style of curing, further processing to rapidly dry the finished meat is required. According to Foxwell (1988), wet cured meats are removed from the brine solution, wiped dry and then hung for a period lasting from several days or even weeks depending on the cut, where they should be exposed to a current of warm, dry air until fully dry. Dry cured meats need to be cleaned of all residual salt on their surface, wiped dry of any fluids which have accumulated and similarly hung where they can be exposed to warm, dry air. In historic times, such an environment was often found in the kitchen, near the hearth. Storage during this period in environments where cold or stagnant air accumulate would lead to the rapid growth of mould on the meat, while excessive warmth would dry out the surface, leading to the formation of a tough crust preventing the interior of the meat from fully drying, causing the preserving process to fail at this late stage. The meat also must be protected from flies during this drying period. Once this phase is completed and the meat is fully dried, it is ready for long term storage. Wrapped in cloth, buried in corn, bran or saw dust and kept in a cool environment in a container covered in cloth, it can be kept for months.

Subsequent to either of these salting processes, the meat can be smoked, thus allowing tarry deposits with naturally antiseptic properties which are carried in the smoke, to settle on the meat. Along with the initial uptake of salt, these deposits inhibit the growth of bacteria which would spoil the food (Erlandson, 2003: 10-11). In the modern world, smoking of meats occurs in specially-designed structures where the salted meats can be hung in a temperature controlled environment for either cold or hot smoking. Cold smoking, requiring temperatures from 10°–29°C (50°–85°F), though 24°–26°C (75°–80°F) is optimal, and does not fully cook meat such as beef fillet. Following a period of cold smoking, which imparts the "smoky" flavour to meat, it can be further treated with a period of hot smoking at temperatures between 82°–115°C (180°–240°F), which will create a fully cooked meat product.

This detailed discussion of the technology of meat preservation through curing and smoking sets the basis for questioning some of the commentary associated with Iron Age food preservation which has previously been presented by archaeologists. It should be clear by now that the technology of salting food for preservation is a complex activity which operates within quite specific parameters, with little about its successful application being either accidental or incidental.

With so many ways for the process to go wrong, and the risk of failure so great, it was likely a focus of ritual behaviour designed to insure a successful outcome. Yet this has not kept some archaeologists from portraying it as an almost insignificant adjunct to Iron Age life. For example, Pope (2007), citing Oliver 1987 and Orchardson and Matson 1961, wrote

> *"Hot air rising from the fire means that an upper floor[in a post ring roundhouse] would be warm and smoky, and food stored there - such as grain, meat or cheese - would become disinfected and well preserved."*

While there is no disagreement with the suggestion about the temperature or smokiness of the upper floor of a roundhouse, there is every reason to believe that untreated or inadequately treated foodstuffs would quickly rot in that environment rather than become properly smoke-treated through the incidental action of trapped smoke. As described by Pope, this upper level would only be a good environment for post-treatment storage of finished dried meat strips, or finished wet or dry cured and smoked meats or cheeses, or previously-dried grains.

The need to store large amounts of finished preserved meat could require supplementary structures for this specific purpose. For the reasons given above, neither subterranean pits nor fogous, both likely to be damp, would be suitable for long term storage of finished salt preserved meats, suggesting that supplemental elevated structures might have been required once roundhouse loft storage capacity had been exceeded. One possibility for such an edifice may be the four post structures commonly found on Iron Age sites, particularly in southern and eastern Britain. While Iron Age four, and even six, post structures are now routinely described as grain stores (e.g. James and Rigby, 1997, Mulville, 2008), such structures could also have been suitable for either meat strip drying or long-term preserved meat storage. As shown in Figure 115, successful frameworks for drying

Figure 114. Dry curing meat in a wooden vat.
Image used courtesy, The Colonial Williamsburg foundation.

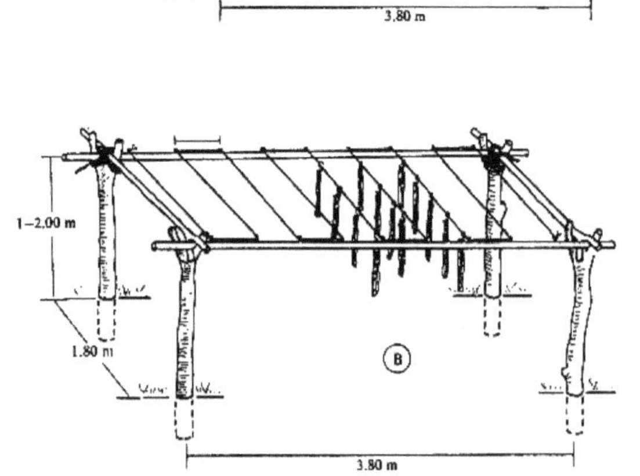

Figure 115. Modern suggested meat drying rack structure.
From http://www.fao.org/docrep/003/x6932e/X6932E02.htm#ch2, [accessed 21 Oct 2008]. Used with the permission of Food and Agriculture Organisation of the United Nations.

meat strips can have remarkably similar dimensions to those recorded for four post structures (Gent, 1983, Cunliffe and Poole, 1991a), though as illustrated here they have much smaller posts and post holes than those associated with the typical prehistoric four posters. However, in the original context of the illustrations, the pictured relatively flimsy racks would have been contained within a more substantial protective structure (FAOUN, 1990).

One site where the interpretation of a four poster as a meat store seems particularly suitable, though at variance with the published explanation, is Brook House Farm, Halewood (Cowell, 2000). Cheshire VCP, suggesting the presence of quantities of salt, was found in various contexts from the Middle to Late Iron Age phase at this high status site. A four poster, 2.5m by 3.5m, had "...small fragments of VCP [in 2 post holes] while the associated north-eastern gulley contained the largest assemblage of VCP from the site [41 sherds, 152.4 g]... " (ibid.: 61). Pollen evidence from the site showed no trace of herbaceous plants in the vicinity (Shimwell and Ogle,

2000), and no grain was found at the site, while butchered pig and cattle bones were present in significant numbers (Cowell 2000: 66). Despite this absence of grain evidence and the presence of animal bones, the four poster, the first found in this region, was described as a grain store and the possibility that it had been a meat store, or had any other possible function, was never explored (ibid.: 61).

The Brook House Farm site is a fairly unique conjunction of a VCP find, a four poster, large quantities of animal bones and a full environmental analysis ideal for illustrating the case for another possible use of the four post structure. However, the more significant underlying point of this example is not the extraordinary nature of this one site but the successive iterations of archaeological thinking over the past 70 years that have turned the *possible* use of four post structures as grain stores suggested by Bersu in his analysis of Little Woodbury (1940) or Cunliffe and Poole at Danebury (1991a: 115) into "four posters *were* grain stores". As with the earlier observation that "salt was used for food preservation," which is often taken as unconsidered fact, this phase has too often become unchallenged wisdom. Bersu mentioned two alternative agricultural uses for four post structures other than grain storage in his article but based his final conclusion on the full range of evidence from the site. Gent (1983: 247), citing Reynolds 1972, suggested an even wider range of alternate agricultural or domestic goods, other than grain, which may have required storage in these structures. Even before Gent's article had appeared, some had noticed the risks of the growing orthodoxy: "There are adequate grounds for caution in accepting that all such rectilinear structures served the same purpose, a generalization that threatens to obscure differences in function, structure, chronology and culture," (Stanford, 1970: 110), though this cautionary voice seems to have been largely ignored. Cunliffe and Poole (1991a: 115) list four commodities other than grain that could be stored in post built structures and point out that a single structure might be used flexibly for storing different products at different points in time. Interpretation of four posters as grain stores makes good sense when charred grain, wheat pollen or querns are in evidence, less so in their total absence, a point made recently by Lambrick and Robinson (2009: 271-72). Needham's recent comment (2007: 55), that "...four- and six-post structures often regarded as granaries..." , should be taken as evidence that the interpretation of these structures is still not always handled with the necessary level of thoughtfulness and finesse. Despite these examples of good practice, the casual explanation of four posters as Iron Age grain stores has become received wisdom that it is now interfering with sound archaeological analysis of site evidence. The preservation of meat using salt was a complex process which would have had required appropriate storage facilities to achieve successful outcomes. It is possible that the Brook House Farm site provides evidence of this activity.

Milk and Dairy Products

Of course meat is not the only product provided by domestic animals. Milk from sheep, goats and cows would have been another valued source of food during the Iron Age, and the predominance of these animals compared to pigs in the Iron Age suggests that milk may have been a dietary staple in that period (Serjeantson, 2007: 92). However, in the absence of refrigeration, most fresh milk would have been processed into cheese, butter or other dairy products to preserve as much as possible for use as food. Bone evidence has been utilised by archaeozoologists to suggest whether herds were being kept for milk production, meat production or both (e.g. Legge, 1981), while lipid analysis on pot sherds (e.g. Dudd et al., 1999) has been used to demonstrate the presence of ancient milk-based fats on cooking pots, suggesting the production of dairy products. Yet while the use of salt in dairy products as a preservative has been noted in archaeological literature (e.g. Serjeantson, 2007: 89), the relationship between milk, salt and dairy products has not been thoroughly reviewed with respect to archaeological interpretation of Iron Age sites. As was the case with meat preservation, few archaeologists have detailed personal knowledge of milk production and/or the making of dairy products, so the role of salt in dairy product production would be even less apparent, a factor presumably contributing to its absence in their writings.

Milk would normally be produced by ruminant animals for a short period after parturition, and herders would have to take the nutritional needs of the young animals into consideration before taking milk for their own use. For example, in the case of sheep, milking would normally commence once lambs begin to graze in late spring and continue until early summer when the ewes' milk dries up (Castell, 2009 pers. comm.). Cows would be a more useful source of milk, as they produce for a longer period, with modern Dexter cows providing 1.5 to 2 US gallons (5.3 to 7.6 litres) of milk for 305 days (Anon., 2000). However, Iron Age cows may have had a shorter productive period due to differing feed regimes and the effect of many generations of controlled breeding since the Iron Age. Killing or removing the calf within days of its birth would end the need to share the milk with the offspring(Legge, 1981: 86). Given these constraints, with natural breeding conditions, fresh milk would not be available throughout the entire year, though it is possible that Iron Age farmers controlled the milk production process by giving rams or bulls access to some females outside of the normal breeding season to insure extended availability of milk. However, if this was an Iron Age practice, it occurred so infrequently that it has not yet been detected and commented upon by archaeozoologists. Given the problems of keeping stored milk fresh, it is probable that some type of a fermented milk drink, as can be found in modern southern and eastern Africa and large parts of Asia (Crawford, 1990), rather than fresh milk, was a beverage in wide use in the Iron Age. In summary, the archaeological evidence to date and what can be determined by both analogy with modern ethnographic parallels and animal physiology suggests that milk was largely a seasonal resource of the Iron Age, and that regular production of dairy products was undertaken to extend its availability as a source of dietary protein. The role of salt, or its absence, in each of the major types of dairy products, sour cream and yoghurt, butter and cheese, will be considered below.

Sour cream and yoghurt

Unpasteurised cream, without any additive, or to which a small quantity of vinegar has been added, will turn into sour cream within 24 hours in a warm environment through bacterial action (Anon., 2006). Though yoghurt has a similar flavour and texture, it results from a different process and is more difficult to produce, as it requires the presence of *Streptococcus thermophilus, Lactobacillus (L.) bulgaricus* and *L. Acidophilus,* all of which occur naturally in milk but develop more slowly than other types of souring bacteria (Ryder, 1983a: 246). Yoghurt production, therefore, requires that milk be boiled to kill these more active bacteria, after which the boiled milk must have live yoghurt culture introduced to initiate the desired form of controlled "spoilage" of the milk. As these live bacterial cultures only live for one to two days (ibid.: 249), production of yoghurt in Iron Age Britain could only have been maintained if herd management practices were advanced enough so as to insure a year-round milk supply, which, as noted above, is not a certainty. It would appear, therefore, that sour cream, though not necessarily yoghurt, could have been widely produced by any pastoralist who had milking animals. Salt is not required for the production of either sour cream or yoghurt.

Butter

Butter production offers a method for extending the useful life of the fat fraction of whole milk found in cream. Unlike the products discussed above, production of butter is so straight-forward that it can occur with little intervention, even being produced accidentally during the transport of full cream milk in a container (Brothwell and Brothwell, 1969: 50-51). Addition of salt to butter, a common practice since at least the Middle Ages in Britain (Cronin et al., 2007: 1019), extends its shelf life. Although butter can be made with the milk of most ruminants, discussion here will focus on cows' milk. Returning to the modern Dexter breed as a model for Iron Age cattle, daily milk yield of these modern animals, as noted earlier, runs from roughly 5.7 to 7.6 litres with roughly a 4% butterfat content (Anon., 2000). Even if Iron Age cows were equally productive and their milk had similar milk to butterfat ratios, which may not be the case, only 0.23 to 0.30 litres of cream per cow would be produced each day, suggesting that it would make sense to accumulate cream over several days before making butter. In doing so, however, the cream would begin a natural fermentation process while in its storage vessel.

Rather surprisingly, Iron Age butter exists in the archaeological record, having been preserved when it was deposited in peat bogs in Ireland and Scotland. Recently nine samples of Iron Age bog butter were subjected to detailed chemical analysis (Cronin et al., 2007) which concluded that salt was absent from the samples. Cronin et al. wrote: "Leaching out of the salt during prolonged storage in the bog may account for its absence from bog butter. However, it may be the case that salt was not widely used in earlier times in the production of butter " (ibid.: 1018). If butter was being made without salt from slightly fermented cream aggregated from milk collected over several days, it would produce a product similar to the unsalted ripened cream butter still widely available in continental Europe, "...southern and eastern Africa, the Middle East - Syria and India and neighbouring countries..." (Crawford, 1990: Sect. 6). The use of slightly fermented cream slows, but does not completely eliminate, the breakdown of fats leading to spoilage of the butter, so while this product would last longer than fresh cream by a matter of weeks, it does not have the same storage life as salted butter, which could have lasted for months. However, as Cronin et al. note (2007: 1019), it is also possible that valuable salt was simply not used in the making of butter which was destined for ritual deposition, but could have been routinely added to butter for normal consumption which is not available for examination.

One further variant of butter, ghee, is worth mentioning as it may have implications for the discussion of the duration of storage. Ghee, or clarified unsalted butter, is made by boiling off the water and milk fat solids from unsalted butter to create almost pure butterfat. Under ideal storage conditions it can last for up to a year even without refrigeration (Mouze, 2008). There is no evidence that this product was in use in Iron Age Britain, but it certainly could have been created from the type of ripened cream butter discussed above, and would have been a highly valued product if it did exist due to its good taste and long term storage potential. Thus, with multiple possible forms of butter, possibly including salted butter, unsalted ripened cream butter, ghee and possibly even a mix of fresh butter with rancid, it is likely that butter was widely used in Iron Age Britain. With the exception of ghee, only addition of salt to butter would have given it an extended storage life.

Cheese

The final class of dairy products which allow for the potential long term storage of milk are cheeses, produced when milk solids coagulate and separate from the watery whey (Foxwell, 1988: 81). Whey should not be seen as not a waste product as it can be used as food for humans (Ryder, 1983a: 244) or pigs (Ryder, 1983a: 244, Kurlansky, 2002: 98). Finds of perforated pot bases which could have served as strainers to separate curds and whey occur in the ceramic assemblage from Phase 3, Late Iron Age Gussage All Saints (Wainwright, 1979: 190), Maiden Castle and Stanwick (Copley et al., 2005: 491), as well as in multiple Iron Age ceramic phases at Danebury where 22 were retrieved (Brown, 1995: 56). These objects were taken as evidence of the production of cheese and other dairy products. Yet unlike the other sites, Gussage All Saints yielded no evidence for the presence of salt. It is possible, therefore, that little salt was available for use in cheese making there, which would have affected the types of cheese produced but would not have made cheese making impossible.

If salt is not available, only two classes of cheese can be produced, those which are preserved through desiccation or the soft cheeses. Desiccated cheese, like the Mongolian byaslag (Mischler and Sosorbaram, 2005-2006), is made from pressed curds, cut into slabs to allow rapid drying, then air dried without the addition of salt. Desiccated cheeses feature in the contemporary cuisine of the Bedouins and the Quashqai as well as the Mongols (Ryder, 1983b: 724). Given the generally higher humidity of Britain than the desert areas where this style of cheese is still produced, it seems unlikely that this type of cheese featured in the Iron Age diet of Britain.

Soft cheeses, however, may well have been dietary staples, as one of the natural by-products of ordinary milk souring is soft cheese (ibid.). Soft cheese formation, however, can be initiated more rapidly through a natural chemical process which occurs when milk is stored in the stomach of a young mammal where rennin, the precursor protein to rennet, occurs naturally, or if rennet is added to the milk. Typically the cheese curds would be separated from the whey, pressed, then placed in a cloth to continue draining for some hours to remove all traces of whey and thereby maximise the storage life of the cheese. However, without addition of salt to the curds, the storage life of soft cheeses in the absence of refrigeration would be a matter of a few days. With salt added to the finished cheese, successful storage of up to a week might be achieved under conditions approximating the temperatures of a modern refrigerator (Foxwell, 1988: 88). Alternatively, if the cheese were stored in strong brine after pressing, similar to modern feta, storage life might be extended to a month or more. Unlike the earlier example of converting butter into ghee, there is no analogous process available to make unsalted soft cheese last for a year.

Hard cheeses, such as cheddar, would last for much longer periods under a range of storage conditions. Yet these cheeses require substantial amounts of salt for their production, on the order of 50gm per 10 litres of milk, averaging the amounts cited in a range of hard cheese recipes examined by the author. The salt is added to the curds before they are pressed (e.g. Bridbury, 1955, Carroll, 1986, GourmetSleuth.com, 2007) about midway through the production process. To estimate how much salt might have been available to the average person in a salt using community in Iron Age Britain, and therefore to theorise about the availability of salt for hard cheese production, a comparison with the Roman diet might be useful. The Roman writer Varro (*DRR*: II, 11, 4) provides a recipe for the production of soft cheese, denegrating the the nutritional value of old, dry cheese. Vergil also mentions the production of both salted and unsalted cheeses (*Georgics*: III, 400-403). Columella (*DRR*: VII, VIII, 4-6) does discuss the production of cheese using salt in several steps of the process, yielding an end product that "...can even be exported beyond the sea," presumably a type of hard cheese. While it is possible that Romans preferred soft cheeses to hard ones, it is also possible that salt was not available in sufficient quantities for the routine production of hard cheese even in this society. If the earlier example of unsalted butter production in Iron Age Britain is part of a wider pattern of limited salt availability for use in dairy products, it seems unlikely that hard cheese was part of the staple diet there. Kurlansky (2002: 188) in fact, dates the earliest English hard cheese production to the medieval period in Cheshire, a salt-rich area, though it is unclear in the text whether he had considered possible prehistoric production in making this comment. While it is impossible to completely rule out production of hard cheeses in salt-rich areas such as Iron Age Essex or Lincolnshire, if hard cheese was being made, it seems it would have been a rare commodity, highly prized for its capacity for extended storage and an ideal trade good.

However, as was the case of meat preservation, detailed study of cheese making processes has not been a focus for many archaeologists, particularly not with respect to a constraint on salt availability. Such gaps in expertise are problematic and should create a degree of caution among archaeologists venturing wide-ranging theories with respect to diet and food supply. For example, the crucial link between salt and hard cheese production was completely overlooked by Legge (2005: 12) when he made the following observation about the prehistoric period in Britain:

> *"Most cheeses from northern latitudes are of the hard type, made to be stored for lengthy periods, and considered by many to be much improved by such storage. A food store of this type is a considerable buffer against famine and stored fats, highly palatable and much desired, are easily moved when free of their incumbent water, and are potentially a powerful social lubricant as items of gift and exchange."*

Nothing in the quote is incorrect as written, and it is correct when applied to the past 800 years or so, however, direct ascription of this behaviour pattern to the past as far back as the Neolithic in Britain, as in the article, is highly questionable. Bone evidence of ruminant herds being raised for milk production and dairy lipid residues on pottery cannot, and should not, be taken as evidence of hard cheese production as has been done by Legge. For similar reasons, Needham's (2007: 56-57) linking of the rise in salt production during the Early Iron Age directly to its use in cheese and butter production is not necessarily supported by the available evidence.

Fruit and Vegetables

As yet, there is no direct evidence for preservation of fruit or vegetables in brine solutions from Iron Age Britain. However, it is possible that such products did exist, that the technology used for wet curing meat may have been extended to cure vegetables as well. Wood (2007: 142 & 144), for example, reports that in the Medieval period rock samphire, *Crithmum maritimum*, harvested in the spring, was stored successfully for a year in strong brine. Certainly the Greeks and Romans used the method to preserve olives, and if the method for brining vegetables had not been found through independent invention in Britain, contact with Rome during the Middle or Late Iron Age could have provided a model to emulate. If they existed, wet cured vegetables in brine could have been easily stored in either subterranean locations or in roundhouses.

6.4 Industrial Uses for Salt

In the modern world, the major uses of salt are not in food preservation but within various commercial processes, including leatherworking and chemical manufacturing (Sherlock, 1921: 1). The question to be addressed here, of course, is whether there was any correspondence between these contemporary uses for salt and its use in the Iron Age. As with the salt preservation of vegetables, the available evidence is often very limited and indirect in its nature.

Leatherworking

Leatherworking would have been an important industry in the Iron Age and preservation of the hide or skin after the death of an animal, a process highly dependent on salt in the modern day (Reed, 1972: 49-50), would have been a routine adjunct of the slaughter process. The significance of the British leatherworking industry in that period is highlighted in an oft-cited quotation from the first century BC Roman writer Strabo (*Geography*, II, v, 2) in prehistoric and Romano-British archaeology texts that lists hides among the products exported from Britain (e.g. Cunliffe, 1982: 41-42). The use of the term "hides" in translation is significant, as in modern leatherworking it applies only to the pelt of larger animals such as horses or cows and the term "skin" is used to refer to smaller creatures such as pigs, sheep and goats (Reed, 1972: 13). The original text, however, uses the Greek word 'dermata' (δέρματα), which is a more generic term for skin (Kamash, 2011 pers. comm.). It is likely that the hides or skins from animals raised here would have produced premium leathers, as animals raised in cooler, wetter regions are said to generally have a superior skin quality to that which is found in hotter, drier climates (Reed, 1972: 37). The large number of cattle bones found at Late Iron Age Hengistbury Head may have related to such exports or indicate a link between local salt production and tanning (Cunliffe, 1987: 339).

There should be no doubt that tanned leather was available in prehistoric Britain, as examples of bovine leather of far earlier Neolithic date have been recovered from Meare, Somerset (Reed, 1972: 259). Yet if there is no evidence for Neolithic salt production in Britain, as was discussed in Chapter 3, then it seems probable that this leather had to have been produced without the use of salt. There are also numerous examples of Iron Age leather found with bog bodies found in Denmark, Germany, the Netherlands and Britain and from within the prehistoric salt mines at Hallstatt, Austria (Groenman-van Waateringe et al., 1999: 884). However, these leather samples have effectively been retanned by the very processes that preserved them, leaving the original tanning process in these cases a mystery. Fortunately tanned skin and leather items were found with the frozen Chalcolithic "Iceman" of Hauslabjoch in the Italian Alps and it was possible to replicate the condition of some of the pollen found on his clothing through hot-smoke tanning of oiled hides, leading to the conclusion that this was the tanning method in use at the time (ibid.). A wide range of natural substances could have provided the fat needed for this process, including "...animal body fat, bone marrow, brains, milk, butter [or] egg-yolk..." (Reed, 1972: 65). Technically this process is properly referred to as currying (ibid.: 67-68) rather than tanning. Unlike leather subjected to other tanning methods, the change in the condition of the leather is not irreversible and such leathers will deteriorate in non-acid waterlogged conditions. This may explain why so little prehistoric leather has been found (Groenman-van Waateringe et al., 1999: 890).

An alternate tanning method which may have been used starting in the Late Iron Age (Groenman-van Waateringe et al., 1999: 886) was described by Reed (1972: 49). The first stage of this tanning process could have involved warm air curing of the hide away from direct sunlight, as prompt water removal from the hide to reduce bacterial putrefaction would have been as important in this instance as it was in the case of meat preservation. By avoiding direct sunlight for this process, the formation of a surface crust on the hide, trapping internal fluid, can be avoided. This could have been followed by hair removal, if desired, and defleshing, a process that could be carried out in a single step by immersion of the dried hide into an acidic liquid created using vegetable-sourced tannins, coming from oak galls among other sources (ibid.: 75-76). Thus, it is, and obviously clearly was, possible to make leather without the use of salt in the past, confirming the earlier supposition that salt, a commodity possibly too rare to even use in all types of food preservation, would not necessarily have been used in leather production (contra Morris, 2007b: 440, and Willis, 2007: 116).

Metallurgy

As in the above cases, the use of salt in metallurgy is generally not directly visible archaeologically. Therefore, this section is rather speculative, and the standard archaeometallurgy texts by Tylecote (1962, 1986, 1992) make no mention of salt usage in any process. Nevertheless, there is reason to believe that salt may have been utilised by prehistoric metal workers as part of their processes based upon evidence in the form of ethnographic and historic analogies. Texts from Qin Dynasty China (221 - 206 BC) report the use of salt as a flux for copper and bronze casting, technologies which were already highly sophisticated there by the much earlier Shang Dynasty (*c.*1600 – 1046 BC) (Qiu, 1993: 92). In European contexts, Sherlock (1921: 1) mentioned the use of salt as a flux for metallurgical applications where the process can be documented back at least to the late medieval/early modern period. MacKinnon and Kepecs (1989: 523) noted an increase in the production of salt in Central America while under Spanish domination to support a concomitant increase in silver production. Salt may have been used for this purpose in the Iron Age as well.

In the contemporary context of the internet (Anon., 2007), salt, along with ammonia and vinegar were listed as ingredients of a solution which would provide a lovely blue patina to new bronze castings. All of these ingredients could have been obtained in prehistory, with the ammonia being derived from urine, thus suggesting the possibility that some metalsmiths may have used this decorative technique in the Iron Age.

The Roman writer Pliny (*NH*, XXXIII, iii, 20, 65) advises that brine could be used as a quenching medium when applying gold leaf to copper. However, the only known examples of gilded metalwork from Iron Age Britain come from the Snettisham treasure and these copper alloy objects were gilded using the cold work mercury gilding technique, so no quenching would have been needed in their production (Northover, 2010 pers. comm.). This single example does not completely eliminate the possibility that brine quenches for hot metalworking techniques, such as forging, might have been utilised, though there is no direct archaeological evidence that this occurred. Healy (1999) draws attention to a report (Hanfmann, 1972) that gold with a pitted surface recovered in excavations near the 6th century BC Artemisium at Sardis, in western Turkey, exhibited the effects of being produced through a salt cementation technique. Another early use of salt in metalworking is revealed in the comment from Diodorus Siculus (*LH*, III, xiv, 3), writing in the first century BC, who reported its use in the refining of gold in Egypt. While such reports are intriguing and open the possibility to similar uses in prehistoric Britain, they are unsupported by archaeological evidence at this time.

At a later date, salt was a very important ingredient in medieval alchemy, whose roots ran back to at least the Roman period, where it

> *"...was considered to be one of the three ultimate elements out of which the seven noble metals were generated. Mercury symbolised the spirit, sulphur the soul, and salt the body; mercury represented the act of illumination, sulphur that of union, and salt that of purification"* (Jones, 1923: 117).

In the discussion of salt production in Chapter 2, the possible link between magic and salt production was introduced. There is reason to believe that prehistoric metalworking was seen to have similar magical overtones (Hingley, 1997), as the refining and smithing processes were complex, difficult to master and subject to unexpected problems related to trace metal impurities and poor temperature control that may have seemed to have a demonic rather than rational origin. Salt, with its ability to be transmuted from solid to liquid and back, just like metals, would seem to be a fairly predictable ingredient for the metal smith's tool kit, even if this cannot be demonstrated.

Dyeing

In the modern world a form of salt with 5 parts per million iron content, referred to as flossy salt, is used as a dye mordant (Kurlansky, 2002: 5, Anon., 2010b), but a similar use in the Iron Age is less certain. As discussed above with respect to leather, occasional finds of prehistoric cloth fibres do occur. North European bogs, for example, have yielded fabric samples dating to the Iron Age. Waterlogged preservation conditions favour woolen and silk fabrics but rapidly break down those of hemp and linen (Ringgaard and Scharff, 2010: 221), so the range of textiles available for study is constrained compared to what may have been in use in the Iron Age, and thus not all possible dyes can be studied. Nevertheless some of the textile finds have been analysed and it was found that they were coloured with vegetable dyes made from a range of lichens, flowering plants and berries (e.g. Vanden Berghe et al., 2010). Such vegetable dyes are more effective and light-fast when used in combination with other chemicals, referred to as mordants (Rudkin, 2007: 11). Salt, vinegar, stale urine, iron and alum have all been used as mordants with natural dyes in antiquity (Rudkin, 2007: vii). Each of these affects the colour tone of the dyes as well as providing them greater durability in the fabrics.

The most common mordants, however, are metallic salts such as alum, or potassium aluminium sulphate, rather than sodium chloride, the "salt" which is the subject of this paper. The use of alum as a mordant can be traced in India as far back as 2,000 BC (Fereday, 2003: 22, 24). Although the English translations of Pliny's first century AD work *Chapters on Chemical Subjects* mention the use of salt in the preparation of *murex*, the highly prized purple dye made from a mollusk (IX, vi, 60, 133), a number of other salts such as alum, "...sulphates and acetate of iron and copper..." (Cardon, 2007: 11-12) are known to have been in use by Roman dyers as mordants.

However, alum as a mineral resource is rare in Britain as it is hygroscopic, and it is possible that chemically similar compounds were found in British plants by dyers, as some 1,600 plant species from 45 botanical families are known to accumulate aluminium in their tissues (Cardon, 2007: 32) and can be used as mordants. The class of plants known as lycopods, which include club mosses, would be likely candidates for this application in Britain (Cardon, 2007: 33-34). Wood ash would also have been useful as a type of mordant and would have been available in abundance in Iron Age Britain. This discussion shows there are many other substances which would have generally been more abundant and more readily available for use in textile dying than sodium chloride. As with other uses discussed above where alternatives existed, salt may have been too valuable for this purpose, although to date the British archaeological record has been silent on this issue.

Soap making

Another significant modern usage of salt is in soap making (Kurlansky, 2002: 5). Chemical analyses of Iron Age preserved human remains found in bogs in Britain,

Ireland and Scandinavia have revealed a range of chemical preparations that had been used on hair (e.g. Mulhall, 2010: 36), while the remains themselves showed that shaving of the face and sometimes the scalp had been practiced. This latter practice had been noted by the Romans, who mentioned the large moustaches of the Celts. While shaving and elaborate hair preparations do not mean that soap was used by Iron Age Britons, they suggest a concern with personal appearance that might have required periodic use of soap-like products. Having said that, the evidentiary chain between the possible use of soap and the potential that salt was used in its production is so attenuated as to be, perhaps, a step too far. Further evidence will be needed to make a case for the use of soap produced using salt during the Iron Age, no matter how common this usage is in the modern world.

6.5 Medicinal Uses of Salt

We have no direct archaeological evidence of the use of salt for medical purposes in the British Iron Age, yet it is difficult to imagine it not being used for this purpose. Even now salt, usually in the form of saline solutions, is still an essential part of the medical arsenal in the modern world. As it is truly efficacious in some situations, and the use of salt for both prevention of, and as a cure for, diseases has been widely documented in societies at all stages of development around the globe (Jones, 1923: 122), it seems quite reasonable to think that there has been continuity in its use in medicine since the Iron Age, if not even earlier.

During the Iron Age, salt would have been one of a small number of natural remedies which were sometimes effective for a wide range of conditions. For example, it would have been a useful, though painful, product for wound cleansing and might, at times, have succeeded in preventing infection. Its use for this purpose by the Romans has been documented (Lane and Morris, 2001: 6), suggesting the possible existence of the practice even earlier in the Iron Age. Pliny states that *flos salis*, a moist, saffron or red coloured salt from mines, is best for medical use (*NH,* XXXI, vii, 40, 90 & 92), the comment serving as a reminder of the possibility that similar distinctions between types of salt from varying sources in alternative containers were perceived as very different products in Iron Age Britain as well. Cato lists salt as an ingredient in several medicinal preparations (*DA*, CLVI, 3, 5 and 7; CLVII, 6 and 9). Other Roman medical uses for salt which may have similarities to British Iron Age practices, for example, include its use in combination with other ingredients as a purgative (Pliny, *NH*, XXXIV, xxii, 106) and as a component of a blemish ointment (ibid.: XXXIV, xxvii, 116). Pliny also advises that a composition of roasted gold, salt and schist can protect from magic potions (*NH*, XXXIII, iv, 25, 84), which were thought to be capable of causing illness. The famous second century AD Roman physician Galen included salt in combination with other ingredients, salt preserved foods and pickled foods in his recommendations for good health (Wilkins and Hill, 2006: 125, 143 & 160). The inclusion of protection from magic potions in this list serves as a reminder that while the practice of medicine in Iron Age Britain was probably more like magic than modern medical activity, it is a reasonable expectation that salt was a prominent factor in many treatment regimes.

6.6 Social salt: salt as a status good

While salt has generally been supplanted as a status good in modern Britain, it is still possible to use salt to impress. For example, a recent article in the travel section of the *Sunday Times* mentioned the appointment of a salt sommelier at an expensive resort in the Maldives which offered guests a choice of 11 types of salt to accompany their meals (Anon., 2008) and a 12 variety gourmet salt set is shown in Figure 116. This continues a long tradition documented in historic times from many different cultures where possessing the right type or quantity of salt was a sign of status. For example, among the mid-twentieth century Baruya of New Guinea, distribution of locally produced salt allowed the salt owner to repay debts for services rendered by the salt maker and local sorcerer, provide salt to his family and to trade for prestige goods such as metal tools (Godelier, 1977). Among the ancient Maya, imported pure white salt from Yucatan was an elite good, while salt from the local lagoons of Belize was not (MacKinnon and Kepecs, 1989: 523). Numerous examples of the use of salt as a status good in widespread locations around the globe during historic times can be found in Jones (1923), Multhauf (1978), Laszlo (2001) and Kurlansky (2002), and thus are not further enumerated here, as there is really no debate about the proposition that salt can act as a status good. It should come as no surprise, then, that the prehistoric past also produces examples. One famous European example of this behaviour in the archaeological record is the apparent exchange of many types of high status goods for salt at Iron Age Hallstatt. In a more directly applicable example, Lambrick and Robinson (2009: 210) suggest that the long distances involved in the procurement of salt in the Upper Thames Valley may have provided a status boost to those who possessed it.

One difficulty with discussions of status in an archaeological context is that "status" is a relative social construct which can only be theorised with respect to prehistory based on modern assumptions about the values assigned to artefacts and through application of ethnographic analogies. As Hodder and Hudson wrote,

> *"Our aim is to interpret the subjective meanings in the sense of the structuring concepts and ideas that were used to organise the recurrent material practices of groups. These social and public ideas were used to constitute subjectivities and they can be examined to see how they were transformed historically through material and social practices and through individual action and interpretation* (2003: 129)."

Status relationships associated with salt would not have been monolithic, but would have operated at different

Figure 116. Twelve types of salt exhibiting a range of colours and textures sold as gourmet seasonings.
Source: author.

levels of society simultaneously. For example, at a macro scale, those groups living in coastal areas with ready access to salt might have been able to exert political or economic power over inland groups reliant on them for salt. These larger relationships would operate in parallel with micro level interactions between the individuals who personally oversaw the production of salt and those who controlled the movement of salt inland, roles possibly filled by a single person though equally possibly done by multiple individuals, and the ultimate salt recipients. Not all of these activity spheres are equally visible archaeologically. On the receiving end, at an individual salt using settlement, as mentioned earlier, there may have been a status advantage accrued by possessing more salt or salt from an unusual source, though this would be impossible to demonstrate archaeologically.

The ability to recognise status symbols in the prehistoric archaeological record is limited due to our partial understanding of these societies. Archaeologists often must resort to simplistic models such as big roundhouses = high(er) status buildings or commodity rarity = high status. While these models often seem to work quite well, too rigorous an application can cause archaeologists to fail to consider more subtle status symbols or ignore them altogether. For example, it is theoretically possible that a smaller structure occupied by a priest or other type of religious leader, having had a highly decorated exterior that is now utterly absent from the archaeological record, was the most powerful and prestigious structure among a group of buildings in its day; and quernstones, while naturally rare in many areas, may have been low status objects due to an association with women. Lambrick and Robinson (2009: 127) have noted the possible disconnection between the physical form of settlements and social status, which may have been much more closely allied to material culture and other considerations. With this cautionary note in mind, it is possible to consider some aspects of salt which may have tempered its social perception and perceived value as a status good.

Renfrew has defined prestige goods as being characterised by "...a restriction in the number of those for whom it is appropriate to have them, and the existence of 'spheres of conveyance' restricting the free exchange of one class of goods for another (1977: 77)." In the preceding chapters and the above text in this chapter, the author has argued that salt would have been required in varying measure by all members of Iron Age British society, a condition which would largely violate the first part of Renfrew's definition. However, as discussed above in the context of the Mayan example, "salt" is not a single monolithic entity; it can have many differing physical characteristics, including crystal size and colour as shown in Figure 116; it can be moulded into shaped forms; it can come from differing sources and it can be transported in a wide variety of containers or possibly without a container. The author would argue that these perceptual distinctions were the key to the differentiation of salt as a status good. For example, Pliny reported that salt made from wood ash residue was always black (*NH*, XXXI, vii, 39, 82) and black salt was often produced due to contamination when straw was used to fuel the brine condensation fires as sparks fell into the solution (Agricola, 1556: 552). Given the wide range of fuel materials used for salt production known in Britain (Gale,

2001), it is probable that black salt was sometimes produced. We cannot know whether such black salt was held to be "good" or "bad", of high or low status, by Iron Age Britons, but we would seriously err in our evaluation of their powers of perception to believe they found no distinction between black salt and pure white salt crystals. It is equally difficult to believe that these two types of salt were considered to be suitable for all purposes, setting up the possibility that these distinctions could become the basis for socially subdividing the broad category of "salt" in a way that allowed some categories of salt to become a status good while permitting all members of society access to other types of salt as required for the multiple uses discussed above.

In Renfrew's definition of prestige goods, the exchange of salt would need to be controlled to create a situation where salt might become a status good rather than a commodity, as described among the Baruya (Godelier, 1977: 136-151). In Chapter 4, the difficulty in finding archaeological evidence of salt distribution in some parts of Iron Age Britain was discussed. Even where briquetage was used for distribution, the activity only becomes widely evident via briquetage finds of the Middle and Later Iron Age; therefore the evidence of the link between salt and status in these archaeological contexts is even more difficult to demonstrate. As noted in earlier chapters, the evidence in Wales, where briquetage finds are usually made in hillforts, certainly suggests this type of pattern. The absence of briquetage away from hillforts suggests that salt may have been decanted there, possibly for redistribution controlled by a local elite caste. In these cases, salt in the original briquetage container may have functioned as a prestige good which was shared out according to some specific social guidelines administered by a group of political or religious leaders. Once removed from the briquetage container, it may have transitioned to the realm of an ordinary commodity. Such a system would have had much in common with that used by the Baruya referenced above. This proposed model of controlled barter for salt in Wales is consistent with the system described by Appadurai, who noted: "Yet wherever evidence is available, the determination of what may be bartered, where, when, and by whom, as well as of what drives the demand for the goods of the 'other,' is a social affair (1986: 11)."

To view "status" archaeologically, we must observe some level of difference in the material culture record when two or more sites are compared. Given the paucity of briquetage evidence in general, it is used to its best advantage as an indicator of status when combined with other types of evidence. An example of this type of application might be the Upper Thames Valley Middle Iron Age site of Watkins Farm, Northmoor, Oxfordshire, where 20% of the animal bone assemblage was composed of horse remains (Lambrick and Robinson, 2009: 243), one of the highest concentrations in the area. Although some horses had been used as food as shown by of the presence of butchery marks on bones (Wilson and Allison, 1990: 59), the age distribution of the horse remains suggested that it was likely that the horses had also been raised for trading (Wilson and Allison, 1990: 61, Lambrick and Robinson, 2009: 247). The concentration of horse remains is a significant find, as horses are generally interpreted as Iron Age status symbols and therefore suggest the potential for the exchange of other high value goods for horses. Watkins Farm, which is 80km from Droitwich, also had 47 sherds of Droitwich briquetage with a total weight of 986g. The Specific Salt Container Index (SSCI), as discussed in Chapter 4, for this site was 0.032, making the assemblage comparable to two hillforts, Bredon Hill, Worcestershire, only 24km from Droitwich, and Dinedor, Herefordshire, 46km from Droitwich. The SSCI is much higher than other sites in the region as shown in Figure 117. For comparison, two other nearby Oxfordshire Iron Age sites often contrasted with Watkins Farm, Mingies Ditch and Farmoor, have SSCI values of 0.004 and 0.0 respectively. Allen noted that the pottery assemblage at Watkins Farm "...was generally more varied and of better quality than that at Mingies Ditch, similar to that from Farmoor..." (1990: 79). The Watkins Farm site also had a single enclosure ditch which surrounded nearly 0.5ha of land, similar in size to the outer enclosure ditch at the Mingies Ditch site, but considerably larger than the enclosed area at other local sites (Lambrick and Robinson, 2009: 121). Despite the presence of significant numbers of horses, better quality and more diverse pottery, a relatively large enclosed area and the high proportion of Droitwich briquetage in the ceramic assemblage compared to other sites in this region, Allen (1990: 79) concluded that there was no indication of greater wealth or status at this site. The significance of the briquetage was discounted by Allen, who pointed out that it might have come from as few as five or six containers and that rapid filling of the roundhouse gullies where the material was found could account for better survival of the material here than at other sites (1990: 52).

Yet given the absolute distance from Droitwich, the relative rarity of briquetage in this region, and the potential widespread use of organic containers for long distance movement of salt as has been argued in Chapter 4, even six containers are not without significance as potential status symbols. The author would argue that while no single item on this artefact list would indicate higher status, the conjunction of the large number of horses, the nature of the ceramic assemblage, the large enclosed area compared to other nearby sites and the presence of a relatively large assemblage of briquetage, considered collectively, do, in fact, suggest that Watkins Farm was of slightly higher status than the other sites in the vicinity. This same conclusion appears in Lambrick and Robinson (2009: 390), albeit stated there based upon the pottery assemblage and later Roman refurbishment of the site after a period of abandonment. Status in the Middle Iron Age could have been expressed as a social continuum, running the full spectrum from low to elite, rather than the great disjunctive blocks of low and high status that are often utilised in archaeological descriptions of sites. Where the evidence permits, as in the case of Watkins Farm, it may be possible to recognise that some

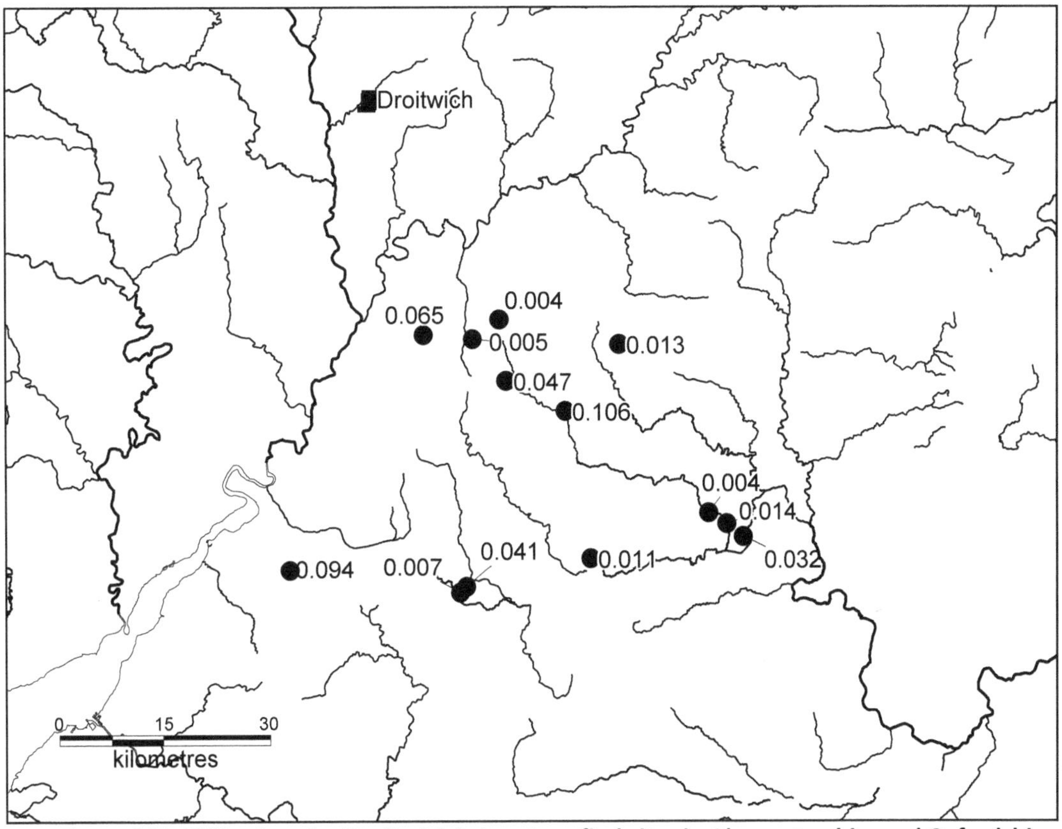

Figure 117. SSCI values for Droitwich briquetage find sites in Gloucestershire and Oxfordshire.
Sites shown on map: Gloucestershire sites – Shenberrow Hill Camp (0.004), Ireley Farm, Hailes (0.005), Spratsgate Lane, Somerford Keynes (0.007), Claydon Pike, Lechlade (0.011), Shorncote Quarry, Somerford Keynes (0.041), Guiting Power (0.047), The Knolls, Oxenton (0.065), Uley Bury (0.094), and Salmonsbury (0.106); Oxfordshire sites – Mingies Ditch (0.004), Chastleton Hill Camp (0.013), Gravelly Guy, Stanton Harcourt (0.014), Watkins Farm, Northmoor (0.032). Underlying map © Crown Copyright Ordinance Survey.

sites seem to sit at the edge between these descriptive terms.

A second example of the evidentiary dilemma presented by some sites with respect to status can be observed at Cowbit 25 and 26, a multi-phase Middle to Late Iron Age salt production site in the Lincolnshire Fens. As shown in Figure 118, multiple possible salt production hearths were accompanied there by the penannular gullies from two structures large enough to have been interpreted as roundhouses, yet these structures were characterised by an absence of domestic refuse. The site did not yield details of either the internal or external structure of the buildings, but the entrances appeared to face the north, an uncommon configuration in domestic structures (Lane and Morris, 2001: 23). Despite its longevity and reuse, we cannot imply or deduce from this site any direct link in terms of status which the production of salt brought to those who operated the salt hearths. In this, Cowbit stands as a representative of virtually all the identified salt production sites of Iron Age Britain and the problems they present.

In Essex, the relationship between salt production and high status seems quite direct in the Late Iron Age. Rodwell (1976: 300) directly linked the wealth of the *oppidum* near Colchester, the regional centre of the Trinovantes, during the late first century BC to the salt trade. Among Rodwell's status indicators were finds of Roman wine amphora sherds at multiple salt production sites in Essex. However, Carver's subsequent study of amphora finds in Britain (2001: 29) concluded that rather than envisioning a two way trade of Essex salt for Roman wine, the finds merely indicated that Roman wine was being distributed inland in the region along the established salt trading patterns. The wine issue may be a moot point; regardless of whether the material wealth of this region during this period was based upon trade of salt with Romans or people living inland in Britain, this region did accumulate material goods, such as imported pottery and metalwork, in the Late Iron Age Belgic period which would have been capable of serving as obvious high status indicators. As these examples show, the presence of salt, as seen through briquetage finds, is a difficult material to link to claims of social status. However, the examples also show that the potential link should not be overlooked.

6.7 Salt in ritual and religion

Salt was likely to have had a role in Iron Age ritual precisely because it is both so commonplace and essential that it was able to easily merge the ritual and mundane aspects of life, a point made by Lambrick and Robinson (2009: 286, citing Hill 1995 and Fitzpatrick 1997).

Finding specific examples of salt rituals in the British Iron Age is difficult, though not impossible, so it may be useful to consider a few Roman comments on religious use of salt and its importance to illustrate practices in use by the Romans in the time period near the end of the Iron Age in Britain. The first century BC writer Varro's ideal farming guidebook mentions that the religious veneration of the goddess Rumina by herdsmen included the offering milk to which salt has been added (*DRR*, II, 11, 5). Slightly later, the poet Ovid, who lived in the late first century BC – early first century AD, wrote that in the past spelt and salt in combination were needed to win the favour of the gods before eastern spices such as saffron and myrrh were available (*Fasti*, I, 337-46). Elsewhere in this text, Ovid also noted the inclusion of salt in special ritual cakes made by the Vestal virgins for the guardians of the city hearth as part of the *Vestalia*. The earlier ritual practice of salting grain offerings was also alluded to slightly later by Pliny (*NH*, XXXI, vii, 40, 89). Providing a further example of salt-mediated religious practice, Jones (1923: 119) states "With the Romans it was a matter of religious principle that no other dish was placed upon the table until the salt was in position," though he does not cite his source text. None of these practices, however vital they were to contemporary Roman religious practice, would be known from archaeological finds, and any similar rituals in Iron Age Britain would be equally opaque in the archaeological record.

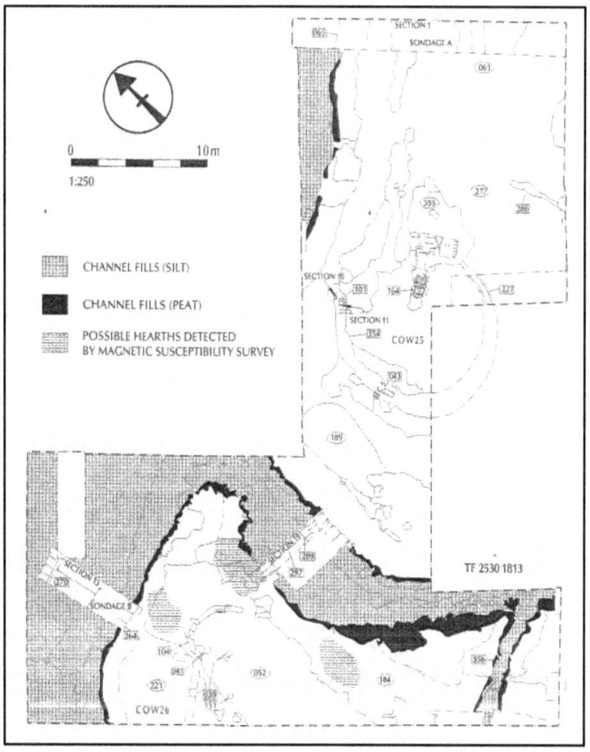

Figure 118. Possible salt production hearth sites near Cowbit 25 and 26 in the Lincolnshire Fens.

Source: Lane and Morris, 2001: 24.

As in the preceding section on salt and status, the use of salt in ritual and religion has numerous examples which can be provided from historical periods and many additional examples from modern societies across the globe which might be enumerated here. The universality of salt in such usage was commented upon by the psychiatrist Ernest Jones early in the twentieth century:

> *"First, that in all ages salt has been invested with a significance far exceeding that inherent in its natural properties, interesting and important though these are.... That this should have been so in all parts of the world and in all times shews [sic] that we are dealing with a general human tendency and not with any local custom, circumstance or notion. Secondly, the idea of salt has in different languages lent itself to a remarkable profusion of metaphorical connotations, so that a study of these suggests itself as being likely to indicate what the idea has essentially stood for in the human mind, and hence perhaps the source of its exaggerated significance* (1923: 113)."

Jones buttresses his argument with examples from contemporary or historical sources, and a range of salt ritual practices gleaned from similar sources are described by Multhauf (1978), Laszlo (2001), Lane and Morris (2001: 6-8) and Kurlansky (2002) among others. Even in our own relatively secular environment in modern Britain salt rituals persist. For example, the inclusion of salt, an ingredient mentioned in multiple Biblical passages, in Christian holy water is an example of the pervasive nature of ritual salt with which many archaeologists would be familiar. In a less formalised ritual, some feel the need to put a pinch of salt over their shoulder to ward off bad luck if salt is spilled. All of these examples can be used to support an assertion that salt would have probably been part to the ritual life of Iron Age Britain, even if most such events are now beyond modern knowledge.

Objects such as the one illustrated in Figure 119 would not survive in the archaeological record of Iron Age Britain. Even if a water-logged site preserved the wooden structure of the artefact, the salt would have dissolved long ago. If placing salt or salt-coated wooden objects into pits or rivers were a common behaviour in Iron Age religious practice in Britain, we are unlikely ever to know it was so through archaeology alone.

Much of the evidence for Iron Age rituals in Britain has been derived from the study of pit fills, particularly those in southern Britain, where pits are a common feature of sites. In a seminal text considering Wessex pit fills, and the distinction between ordinary rubbish and ritually deposited materials, Hill concluded "Depositing material in pits was not a part of the daily refuse maintenance strategies (1995: 100)," suggesting instead that most domestic refuse would have been burned (ibid.: 4). There is reason to believe that Hill's conclusion that "The deposition of certain types of pottery in certain types of contexts and associations suggests their use was ritual, ideally restricted to certain types of people, activities or spheres of practice (ibid.: 109)," is equally valid for briquetage. Although the author would suggest that the actual deposition of salt would seem a greater gift to provide to the gods than just the briquetage in which it was contained, there is no evidentiary basis as yet to support this belief that salt was used for ritual deposition. Therefore, in the absence of salt itself, it is necessary to consider finds of briquetage whose context suggests that they may relate to ritual practices.

Figure 119. A salt encrusted star from Salinas La Concordia, Chiapas, Mexico.
Photograph by Frans Blom *c.* 1945. From Andrews, 1980: 31. Copyright The Archaeological Institute of America, 1980.

Despite the fact that far more briquetage has been found at salt production sites than at salt using locations in Britain, all of the examples of possible ritual use of salt come from the salt using sites. This may be due to the inability of archaeologists to differentiate ritual accumulations of briquetage from that which was ordinarily-discarded working briquetage at production sites, the failure of archaeologists to look for possible signs of ritual amid accumulations of briquetage at production sites or the real absence of ritual activities involving briquetage at salt production sites; whatever the cause, the problem seems to be endemic within British archaeology. The total absence of ritual associated with Iron Age salt production seems improbable; as Hill has pointed out (1995: 112), there may have been no dichotomy between ritual behaviour and routine behaviour in the Iron Age. The selected shape of the salt production and drying vessels and the resulting salt cakes, which differed by region as has been discussed earlier, might have been linked to the cosmology of the salt producers (Miller, 1987: 113) as well as the functional requirements of the process. In the view of Gell (1992: 58-59), magic was a reverse comparator of technological processes, the labour-free, risk-free opposite of the hard work needed for technological production where the process was fraught with uncertainty. Although his example applied to the manufacture of Melanesian canoe prow boards which he had studied, the same production uncertainties would also have applied to salt production in Iron Age Britain, thus setting the stage for an association of the process with magic. This potential link between prehistoric salt production and magic was made explicit in the title of a recent paper, “Making magic: later prehistoric and early Roman salt production in the Lincolnshire fenland” (Morris, 2007b). It is not difficult to imagine appropriate rituals for salt production; for example, there may have been specific “closure” rituals associated with the change in status of a saltmaking hearth later converted for use as a saltmaking waste dump, a pattern seen at Billingborough, Lincs., (Chowne et al., 2001) among many other sites, where 3 hearths were later filled with briquetage and other waste material. Consideration of the possible ritual life of salt production sites is a notable absence in archaeological literature describing these sites to date.

It is possible that the most obviously archaeologically visible ritual magic associated with salt production sites may have occurred at a landscape scale, with a taboo status for the production areas (Morris, 1994b: 385), as most production sites are separate from habitation and potential defensive structures or fortifications are largely unknown at these sites despite the value of the salt being produced. It is also conceivable that there was a different sort of ritual relationship between the more numerous sea salt production sites, whose locations were somewhat variable over time due to coastal erosion and sea level changes, and the inland salt production which took place at specific salt springs. Perhaps the sea salt production sites might have been annually consecrated while the salt springs had associated deities, not unlike the goddess Sulis, venerated at the hot springs of Bath (Green, 2005: 112-14). Although the most widely promoted models of the salt trade in Iron Age Britain are drawn from economics, some salt exchange may have been conducted as part of ritual activities. As discussed in Chapters 4 and 5, such behaviour could, for instance, account for the atypical inland finds of Essex briquetage far from coastal and tidal salt production sites.

Trying to distinguish briquetage disposed of as rubbish from that which entered the archaeological record as a result of ritual activity is not straightforward. Schiffer (1987: 75) summarised the difficulty: "In order to demonstrate symbolic causality, one must show that the behaviors [sic] in question depart from expectations generated by utilitarian considerations." Of course the problem with this definition is that the “behaviors in question” cannot be directly observed and can only be imputed through the artefact record which has survived all taphonomic events subsequent to the entry of the object into the archaeological record, making Schiffer's definition circular. Nevertheless, this observation encapsulates much of the subsequent work wherein archaeologists have attempted to differentiate "rubbish" from symbolically meaningful disposal of artefacts. While salt has many potential symbolic meanings, clearly the presence of briquetage in a pit is not sufficient in itself to lead to a definition of the pit fill as a ritual deposition. Such is the case, for example, with the briquetage finds from the two pits at Madeley Heath, Belbroughton, Worcs., mentioned previously in Chapter 4, where the fills included burnt stone "pot boilers", charcoal, animal bones and iron slag along with 156 sherds of briquetage with a total weight of 2.825kg (Hurst and Pearson, 1996). Although this is a large quantity of Droitwich briquetage compared to other sites in the area and the assemblage includes briquetage vessel base forms rarely found away from Droitwich , the overall placement of the finds in the pits did not lead to any interpretation other than possibly as domestic rubbish. This site was excavated during the construction of a gas pipeline; it was suspected that there may have been a nearby domestic site, unknown from cropmarks (ibid.: 132), a presumption integral to the logic of the interpretation as domestic rubbish. Had the assumption been that the absence of cropmark evidence meant these were isolated pits, an interpretation as ritual deposits would have been more likely. As Lambrick and Robinson recently noted (2009: 282), the archaeological difference between defining something as refuse and a special deposit is often contextual and is linked as well to the subsequent fate of the material after it entered the context in which it was deposited.

The definition used by Schiffer (1987: 79) of a ritual deposit, which he termed a ritual cache, can also be difficult to reconcile with the evidence provided by many briquetage finds; therefore, linking these finds to symbolic behaviour can be difficult. For Schiffer, there was a need for "...a reasonably discrete concentration of artifacts [sic], usually not found in secondary refuse deposit; in addition, ritual caches generally contain complete artifacts [sic], sometimes unused, that are intact or easily restored." The extremely fragmentary nature of briquetage finds at most salt production or salt usage sites and its frequent inclusion in contexts in which it has been construed as ordinary household rubbish seem to violate key components of this definition and would imply it is unlikely to be used in rituals. However, Schiffer then proceeded to define two types of contextually-referenced caches, the dedicatory ones from construction sites and "offertory or votive caches (1987: 80)" typified by "the (often periodic) placement of artifacts [sic] in a special location..." which are more representative of those briquetage finds which have been construed as ritual in nature. The largely unpublished site of Poulton, Cheshire, may be an example of a site with an offertory or votive cache. Substantial quantities of Cheshire VCP were recovered from a prehistoric ring ditch that had once been a Neolithic wooden henge that was later converted into a ring ditch with an internal circle of posts. The VCP had been broken up and intentionally deposited in the southern portion of the ring ditch (K. Cootes 2010 pers. comm.). This assemblage differed from VCP finds in two further ring ditches at the site and from an area near the site's medieval chapel which appeared to have resulted from routine disposal of occupation debris. It may be significant that Poulton could have acted as a location where salt was transferred from land based to water based transportation as it headed toward the port of Meols, an activity one might suspect was likely to require the blessing of the gods.

The potential symbolic significance of artefact finds in pits, as well as some finds in ditches, was emphasized by Hill who recognised the importance of pit and boundary deposition during the Iron Age as a ritual activity which "...served to presence and renegotiate a cluster of symbolic resources...through which the implicit meanings of everyday life were articulated (1993: 64)." The inclusion of salt or at least the containers for salt, in these symbolic spaces could, therefore, be interpreted as linking the imputed characteristics of salt, such as its action as a preservative or an antiseptic, to the ritual being undertaken and its desired outcome. Such a linkage is certainly clearly possible in the find of Essex briquetage in Stanway mortuary ditches (Crummy, 2007: 377) which was discussed in Chapter 4 and 5. A similar meaning could be inferred for the briquetage found with the Bronze Age cremation at Easington, East Yorkshire (Faulkner, 2008: 41). An alternate explanation for the presence of briquetage in funeral contexts, also provided by Crummy (2007), would suggest the use of salt as part of a funeral feast, with the briquetage left behind as rubbish along with other waste, such as animal bones.

As shown in Chapter 5, although relatively rare at salt using sites, briquetage has frequently been found in pits at these locations. A relatively recent find of Late Bronze Age date from a pit fill at Carshalton was composed of a single 2gm briquetage sherd along with an antler and a deer skull (Macpherson-Grant, 2002: 81), though its presence was discounted as part of the offering by the excavator due to the small size of the sherd and its worn condition. As this is possibly the only sherd of briquetage known from Surrey of either Bronze or Iron Age date, this may not be an accurate interpretation.

A far better known and more substantial example of pit finds of briquetage is Danebury hillfort, as was discussed in depth in Chapter 5. The detailed analysis of the contents of the 1,707 pits excavated there over a-20 year period (Cunliffe and Poole, 1991a: 153) provides some evidence for the argument that salt deposition may have been part of the rituals played out at that location. Relatively large amounts of south coast briquetage (Cunliffe, 2000: 45) were found at this hillfort, although not all came from pits (e.g. Cunliffe, 1995: 261) leading the excavators to conclude that salt was an important commodity there (Cunliffe and Poole, 1991b: 486). Placement of briquetage into pits seemed to be an unusual event at Danebury, with the vast majority of excavated pits containing no briquetage, but where it was present, it appeared in considerable amounts. This pattern was initially noted with respect to the nearly 5kg of briquetage finds made between 1969 and 1978 (Poole, 1984b: 426). In the later years of excavation between 1979 and 1988, 65% of the more than 2.6kg of briquetage found was recovered from only 91 pits (Cunliffe and Poole, 1991b). In most of these cases we cannot know whether the briquetage was physically associated with salt at the time of its deposition, or if only the ceramic material went into the pit. However, in the second decade of work on the site a pit was located with the possible remains of one or two nearly complete containers (Cunliffe and Poole, 1991b: 404) from ceramic phase 4, the mid-fourth century BC (Cunliffe, 1995: 17), which might have contained salt at the time of deposition. Despite this find, none of the pit finds of briquetage were considered by the excavators to represent special deposition of the material.

A second site with a similar example of a nearly complete container, this one from Droitwich, is known from Conderton Camp, a Middle Iron Age hillfort in Worcestershire, where a single pit, Pit A, held more than a kilo of briquetage, 16% of the total briquetage found at the site. It was suggested that the briquetage might represent a special deposit of the majority of a single salt container (Morris et al., 2005: 120 & 145). These instances seem to meet Schiffer's definition of a ritual cache object in terms of physical completeness, as discussed above.

An unusual pit assemblage was also recovered from the Middle Iron Age site at Aston Mill Farm, Kemerton, Worcestershire. Of the 55 storage pits which were either partially or fully excavated, only 4 had significant quantities of pottery or briquetage (Dinn and Evans, 1990: 21-22), a pattern similar to the Danebury example. Pit 432 had the largest assemblage; it held 201 sherds of pot weighing 4.5kg, a single 60gm briquetage rim sherd, 158gm of oven sherds and 674gm of other fired clay. Although not discussed as such in the original site report, this unusual assemblage of material within a single pit is recognisable as at least a "structured deposit", which Lambrick and Robinson differentiate from "special deposits," attributing to them a symbolic or spiritual dimension (2009: 278). It would seem likely that some social imperative led to the selection of material for inclusion in this pit that did not apply to filling other pits at the site; this unknown social imperative could therefore be defined as a ritual, even if the practice was short-lived and devoid of deep spiritual meaning.

A final Middle Iron Age example is a pit, this time containing Cheshire VCP, located at the Coventry Road, Hinckley, Leicestershire, site (Chapman, 2004: 51 & 53), a rare example of this material in the central Midlands region. Pit 37, which contained the VCP, although only partially excavated due to flooding, was interpreted as containing special deposits.

The examples discussed here only serve to hint at the potential range of ritual uses for salt in Iron Age Britain. Unfortunately, the probable majority of such usage is forever beyond the reach of archaeology.

6.8 Gender and its relationship to the experience of salt

Knowledge of the relationship between the genders is one fundamental dimension needed to understand how societies function, yet prehistoric archaeology usually leaves more to the imagination in this area than can be demonstrated. There is evidence in the Roman description of the Boudiccan uprising (Tacitus, Annals: XIV, 29-38) suggesting that during the Late Iron Age some British women could be equal to their male counterparts in aspects of life which were shocking to the Romans. However, the archaeological evidence from inhumations and cremations of the period, limited though it may be, suggests that there were clear differences between the sexes that translated into the material culture of the period. Incorporating these perceived gender differences into the "habitus" of daily life (Hodder and Hudson, 2003: 90-93), would help to define a wide range of behaviours as diverse as household space utilisation, speech patterns, clothing and task assignment. We can theorise that there would have been socially constrained "right" and "wrong" ways for men and women to interact with each other, and possibly with salt, in Iron Age Britain. To date, however, discussion of gender and salt has largely focused on the question of whether production of briquetage and salt was accomplished by men or women, a topic touched upon in Chapter 3.

The discussion of gender and briquetage production has followed essentially the same line of reasoning as the discussion of gender and pottery production. Murdock and Prevost (1973) compiled data on the role of gender and task assignment across 50 different technical activities, including pottery production, across 185 societies using data from published reports. In all societies examined, gender was a statistically significant determinant with respect to assignment of most of the activities studied. Of the 105 societies for which pottery production was studied, 19 had either exclusively male or predominantly male pottery production (18%), with only 6 groups showing no gender bias (6%); in the remaining 76% of the cases, the task was either predominantly (6 groups) or exclusively female (74 groups) (Murdock and Prevost, 1973: 207). Of the 50 tasks studied, only 7 were found to be more highly feminised than pottery making. Quoting Judith Brown, Murdock and Prevost (ibid.: 211) link feminised tasks to their compatibility with childcare, ability to be performed near home, inherent task monotony and limited requirement for acute mental focus, lack of danger and easy ability to be resumed after interruptions. The encyclopdic nature of the Murdock and Prevost study meant that it was still being cited late in the twentieth century (Hays-Gilpin and Whitley, 1998: 139), although by that date the rationale for the range of tasks performed by women put forward by Brown was under attack by some archaeologists interested in the issue of gender (e.g. Nelson, 1997: 85-88).

Brown's logical framework does fit well with the writings of Peacock (1981), who assigned low volume, household level ceramic production, whether for a single domestic unit or a local market, to women, but theorised that higher volume production in workshop industries with an investment in technology would be masculinised. Arnold (1985: 100-101) drew upon these earlier studies when he stated that three factors "...(1) the compatibility of pottery production with household tasks, (2) the subsequent female advantage in making pottery, and (3) the male involvement in subsistence activities which may require time away from his residence," determined the gender of pottery makers. He also concluded that seasonal pottery production when undertaken as a part-time household craft was generally a feminine occupation. Gell (1998: 73) also included pottery production as primarily a women's decorative art in his list of agrarian/subsistence society roles. Morris (2007b: 440-41) likewise saw low level briquetage production as being associated with women. Support for this model is not universal; recently the American archaeologist Weedman (2006) has challenged this association of small scale pottery production with women and large scale with men as being derived from Western industrial model bias, and suggesting that craft assignment by gender is highly idiosyncratic.

If the gender of prehistoric pottery producers is uncertain, the gender of briquetage producers should be seen as equally ambiguous. The finger marks sometimes found on briquetage often occur in sizes that could belong to either men or women, so they are of little help in this debate. The same study cited above which showed pottery production to be highly feminised found ore smelting, studied in 37 societies, and metalworking in 85 societies to be virtually 100% masculine occupations (Murdock and Provost, 1973). These technologies are often accompanied by the development of ceramic crucibles and ceramic metal moulds, yet the production of these objects is never attributed to women in archaeological or ethnographic literature. Why should it follow that the industrial ceramics associated with salt production are made by women for hundreds of years, then by men, but only when larger quantities are needed? Is it not possible that all industrial ceramics, including briquetage, and pottery were never seen as equivalent goods in Late Bronze Age and Iron Age Britain despite the shared aspects of the technology?

It might appear that the above paragraph is contradicting ideas put forward by the author in Chapter 3 relative to the gender of briquetage and salt producers, but this is not the case. In the earlier chapter, the issue was the potential changing participation of women in briquetage and salt production, but accepting the assumption that women are initially involved in making small quantities of briquetage and salt as a starting point. The discussionbove is restricted solely to the issue of the gender of the briquetage producers, which may or may not be the same as that of the producers of salt, and questions the widely accepted initial assumption about the sex of briquetage producers. While the two activities are closely related, they need not be performed by the same people. Even if all of the briquetage was made by men, there is no inherent reason why women could not have made the salt. However, as was pointed out in the earlier discussion of gender and salt production, social custom, rather than any physical necessity, may well have been the actual determinant of the gender of salt producers. It is unlikely that archaeology will be able to move this issue beyond the realm of theory based on ethnographic parallels.

There is yet another aspect to salt and gender which would relate to its common usage in tasks which usually were assigned based upon gender. As was discussed earlier, this concept ventures into the realm of the perception of salt, and how many "types" of salt might have been thought to exist in Iron Age Britain. Did the salt used for cooking, a traditional female occupation, arrive in a different container or consist of a different crystal size than the salt which might be used for payment of a bride price by a man, if such a practice occurred in Iron Age Britain? In an example of a gender-mediated interaction with salt, in Godelier's study of the Baruya (1977), salt bars were owned and traded only by men, though salt was used by both men and women. This modern example shows that gender could be a determinant in utilisation of some forms of salt, and suggests that such behaviour might also have existed in the past, though it cannot be readily demonstrated archaeologically.

6.9 Using salt in the Iron Age

The great effort associated with salt production and distribution in the British Iron Age was necessary and worthwhile, because salt was essential for numerous aspects of daily life, far beyond simple usage for food preservation. This chapter has looked at salt and animal welfare, technology and the social characteristics of salt. Although most aspects of salt usage are archaeologically invisible, hints of their existence can be found by drawing parallels with salt usage in other societies. By fully considering this information, new insights can be gained with which to interpret Iron Age diets, structures and social practices. As in cooking, adding salt can lead to just the right result.

Chapter 7
Conclusions: making sense of salt

7.1 Introduction

The preceding chapters have sought to comprehensively address four questions related to the production, distribution and use of salt during the Iron Age in Britain and to put this activity into a wider Iron Age context. As preparation for addressing the primary research questions, the history of the development of the study of prehistoric salt by British antiquarians and archaeologists has also been reviewed. The research has utilised information drawn from what might be termed traditional archaeological resources or Classical literature as well as less traditionally utilised information resources from other disciplines including, but not limited to, food science, veterinary medicine and craft working, that have been combined to create a wide ranging view of these associated topics.

7.2 The questions answered

One of the initial research questions required an assessment of the current state of knowledge about production of salt in Iron Age Britain. This question was answered by compiling information obtained through the review of primary source documents, particularly the journals of archaeological societies or field clubs, the publications of archaeological organizations and excavation reports, public records held in county Historic Environment Records, English Heritage National Monuments Records and archaeological reports made available through commercial archaeology firms, the "grey literature" of the past twenty years. Beginning from what was a complete lack of comprehension of the artefactual evidence approximately 160 years ago (Franks, 1850), archaeologists now recognize that salt was being produced during the Iron Age using briquetage in the Lincolnshire and Cambridgeshire Fenlands, the Norfolk Broads, coastal Essex and Kent, along the south coast from Dorset to East Sussex and on the Isle of Wight, in the Tyne, Tees and Humber coastal wetlands, and at natural salt springs in Droitwich and either Nantwich or Middlewich, or possibly both locales in Cheshire. Not all of these salt production areas have yielded archaeological evidence suggesting that they were of the same scale, with Lincolnshire, Essex, the south coast and Droitwich the most extensive and best understood production areas. Production output levels also varied over time, with the Late Iron Age being a period of almost industrial scale production in some areas, such as Essex, the south coast, the Fenlands, and the Droitwich and Cheshire salt springs. More fieldwork is needed in the future to clarify whether this reflects an actual hierarchy of the economic importance of these production activities, differences in production methods which make it archaeologically difficult to correctly evaluate the scale of the regional production or if this is an artefact of the lack of knowledgeable archaeological attention in the differing locales. It would seem highly probable that there is much missing archaeological evidence, as wide areas of Britain have, as yet, provided no archaeological clues to prehistoric salt production. Some areas, such as the Somerset Levels and Cornwall, were highly productive salt sources in the Roman period (Bradley, 1992: 44), making the present virtual absence of Iron Age salt production evidence in these areas puzzling. Similarly, the unmarked areas on the briquetage finds distribution map which appears in Chapter 4, Figure 42, showing that the material is almost unknown north of the river Mersey with the exception of a finds cluster in the Tyne-Tees region, demonstrates clearly how much salt production information may yet be discovered.

Chapter 3 considered the production evidence now available, allowing certain conclusions to be drawn. First, there is only a weak claim for evidence of salt production in Britain during the Neolithic (Brothwell and Brothwell, 1969: 160, KCC, 2009) although it had already begun at salt springs in Poland at that date (Hopkinson, 1975: 11, Ryka and Werner, 1985: 169). Second, a Bronze Age date for the use of briquetage in Britain is now well established. As was discussed in Chapter 3, 31 Bronze Age sites in Britain have now been confirmed as having briquetage in their ceramic assemblages, and 13 of these are thought to have been salt production sites. The earliest unchallenged evidence for British salt production with briquetage presently known comes from the Middle Bronze Age sites at Brean Down, Somerset (Foster, 1990: 172), where a date of *c.*1400 BC is likely, Pode Hole Quarry (Morris, 2009) and Northey, Cambridgeshire (Gurney, 1980), thought to be of similar date. Pode Hole Quarry differs slightly from the other two sites, in that the excavated site was not a saltern, but yielded more than 14kg of production briquetage that had to result from nearby salt production in a more saline portion of the Fens. Production may have occurred at Billingborough, Lincolnshire, in the same period (Morris, 2007b: 435). The Bronze Age briquetage find at Knight's Farm, Burghfield, Berkshire, (Bradley et al., 1980: 275) on the Kennet gravels, opens the possibility of river transport of salt in briquetage containers being conducted from an early date relative to the start of production utilising briquetage, though land transport from the south is also a possibility for this find, which has never been subjected to petrographic examination. Whether this salt briquetage had moved as part of a gift exchange or as part of a trade network is a topic open to debate.

The second research question considered the state of knowledge about Iron Age salt distribution and was the subject of Chapter 4. The answer to the question necessitated an assessment of the current information available about salt distribution during the period. In the absence of salt, the most direct archaeological evidence for salt distribution is the presence of briquetage at a site remote from salt production, and in this book all sites not considered to be salterns have been counted as being related to distribution. Yet there is some ambiguity in using this classification system, as some finds, such as that at Mucking, Essex, or the just mentioned Pode Hole Quarry site, have been so close to production areas that they have been interpreted as perhaps representing the location of some salt production process step, such as final salt crystal drying by the producers or being otherwise integral to the distribution process (e.g. Jones, 1977: 318), thus blurring the distinction between production and distribution as separate classifications. Discussing the south coast industry, Bradley (1975: 24) raised similar concerns with inland finds, noting that they might occur at the settlements of the salt producers and represented the evidence of locations where briquetage had been made for later use at the shore. Another possible explanation is that these find sites may have resulted from secondary processing operations such as repackaging of crystals in other containers which could be performed more readily away from the saline environment of the salt production site. These theories would suggest that there is some unquantified spatial distance between finds at definite production sites and sites where briquetage can only be interpreted as occurring as the residue of salt which had been distributed in briquetage containers. While it may seem obvious that the find of Droitwich briquetage at Groundwell Farm, Wiltshire (Morris, 1985: 350), approximately 80km from its source, is evidence of salt distribution, any finds made within an hours' walk of a production source are much less clearly so. Further, the possibility exists that even more remote locations might have acted as fixed points in a trading network served by itinerant salt traders who were part of the social group producing the salt, further blurring the distinction between production and distribution in another fashion. Perhaps the best approach is to consider salt production and salt distribution as a continuum of activity with many possible permutations, most of which are beyond the reach of archaeology. Beyond these issues, the reliance on briquetage evidence to demonstrate the movement of salt in the Iron Age is an imperfect substitute for salt in the British archaeological record, as not all salt production areas seem to have utilised the material for transportation of the finished crystals. The Lincolnshire Fens, for example, although having abundant briquetage evidence for Iron Age salt production, have no definitive evidence for the movement of salt away from the production sites (Lane and Morris, 2001: 460), yet the scale of production is such as to imply surplus existed for use in trade or other types of socially mediated exchange.

As noted above, a bulk commodity such as salt would have been a challenge to distribute during the Iron Age due to the absence of paved roads. The archaeological evidence for the probable use of boats for water transport of this heavy commodity has been presented, and it would seem possible, given the pattern of briquetage find sites which has emerged, that water transport was a factor in some regional distribution networks. The Fenlands (Lane, 2001: 460), Essex, north Kent and Droitwich all would have had ready access to riverine networks; access via reasonably short distances overland to points where water transport became feasible, as at the port of Meols or the river Severn, seems to fit the evidence in the case of Cheshire salt (Matthews, 1999: 181). The salt distribution evidence in the Severn-Cotswolds region, for example, though more limited, dovetails with similar information on the long distance movement of quern stones (Moore, 2007: 48-49), a fortunate conjunction as that pattern rests on more secure archaeological evidence and which can, therefore, be cited as an analogy for the movement of other bulk commodities such as salt.

In Chapter 5, the salt production and distribution evidence was placed alongside evidence from excavations of Iron Age sites without briquetage in four study areas in southern Britain. This exercise demonstrated that the majority of these Iron Age sites did not provide archaeological evidence of the presence of salt when excavated. In most cases, there was little difference in a region between the types of sites where salt evidence was found and locations where it was absent. On one hand, this might suggest that salt was widely present on many types of sites even though it is largely archaeologically invisible now (Lane, 2001). On the other hand, a strict interpretation of the evidence might suggest that salt availability was highly restricted during the Iron Age. This latter position raises serious questions about utilisation of milk from flocks of sheep or herds of cows and the quantities of meat resulting from autumn herd culls as theorised from animal bone evidence (e.g. Serjeantson, 2007) particularly at sites with no evidence of the presence of salt. These issues were explored at length in Chapter 6. If the argument presented in this study and elsewhere for the widespread existence of archaeologically invisible salt is accepted, this interpretive problem is avoided.

Chapter 5 also provided a retrospective attempt to utilise published information on briquetage finds in the four study areas to develop detailed site context level distribution plans for these finds. The exercise demonstrated that the availability of the relevant data in these instances often seemed to be in inverse proportion to the size of the briquetage assemblage for the sites used in the exercise. While no single site plan proved statistically meaningful, the data sets from which they were developed suggested collectively that special deposition of briquetage was not an activity practiced in these study areas, though the evidence is insufficient to offer an opinion on the special deposition of salt without a briquetage container. The diffuse presence of the briquetage that had been illustrated rarely gave a suggestion of concentrated salt usage in any part of any of the studied sites, though Pollard (2002: 22) has cautioned that seeking such "...a direct correlation between the discard of material and the spatial presence and organisation of productive activities..." may not yield the anticipated results if left uncoupled with consideration of special types of depositional activity, an impossibility in the case of briquetage and salt.

The third research question required the exploration of the various aspects of the utilisation of salt during the

Iron Age which might be demonstrated or theorised. The role of salt in the British Iron Age, touched upon only occasionally by archaeologists, was the focus of Chapter 6. Development of this chapter required the utilisation of many sources for information beyond those which traditionally compose the knowledge base routinely incorporated for archaeological interpretation of the relatively limited physical evidence available. It had become clear to the writer in the course of researching this chapter that the absence of personal experience in animal management, agriculture, food production and preservation, and other technologies associated with Iron Age daily life was a problem of considerable magnitude in the development of some contemporary archaeological site interpretations. For example, the potential role of salt with respect to domestic animal welfare and management was discussed in this chapter, though it is rarely incorporated into discussion of the rationale for salt exchange elsewhere.

The short but oft-repeated phase that "salt was used for food preservation" has been expanded in Chapter 6 to include a detailed discussion of the physical and chemical aspects of food preservation technology. While earlier generations of archaeologists may well have been involved in the slaughter of animals and salting of their own meats, few now are likely to have had this personal experience to add to their consideration of the space and salt requirements for this task. They are equally unlikely to have attempted any of the traditional methods of milk preservation at first hand, electronic home yoghurt makers excepted, particularly so as to replicate Iron Age production conditions. This leaves archaeologists potentially unable to connect an artefact, such a ceramic strainer that might have had a role in cheese making, with structures that would have been suitable for cheese storage even when they are evidenced, as the requirements for cheese production and storage cannot be integrated into the analysis of the site from their range of personal experience.

Industrial uses of salt were also considered in Chapter 6, again with a focus on the technologies which would have been available in the Iron Age. It is easy to suggest that salt would have been used for the production of leather, for example, based upon modern practices, but this suggestion does not always stand up to close scrutiny when the archaeological evidence is considered. Bradley (1975: 22) thought brine might have had a role in early leather working on the Hampshire – Sussex shoreline, but little evidence of herd culls in the area has accumulated since he wrote this to support the idea. The evidence that has accumulated has suggested that leather tanning was accomplished through hot smoke tanning of oiled hides (Groenman-van Waateringe et al., 1999: 886, 889) or processing with locally available vegetable materials, such as tannin from oak galls, rather than with brine or salt (Reed, 1972: 51 & 53).

The potential roles of salt in social activities such as gift exchange, ritual and religion, which can rarely be demonstrated in Britain, were also reviewed in Chapter 6. Much of this section relied upon the use of analogies from other sources such as ethnography or anthropology, along with Classical literature, all of which have become routine data sources for archaeologists. Even though this section is largely speculative in nature, it was grounded in archaeological evidence. The lack of a large body of archaeological evidence for these uses is not sufficient justification to believe salt was absent or unimportant in the wider social structure of the Iron Age.

In support of this research, a searchable Access database of British archaeological sites with more than 515 briquetage finds has been compiled and is included on a CD with this book. Additionally, a number of appendices have been prepared with detailed information to supplement material included in the various chapters.

7.3 Summation

One hundred years ago archaeologists were still debating the date and purpose of the Red Hills of Essex. Fifty years ago, an archaeologist raising the prospect of Bronze Age salt production in Britain involving the use of briquetage containers would have been considered to be a fantasist. It is only in the past 40 years that the awareness of briquetage and the number of sites where it was identified has grown exponentially. This expansion of site numbers has resulted in no small part from the work done by Elaine Morris (e.g., 1983, 1985) which raised the profile of this material. Her continued efforts in studying and publishing assemblages of briquetage and VCP have created a widespread archaeological appreciation that briquetage could be found at sites 100km or more from salt production locations in Britain that had largely been lacking before her work. During this same period, landscape surveys conducted in East Anglia (e.g. Wilkinson and Murphy, 1986, Wilkinson and Murphy, 1995) and at various points along the south coast (e.g. Allen and Gardiner, 2000) have greatly increased the number of known production sites. Archaeological work during gravel quarrying and the redevelopment of towns and cities, paid for by site developers under the PPG 16/PPS 5 regime, added greatly to this area of knowledge as well, particularly with respect to Iron Age Droitwich production and the distribution patterns of both Droitwich and Cheshire briquetage. This research has sought to bring together the information about Iron Age salt production, distribution and use that has previously existed in many disparate sources with regional scope to produce a coherent view of the role of this mineral during later prehistory in Britain.

When beginning to compile the background research for this book, the author had the impression that briquetage was largely an artefact found at Iron Age sites, with perhaps one or two sites where it had been found in Bronze Age contexts. As was stated above, there are now 31 known sites of Bronze Age date where finds have been made, representing approximately 6% of the prehistoric briquetage finds in the author's database. The Cambridgeshire and Lincolnshire Fenlands, Essex and Kent can now be identified as active salt production centres in this period. No briquetage find sites dated to this period were known to exist prior to the late 1970s (Jones, 1977). A number of the sites have been found unexpectedly, being covered with thick layers of clay or silt deposited due to sea level rises since the Bronze Age. The pace of discovery seems to have accelerated sharply: the two Mucking sites were the only Bronze Age briquetage find site identified in the 1970s, six were published during the 1980s, six more in the 1990s and seventeen between the years 2001 and 2009. In the years since the Mucking excavations, the shock and disbelief associated with the initial few discoveries has been replaced by a nearly continuous stream of new find site publications to extend the list and the writer has no doubt that more Bronze Age sites with salt evidence will be identified in the future.

Perhaps the most controversial idea which has been put forward in this book is that the movement of salt across the British landscape during the Bronze and Iron Ages took place primarily in containers which are archaeologically invisible. As discussed in Chapter 4, the number of finds of briquetage from salt production areas such as Dorset, Essex, Kent and Lincolnshire greatly exceeds the number of other sites where briquetage from these sources has been found. Archaeology has thus far provided no answer to the question as to where the salt produced at these many sites went. Even in areas such as Oxfordshire and the Danebury region of Hampshire and Wiltshire, where briquetage seems to have been used for distribution on a regular basis, the case studies in Chapter 5 demonstrated that it was absent at the majority of Iron Age sites. In noticing the correspondence of the distribution of Malvernian pottery and Droitwich briquetage, Morris (1983, 1985) had indirectly raised the question which is yet to be answered as to whether the Malvernian pots moved along their common distribution route empty or filled with salt. There is no reason to believe this issue is restricted to this single region in light of the imbalance between production and distribution briquetage. The find in the Mendips of a pot from the Lincolnshire and Northamptonshire region (Todd, 2001), teases the possibility that these pots might have been filled with salt in Lincolnshire and used for transport of salt to Northamptonshire, the source of most archaeological finds of this particular pot type. Had this example reached the Mendips as part of an exchange of salt and other goods for silver to be used in coin or jewellery production?

There is still ample scope for further research on the subject of prehistoric salt production. Without a doubt, the number of sites with recorded briquetage finds will continue to expand during the twenty-first century, though whether at the same pace under the new PPS 5 (Anon., 2010a) is yet to be seen, thus creating the possibility for later scholars to revisit the subjects considered here and to draw new conclusions from the data. Evidence for production before the Middle Bronze Age is still completely absent in Britain, though salt would have had all of the same potential uses discussed in Chapter 6 if it were available. Information about the production of salt in areas yet to produce finds of Iron Age briquetage is a topic which needs further work, including assessment of previously collected pottery assemblages to look for missed finds, similar to the work of Willis (1995) with respect to Teeside Iron Age salt production. It is also possible that additional Bronze Age sites might be identified through reassessment of ceramic assemblages from appropriate locations in the Fens or Kent. Landscape surveys in Cornwall might locate Iron Age production sites there, and it seems likely that further work on the Somerset Levels will identify more Iron Age salt production. This is clearly not the final word to be published on the subject of Iron Age salt in Britain.

Appendix 1 – Figure 6 Essex salt production sites

NB: In Appendices 1-7 Ordinance Survey National Grid References in the Location column shown in boldface were estimated by the author as they were not provided in original source texts.

Bronze Age and Iron Age Sites

Site Name	Site Type	Location	Period
Abbotts Hall Farm, Great Wigborough	Red Hill	TL 96444 13604	M-LIA or RB
Ashingdon	Red Hill	TQ 862 958	IA,RB
Barling Marsh Trench C, Barling Magna	Red Hill	TQ 934 902	IA, RB
Bradwell on Sea	Red Hill	TM 00 00	IA
Breadward Farm, Southminster	Red Hill	TQ 9815 9837	IA, RB
Breadward Farm, Southminster	Red Hill	TQ 9820 9836	IA, RB
Breadward Farm, Southminster	Red Hill	TQ 9828 9855	IA, RB
Breadward Farm, Southminster	Red Hill	TQ 9833 9858	IA, RB
Blackmore Ave., Canvey Island	Red Hill	TQ 798 825	LIA,RB
Blackmore Ave., Canvey Island (RH 274)	Red Hill	TQ 7980 8263	LIA, RB
Bower Hall Farm, East Mersea	Red Hill	TM 0220 1541	IA, RB
Bradwell on Sea (Blackwater 35, RH 191)	Red Hill	TM 009 088	LIA
Burnham-on-Crouch	Red Hill	TQ 9359 9562	IA or RB
Canewdon	Red Hill	TQ 8878 9523	IA, RB
Canewdon	Red Hill	TQ 8880 9546	IA, RB
Canewdon I (RH 239)	Red Hill	TQ 8885 9535	LIA,
Canewdon	Red Hill	TQ 8973 9558	IA, RB
Canvey Island (RH XIII)	Red Hill	TQ 77 84	IA, RB
Canvey Island (RH XII)	Red Hill	TQ 7878 8208	IA, RB
Canvey Island (RH 278)	Red Hill	TQ 7881 8208	IA, RB
Canvey Island	Red Hill	TQ 790 820	LIA, RB
Canvey Island (RH 277)	Red Hill	TQ 794 819	IA, RB
Canvey Island	Red Hill	TQ 795 819	IA to Medieval
Canvey Island	Red Hill	TQ 8225 8331	IA, RB
Clacton (RH 34)	Red Hill	TM 160 132	LIA
Crouch Site 2 (Fenn Creek, Woodham Ferrers)	Red Hill	TQ 8020 9646	LBA
Crouch Site 60	Red Hill	TQ 869 973	IA
Cudmore Grove, East Mersea	Red Hill	TM 069 151	IA
East Mersea (RH 61)	Red Hill	TM 05721380	LIA, RB
Ferndale Crescent, Canvey Island	Red Hill	TL 9014 0725	IA, RB
Goldhanger VIII	Red Hill	TL 9082 0820	IA
Goldhanger X (RH 169)	Red Hill	TL 9259 0870	LIA, RB
Great Oakley	Red Hill	TM 208 270	IA, RB
Great Wakering	Red Hill	TQ 951 897	IA
Gt & Lt Wigborough (Blackwater 16, RH 306)	Red Hill	TL 994 140	IA
Great Wigborough (RH 147)	Red Hill	TL 975 135	LIA, RB
Kirby-le-Soken (RH 30)	Red Hill	TM 2273 2255	IA, RB
Langenhoe I (RH 95)	Red Hill	TM 0138 1667	LIA, RB
Langenhoe III (RH 89)	Red Hill	TM 0166 1674	LIA, RB
Langenhoe V (RH 78)	Red Hill	TM 0294 1666	LIA, ERB
Langenhoe	Red Hill	TM 0294 1787	IA, RB
Langenhoe VI (RH 70)	Red Hill	TM 0352 1718	MIA
Lauriston Farm, Tolleshunt D'Arcy	Red Hill	TL 926 088	IA
Leigh Beck	Red Hill	TQ 820 830	IA, RB
Little Wakering (RH 301)	Red Hill	TQ 947895	IA, RB

Maldon Landfill Site	settlement	TL 863 057	LIA
May Avenue, Canvey Island	Red Hill	TQ 806 827	LIA, RB
May Avenue, Canvey Island	Red Hill	TQ 8048 8283	LIA, RB
Maydays Marsh 1	Red Hill	TM 0315 1545	IA
Maydays Marsh 2	Red Hill	TM 034 154	IA
Mucking North Ring	settlement	TQ 6755 8112	LBA, EIA
Mundon (RH 295)	Red Hill	TL 8969 0487	IA, RB
Old Moze Dock Area	Red Hill, pit	TM 201 247	LIA
Osea Island (Blackwater 29)	Red Hill	TL 912 066	IA
Osea Road, Maldon (RH 184)	Red Hill	TL 8873 0743	IA, RB
Peldon (RH 129)	Red Hill	TM 0000 1562	IA
Peldon	Red Hill	TM 005 157	LIA
Peldon (RH 118)	Red Hill	TM 0051 1537	IA
Peldon (RH 117) (Bonner's Saltings Nature Reserve)	Red Hill	TM 0055 1560	M-LIA
Peldon	Red Hill	TM 006 155	LIA
Peldon (RH 106)	Red Hill	TM 0112 1605	IA, RB
Peldon	Red Hill	TM 0125 1595	IA, RB
Salcott	Red Hill	TL 957 128	IA
Salcott (RH 157)	Red Hill	TL 9579 1300	LIA, RB
Sea Road, Maldon	Red Hill	TL 855 065	LIA
South Benfleet (RH 283)	Red Hill	TQ 765 863	IA, RB
Southminster (RH 216)	Red Hill	TQ 993 984	IA
St Lawrence (RH 189, Blackwater 6)	Red Hill	TL 9598 0553	IA
Stanford-le-Hope (RH 289)	Red Hill	TQ 695809	LIA, RB
Stanford-le-Hope	Red Hill	**TQ 698 809**	IA, RB
Thorpe-le-Soken	Red Hill	TM 195 245	IA, RB
Tollesbury	Red Hill	TL 95920 11295	MIA
Tollesbury (RH 152)	Red Hill	TL 9623 1113	IA, RB
Tollesbury (Blackwater 12)	Red Hill	TL 9626 1111	LIA, RB
Tollesbury	Red Hill	TL 964 128	LIA
Tolleshunt d'Arcy (RH 168, Blackwater 11)	Red Hill	TL 9390 0840	IA, RB
Wallasea Island, Canewdon	Red Hill	TQ 98 93	IA,RB
Walton-on-the-Naze	Red Hill	TM 243 228	IA,RB
Walton on the Naze (RH 23)	Red Hill	TM 2675 2459	LIA
West Mersea	Red Hill	TM 0085 1664	IA,RB
West Mersea	Red Hill	TM 0244 1240	IA,RB

Romano-British Sites

Site Name	Site Type	Location	Period
Canvey Island near Russell Head	Red Hill	TQ 7795 8425	RB, Medieval
East of Ware Farm, Dengie	Red Hill	TM 005 014	RB
Fobing Marsh	Red Hill	TQ 724 836	RB
Foulness Island	Red Hill	TQ 980 904	RB ?
Goldhanger	Red Hill	TL 8988 0774	RB
Havengore Island Site HG/1	Red Hill	TQ 965 901	RB
Havengore Island Site HG/2	Red Hill	TQ 969 902	RB
Havengore Island Site HG/3	Red Hill	TQ 900 968	RB
Havengore Island Site HG/4	Red Hill	TQ 968 896	RB
Layer de la Haye	Red Hill	TL 9641 1282	RB
Leigh Beck, Canvey Island (RH 111, Thames Site 1)	Red Hill	TQ 7966 8275	RB

Leigh Beck, Canvey Island (RH 270 & 271 ?)	Red Hill	TQ 823 832	RB
Leigh Beck, Canvey Island (RH 270 & 271 ?)	Red Hill	TQ 823 8330	RB
Leigh Beck, Canvey Island	Red Hill	TQ 824 832	RB
New England Island Site NE/1	Red Hill	TQ 979 897	RB
The Old Ranges, Shoeburyness	Red Hill	TQ 93 84	RB
Oxenham Farm, Great Wakering	Red Hill	TQ 960 880	RB
Tollesbury	Red Hill	TL 955 087	RB or Medieval
West Mersea	Red Hill	TM 022 154	RB

Undated Essex Red Hill Sites

Site Name	Site Type	Location
Ashingdon	Red Hill	**TQ 857 960**
Barling Gravel Quarry	Red Hill	TQ 928 899
Beaumont cum Moze	Red Hill	TM 193 245
Beaumont cum Moze	Red Hill	TM 195 244
Beaumont cum Moze	Red Hill	TM 203 250
Beckney Farm, Ashingdon	Red Hill	TQ 84 95
Birch Hall Farm, Thorpe-le-Soken	Red Hill	TM 221 235
Bradwell Marshes	Red Hill	TM 0130 0635
Bradwell Marshes	Red Hill	TM 0230 0650
Bradwell on Sea	Red Hill	TL 998 083
Bradwell on Sea	Red Hill	TM 0110 0620
Bradwell on Sea	Red Hill	TM 016 081
Brightlingsea Rising Main	Red Hill	TM 072 178
Burnham on Crouch	Red Hill	TQ 9672 9752
Canewdon	Red Hill	TQ 8765 9584
Canewdon	Red Hill	TQ 8835 9563
Canewdon	Red Hill	TQ 8891 9534
Canewdon (Crouch 44, RH 297)	Red Hill	TQ 8977 9600
Canewdon	Red Hill	TQ 8988 9539
Canewdon	Red Hill	TQ 9065 9529
Canewdon	Red Hill	TQ 9149 9526
Canewdon	Red Hill	TQ 956 928
Canvey Island, Denham Road	Red Hill	TQ 791 836
Canvey Point, Canvey Island	Oven-like structures	TQ 823 832
Coleward Farm, Burnham on Crouch	Red Hill	TQ 97 95
Crouch 14, Rettendon	Red Hill	**TQ 7910 9470**
Dengie	Red Hill	TM 0047 0166
Dengie	Red Hill	TM 0053 0145
East Mersea	Red Hill	TM 0460 1555
East Mersea	Red Hill	TM 059 156
East Mersea	Red Hill	TM 066 155
East Mersea (Blackwater 27, RH 56)	Red Hill	TQ 8326 9643
Fen Farm, East Mersea	Red Hill	TM 0600 1410
Goldhanger	Red Hill	TL 8897 0771
Goldhanger (Blackwater 28)	Red Hill	TL 8969 0777
Goldhanger	Red Hill	TL 8971 0780
Goldhanger	Red Hill	TL 8984 0780
Goldhanger	Red Hill	TL 8999 0778
Goldhanger	Red Hill	TL 9052 0793
Goldhanger	Red Hill	TL 9083 0837

Goldhanger	Red Hill	TL 9133 0847
Goldhanger (Blackwater 19, G. III, RH 180)	Red Hill	TM 0607 1401
Goldhanger, Lee Wick	Red Hill	**TM 108 143**
Great Oakley	Red Hill	TM 202 248
Great Oakley	Red Hill	TM 204 250
Gt & Lt Wigborough	Red Hill	TL 974 136
Gt & Lt Wigborough (Blackwater 16, RH 306)	Red Hill	TL 9770 1395
Gt & Lt Wigborough	Red Hill	TL 9772 1413
Gt & Lt Wigborough	Red Hill	TL 9780 1409
Gt & Lt Wigborough	Red Hill	TL 980 140
Gt & Lt Wigborough	Red Hill	TL 9822 1385
Gt & Lt Wigborough	Red Hill	TL 9835 1391
Gt & Lt Wigborough	Red Hill	TL 9849 1391
Gt & Lt Wigborough	Red Hill	TL 9858 1427
Gt & Lt Wigborough	Red Hill	TL 9860 1388
Gt & Lt Wigborough	Red Hill	TL 9863 1415
Gt & Lt Wigborough	Red Hill	TL 9865 1441
Gt & Lt Wigborough	Red Hill	TL 9929 1499
Horsey Island	Red Hill	TM 235 245
Kirby le Soken	Red Hill	TM 216 238
Kirby le Soken	Red Hill	TM 231 228
Langenhoe	Red Hill	TM 0129 1686
Langenhoe (RH IV)	Red Hill	TM 0208 1654
Langenhoe	Red Hill	TM 0288 1776
Latchingdon (Crouch 60, RH 296)	Red Hill	TQ 8691 9734
Lee Wick Farm, Barnfield (?), St Osyth	find site	**TM 107 141**
Little Oakley	Red Hill	TM 223 279
Little Oakley	Red Hill	TM 231 283
Little Oakley	Red Hill	TM 231 284
Little Oakley Group	Red Hills	TM 232 279
Little Oakley	Red Hill	TM 233 277
Little Oakley	Red Hill	TM 233 282
Little Oakley	Red Hill	TM 234 286
Little Oakley	Red Hill	TM 235 285
Little Oakley	Red Hill	TM 238 281
Little West Wick, Burnham (RH 219)	Red Hill	TQ 9814 9743
Maldon (Blackwater 32)	Red Hill	Not available
Maldon	Red Hill	TL 8661 0515
Maydays Farm, West Mersea	Red Hill	TM 0288 1532
Mundon (Blackwater 1)	Red Hill	TL 881 052
The Naze Tower, Frinton and Walton	Red Hill	TM 2661 2362
New Hall Farm Group A, Little Wigborough	Red Hills	**TL 990 151**
New Hall Farm Group B, Little Wigborough	Red Hills	**TL 989 148**
New Hall Farm Group C, Little Wigborough	Red Hills	**TL 993 139**
One Tree Marsh, Jaywick, Clacton	Red Hill	TL 8810 0526
Osea Island (Blackwater 29, RH 294)	Red Hill	TL 9113 0664
Peldon	Red Hill	TL 9954 1474
Peldon	Red Hill	TL 996 147
Peldon	Red Hill	TL 997 147
Peldon	Red Hill	TL 998 148
Peldon	Red Hill	TL 998 157

	Red Hill	TL 999 149
Peldon	Red Hill	TM 000 149
Peldon	Red Hill	TM 000 156
Peldon	Red Hill	TM 0079 1617
Peldon	Red Hill	TM 012 155
Rawreth (Crouch 14)	Red Hill	Not available
Rochford (Crouch 37, RH 315)	Red Hill	TQ 8295 9566
Rochford (Crouch 39)	Red Hill	TQ 8332 9511
Rushley Island	Red Hill	**TQ 96 89**
Salcott	Red Hill	TL 9560 1313
Salcott	Red Hill	TL 9567 1357
Salcott	Red Hill	TL 9575 1293
Salcott	Red Hill	TL 9576 1302
Salcott	Red Hill	TL 9578 1291
Salcott	Red Hill	TL 958 130
Salcott	Red Hill	TL 95961 13112
Sampson's Lower Barn Field	Red Hill	Not available
Skippers Island	Red Hill	TM 215 242
Southminster	Red Hill	TQ 976 993
Southminster	Red Hill	TQ 9892 9864
Southminster	Red Hill	TQ 9912 9925
Stambridge	Red Hill	TQ 924 915
St Lawrence	Red Hill	TL 9591 0551
Steeple	Red Hill	TL 93 03
Steeple	Red Hill	TL 9479 0566
Thorpe-le-Soken	Red Hill	TM 206 236
Thorpe-le-Soken	Red Hill	TM 214 239
Tillingham	Red Hill	TM 004 054
Tillingham	Red Hill	TM 006 047
Tillingham	Red Hill	TM 007 048
Tillingham	Red Hill	TM 007 0497
Tillingham	Red Hill	TM 007 051
Tillingham	Red Hill	TM 0088 0451
Tillingham	Red Hill	TM 010 043
Tillingham	Red Hill	TM 011 044
Tillingham	Red Hill	TM 012 028
Tillingham	Red Hill	TM 015 028
Tillingham	Red Hill	TM 029 025
Tollesbury	Red Hill	TL 9425 0837
Tollesbury	Red Hill	TL 9457 0841
Tollesbury	Red Hill	TL 9485 0856
Tollesbury	Red Hill	TL 9550 1176
Tollesbury (RH 156)	Red Hill	TL 9585 1134
Tollesbury RH A	Red Hill	TL 959 112
Tollesbury RH B	Red Hill	TL 959 112
Tollesbury, Old Hall Marshes	Red Hill	TL 9610 1260
Tollesbury	Red Hill	TL 96375 12465
Tollesbury	Red Hill	TL 9652 1020
Tollesbury	Red Hill	TL 96584 12995
Tollesbury, Old Hall Marshes	Red Hill	TL 96823 13942
Tollesbury	Red Hill	TL 96981 12369

Tollesbury	Red Hill	TL 97248 12813
Tollesbury, Old Hall Marshes	Red Hill	TL 97957 12852
Tolleshunt Major	Red Hill	TL 9188 0869
Walton on the Naze	Red Hill	TM 246 228
Walton on the Naze	Red Hill	TM 248 229
Walton on the Naze (Clacton 2, RH 11)	Red Hill	TM 2645 2475
West Mersea	Red Hill	TM 009 154
West Mersea	Red Hill	TM 0093 1628
West Mersea	Red Hill	TM 010 167
West Mersea	Red Hill	TM 0115 1421
West Mersea	Red Hill	TM 013 157
West Mersea	Red Hill	TM 02 14
West Mersea	Red Hill	TM 0355 1524

Appendix 2 – Figure 10 south coast salt production sites

Site Name	County	Site Type	OS Grid Ref	Date
East of Corfe River	Dorset	settlement	**SY 968 853**	LIA
Furzey Island, Poole Harbour	Dorset	enclosure	SZ 0091 8694	IA,RB
Gaulter Gap, Kimmeridge	Dorset	saltern	SY 907 791	IA "A"
Godlingston Heath, Studland	Dorset	saltern	SZ 014 821	IA, RB
Hengistbury Head, nr. Christchurch	Dorset	*oppidum*, port	**SZ 184 904**	IA, RB
Hobarrow, Isle of Purbeck	Dorset	saltern	SY 89540 799090	EIA, RB
Kimmeridge, Isle of Purbeck	Dorset	saltern	SY 90805 78710	EIA
Middlebere Marsh, Wytch Farm	Dorset	industrial area	SY 9660 8530	IA
Ower, Isle of Purbeck	Dorset	saltern & enclosure	**SZ 000 860**	IA, RB
Poole Foundry (PM46)	Dorset	ditch	SZ 00730 90360	LIA
Poole Foundry	Dorset	saltern	SZ 0075 9038	IA
Rope Lake Hole	Dorset	saltern	SY 9253 7798	IA
Worbarrow Bay	Dorset	saltern	SY 8954 7909	IA-RB
Wyke Regis	Dorset	saltern	SY 65690 76950	LIA, RB
Bishopstone	East Sussex	settlement	TQ 4675 0072	EIA
Brownwich Farm, Titchfield	Hampshire	saltern & ditch or gully	SU 5205 0324	IA
East Park Terrace/St Andrew's Road, Southampton	Hampshire	settlement	SU 4230 1229	LIA
Fareham	Hampshire	saltern	SU 4911 0463	LIA
Fareham	Hampshire	saltern	SU 5204 0322	IA
Havant	Hampshire	saltern	SU 7005 0565	IA
Havant (Ha.74)	Hampshire	saltern	SU 7012 0575	IA, RB
Havant	Hampshire	saltern	SU 715 049	LIA, ERB
Havant	Hampshire	saltern	SU 7153 0260	IA
Havant	Hampshire	saltern	SU 7215 0259	IA, RB
Havant	Hampshire	saltern	SU 7216 0269	IA, RB
Hook, Warsash	Hampshire	saltern	**SU 505 055**	IA, RB
Langstone Harbour-Baker's Island (BI.87)	Hampshire	salterns	SU 6950 0350	LIA,RB
Langstone Harbour - Creek Field (HI.72)	Hampshire	saltern	SU 7180 0355	IA
Langstone Harbour - Farlington Marshes (FM.57)	Hampshire	find site	SU 6840 0350	IA

Langstone Harbour - Farlington Marshes (FM.84)	Hampshire	find site	SU 6830 0340	IA, RB
Langstone Harbour - Hayling Island (HI.69)	Hampshire	find site	SU 7139 0314	IA
Langstone Harbour - Hayling Island (HI.75)	Hampshire	find site	SU 7152 0321	IA, RB
Langstone Harbour - Hayling Island (HI.76)	Hampshire	saltern	SU 7160 0310	LIA
Langstone Harbour - Hayling Island (HI.77)	Hampshire	find site	SU 7166 0326	IA, RB
Lymington (Manor Farm, Pennington)	Hampshire	settlement	**SZ 316 925**	LIA
Paulsgrove, Portsmouth	Hampshire	saltern	SU 6392 0558	EIA
Poole Harbour	Hampshire	salterns	**SZ 012 900**	LIA
Portsmouth	Hampshire	saltern	**SU 63 05**	IA
Grange Chine	Isle of Wight	saltern	SZ 410 810	LIA
Knighton	Isle of Wight	saltern	**SZ 566 866**	LIA
Redcliff, Sandown	Isle of Wight	saltern	**SZ 618 853**	M-LIA
Chichester Harbour	West Sussex	saltern	**SU 788 024**	LIA
Chidham, Chichester Harbour, Site A	West Sussex	saltern	SU 7780 0422	M-LIA, RB
Chidham, Chichester Harbour, Site B	West Sussex	saltern	SU 7798 0348	M-LIA, RB

Appendix 3 – Droitwich non-production briquetage find sites

Site Name	County	Site Type	Location	Dating	Morris 85
Bagendon	Gloucestershire	find site	SP 018 064	M-LIA	*
Bourton-on-the-Water Primary School	Gloucestershire	find site	SP 1666 2100	EIA	
Claydon Pike, Lechlade	Gloucestershire	enclosure	SP 210 005	M-LIA	*
Crickley Hill	Gloucestershire	hillfort	SO 937 165	EIA	*
Guiting Power	Gloucestershire	settlement	SP 0892 2497	M-LIA	*
Highgate House	Gloucestershire	enclosure	**SO 952 129**	LIA	
Horcott Pit, Fairford	Gloucestershire	enclosure ditch and pits	SU 1432 9875	MIA	
Ireley Farm, Hailes	Gloucestershire	find site	SP 042 307	M-LIA	*
The Knolls, Oxenton	Gloucestershire	unenclosed settlement	SO 973 313	M-LIA	*
Lydney Park	Gloucestershire	find site	SO 616 027	M-LIA	*
Oxenton 2	Gloucestershire	unenclosed settlement	SO 972 313	MIA	
Salmonsbury	Gloucestershire	large enclosure	SP 173 208	M-LIA	*
Shenberrow Hill Camp	Gloucestershire	hillfort	SP 080 334	EIA	*
Shorncote Quarry, Somerford Keynes	Gloucestershire	roundhouse gulley	SU 033 966	MIA	
Spratsgate Lane, Somerford Keynes	Gloucestershire	partially enclosed settlement	SU 0242 9579	MIA	
Thornhill Farm	Gloucestershire	enclosed settlement	SU 183 997	MIA-RB	
Uley Bury	Gloucestershire	hillfort	ST 784 989	M-LIA	*
Warrens Field Tr 6, Claydon Pike, Lechlade	Gloucestershire	settlement	SU 19335 99730	MIA	
Warrens Field Tr 12, Claydon Pike, Lechlade	Gloucestershire	settlement	SU 19220 99785	MIA	
Warrens Field Tr 14, Claydon Pike, Lechlade	Gloucestershire	settlement	SU 19070 93830	MIA	
Winson	Gloucestershire	find site	**SP 090 086**	E-MIA	
Lodge Hill Camp	Gwent	hillfort	ST 323 914	M-LIA	
Credenhill Camp	Herefordshire	hillfort	SO 450 455	M-LIA	*
Croft Ambrey	Herefordshire	hillfort	SO 443 667	EIA	*
Dinedor	Herefordshire	hillfort	SO 524 364	M-LIA	*

Ivington Camp	Herefordshire	hillfort	SO 4850 5470	M-LIA	
Kenchester	Herefordshire	enclosure	SO 448 426	M-LIA	*
Sutton Walls	Herefordshire	hillfort	SO 525 465	M-LIA	*
Wellington Quarry	Herefordshire	ditched enclosure	SO 508 479	M-LIA	
Portskewett New School	Monmouthshire	ditch	SO 4983 8833	M-LIA	
Sudbrook	Monmouthshire	find site	ST 500 868	M-LIA	*
Twyn-y-Gaer	Monmouthshire	hillfort	SO 294 219	M-LIA	*
Bicester Fields Farm	Oxfordshire	enclosed settlement	SP 592 222	M-LIA	
Chastleton Hill Camp	Oxfordshire	hillfort	SP 250 300	EIA	*
Chinnor	Oxfordshire	settlement	**SP 767 006**	IA	
Coxwell Road, Faringdon	Oxfordshire	settlement	SU 2812 9464	MIA	
Gravelly Guy, Stanton Harcourt	Oxfordshire	settlement	SP 403 053	EIA-RB	
Mingies Ditch, Hardwick-with-Yelford	Oxfordshire	enclosed settlement	SP 377 068	MIA	*
Mount Farm, Dorchester	Oxfordshire	enclosure and pits	**SU 579 961**	EIA	
Purwell Farm, Cassington	Oxfordshire	settlement	**SP 443 118**	EIA	
Rollright Stones	Oxfordshire	enclosure	**SP 296 308**	MIA	
Uffington	Oxfordshire	enclosed settlement	SU 295 899	EIA	
Watkins Farm, Northmoor	Oxfordshire	enclosed settlement	SP 426 035	MIA, RB	
Yarnton	Oxfordshire	find site	**SP 475 125**	EIA	
Bromfield	Shropshire	enclosure	SO 485 765	MIA	
Bromfield Site 2	Shropshire	enclosure	SO 483 766	M-LIA	*
Sharpstones Hill site A, Weeping Cross	Shropshire	enclosed farmstead	SJ 5085 1042	EIA	*
The Wrekin	Shropshire	hillfort	SJ 630 083	EIA	*
Eckweek	Somerset	ditch & post holes	ST 713 574	E-MIA	
Long Compton	Warwickshire	settlement	SP 2972 3105	MIA	
Marsh Farm Quarry, Salford Priors	Warwickshire	linear trackways	SP 0791 5204	IA and/or RB	
Stretton-on-Fosse	Warwickshire	find site	SP 222 384	M-LIA	*
Tiddington	Warwickshire	find site	SP 2168 5564	LIA or RB	
Meriden Quarry, Solihull	West Midlands	enclosure and roundhouses	SP 228 821	MIA	
Groundwell Farm, Blunsdon St Andrew	Wiltshire	enclosure	SU 157 889	MIA	*
Aston Mill Farm	Worcestershire	enclosures	SO 952 354	MIA	
Badsey	Worcestershire	find site	SP 070 435	M-LIA	*
Beckford	Worcestershire	settlement	SO 984 364	M-LIA	
Beckford I	Worcestershire	settlement	SO 980 361	MIA	*
Beckford II	Worcestershire	find site	SO 980 365	M-LIA	*
Blackstone	Worcestershire	enclosure	SO 790 750	M-LIA	*
Bredon Hill	Worcestershire	hillfort	SO 958 402	M-LIA	*
Broadway	Worcestershire	enclosure	SP 090 380	M-LIA	*
Conderton Camp (formerly Danes' Camp)	Worcestershire	hillfort	SO 973 384	MIA	*
Clifton Quarry, Severn Stoke	Worcestershire	unenclosed settlement	SO 8460 4670	MIA	
Deansway, Worcester	Worcestershire	roundhouse & pit; RB farm	SO 849 548	LIA-RB	
Droitwich, 97 Friar Street	Worcestershire	find site	SO 8988 6345	IA, ERB	
Droitwich, Salter's Shopping Centre	Worcestershire	residual find site	**SO 897 633**	IA,RB	
Droitwich, Whittingham House	Worcestershire	RB ditches	SO 9003 6328	IA,RB	
Elm Farm, Beckford	Worcestershire	settlement	**SO 969 343**	M-LIA	

Evesham, 93-97 High Street	Worcestershire	domestic site & ditches	SP 0368 4415	MIA	
Fladbury, Evesham Road	Worcestershire	settlement	SO 9887 4703	IA	
Hampton Lovett and Westwood	Worcestershire	settlement	SO 8830 6475	IA-RB	
Hanbury, St Mary's Church	Worcestershire	hillfort	SO 9540 6430	IA	
Impney Farm, Dodderhill	Worcestershire	ditch	SO 910 634	IA	
Madeley Heath, Belbroughton	Worcestershire	pits	SO 9527 7702	IA	
Midsummer Hill	Worcestershire	hillfort	SO 760 375	MIA	
Midsummer Hill	Worcestershire	hillfort	SO 763 372	M-LIA	*
Newdix Court, Worcester	Worcestershire	bank & ditch	SO 851 547	LIA	*
Perrin's Farm, Childswickham	Worcestershire	ditches & pits	**SP 076 389**	M-LIA	
Sidbury	Worcestershire	find site	SO 848 547	M-LIA	*
Stoke Lane, Wychbold	Worcestershire	enclosure	SO 9215 6560	MIA	
Stonebridge Cross, Westwood	Worcestershire	farmstead	SO 87648 64647	IA	
Throckmorton Airfield	Worcestershire	enclosed settlement	SO 9750 4940	MIA	
Wormington Farm, Aston Somerville	Worcestershire	trackways & enclosure	**SP 039 368**	M-LIA	
Wyre Piddle	Worcestershire	defended enclosure	**SO 975 477**	IA	

Appendix 4 – Cheshire non-production briquetage find sites

Site Name	County	Site Type	Location	Dating	Morris '85
Abbey Green, Chester	Cheshire	residual find site	SJ 4048 6667	EIA	
Beeston Castle	Cheshire	hillfort	SJ 538 592	E-LIA	*
Brook House Farm, Bruen Stapleford	Cheshire	settlement	SJ 4975 6385	LBA, M-LIA	
Chester Amphitheatre	Cheshire	amphitheatre	SJ 408 661	LIA,RB	
Eddisbury Castle Ditch	Cheshire	hillfort	SJ 5530 6930	IA	
Mellor	Cheshire	settlement	**SJ 985 895**	EIA-RB	
Poulton	Cheshire	ring ditches	SJ 403 584	IA, RB	
Tarvin	Cheshire	ditch	SJ 4857 6705	LIA-ERB	
Braich-y-Ddinas	Conwy	hillfort	SH 701 753	M-LIA	*
Dinorben	Conwy	hillfort	SH 968 757	M-LIA	*
Moel Hiraddug	Denbighshire	hillfort	SJ 063 785	IA	*
Prestatyn	Denbighshire	farmstead	SJ 062 817	M-LIA	
Foxcovert Farm, Aston-on-Trent	Derbyshire	enclosures	SK 417 305	IA	
Castell Odo	Gwynedd	hillfort	SH 1870 2847	IA	
Garn Boduan	Gwynedd	hillfort	SH 310 390	M-LIA	*
Pen Dinas, Llanaber	Gwynedd	hillfort	SH 606 208	M-LIA	*
Credenhill Camp	Herefordshire	hillfort	SO 450 455	M-LIA	*
Croft Ambrey	Herefordshire	hillfort	SO 443 667	M-LIA, RB	*
Dinedor	Herefordshire	hillfort	SO 524 364	M-LIA	*
Ivington Camp	Herefordshire	hillfort	SO 4850 5470	M-LIA	
Kenchester	Herefordshire	enclosure	SO 448 426	M-LIA	*
Sutton Walls	Herefordshire	hillfort	SO 525 465	M-LIA	*
Wellington Quarry	Herefordshire	ditched enclosure	SO 508 479	M-LIA	
Pant-y-Saer	Isle of Anglesey	enclosed settlement	SH 495 847	MIA	*

Breedon-on-the-Hill	Leicestershire	hillfort	SK 400 220	LIA	
Coventry Rd., Hinckley	Leicestershire	enclosed settlement	SP 405 930	MIA	
Enderby II	Leicestershire	enclosure	SP 5496 9988	LIA	
Grove Farm, Enderby	Leicestershire	farmstead	SK 551 001	LIA	
Huncote	Leicestershire	enclosure	**SP 516 984**	LIA-ERB	
Kirby Muxloe	Leicestershire	enclosures	SK 530 050	IA	
Leicester Lane, Enderby I	Leicestershire	enclosure	SK 5505 0022	LIA	
Normanton-le-Heath	Leicestershire	enclosure	SK 389 118	LIA	
Great Woolden Hall Farm, Glazebrook	Manchester	enclosed settlement	SJ 6910 9360	LIA	
Brook House Farm, Halewood	Merseyside	enclosure, ditches	SJ 473 850	M-LIA, RB?	
Irby on the Wirral	Merseyside	oval enclosure	SJ 2544 8375	IA	
Mill Hill Road, Irby	Merseyside	farm & enclosures	SJ 2520 8520	IA	
Twyn-y-Gaer	Monmouthshire	hillfort	SO 294 219	M-LIA	*
Crick Covert Farm	Northamptonshire	settlement	**SP 571 732**	M-LIA	
Gamston	Nottinghamshire	settlement	SK 602 369	LIA	
Arddleen	Powys	enclosure	SJ 260 160	M-LIA	*
the Breiddin	Powys	hillfort	SJ 292 144	MIA	*
Cefn Carnedd	Powys	hillfort	SO 017 900	M-LIA	*
Collfryn	Powys	hillslope enclosure	SJ 222 173	M-LIA, RB	*
Ffridd Faldwyn	Powys	hillfort	SO 217 969	IA	
Llwyn Bryn-Dinas	Powys	hillfort	SJ 172 247	M-LIA	*
Twyn Llechfaen	Powys	hillfort	SO 082 291	M-LIA	*
the Berth	Shropshire	hillfort	SJ 429 237	M-LIA	*
Bromfield	Shropshire	enclosure	SO 485 765	MIA	*
Burrow Hill Camp	Shropshire	hillfort	SO 382 831	M-LIA	*
Bury Walls	Shropshire	hillfort	SJ 575 270	EIA	*
Castle Farm, nr. Shifnal	Shropshire	minor fort or enclosure	SJ 724 095	M-LIA	*
Ebury Hill	Shropshire	hillfort	SJ 546 164	M-LIA	*
Old Oswestry	Shropshire	hillfort	SJ 295 310	EIA	*
Sharpstones Hill site A, Weeping Cross	Shropshire	enclosed farmstead	SJ 5085 1042	EIA	*
Sharpstones Hill site E, Weeping Cross	Shropshire	enclosed farmstead	SJ 496 104	M-LIA, RB?	*
Wall Camp, Kynnersley	Shropshire	enclosure	SJ 681 178	M-LIA	*
the Wrekin	Shropshire	hillfort	SJ 630 083	EIA	*
Fisherwick 1974-1975	Staffordshire	buildings	SK 187 082	M-LIA	*
Fisherwick 1976	Staffordshire	enclosure	SK 187 080	MIA	
Whitemoor Haye Quarry	Staffordshire	pits, ditches	SK 180 130	M-LIA	
Meriden Quarry, Solihull	West Midlands	enclosure and roundhouses	SP 228 821	MIA	
Beckford II	Worcestershire	find site	SO 980 365	M-LIA	*
Blackstone	Worcestershire	enclosure	SO 790 750	M-LIA	*
Conderton Camp (formerly Danes' Camp)	Worcestershire	hillfort	SO 973 384	MIA	*
Midsummer Hill	Worcestershire	hillfort	SO 760 375	MIA	
Midsummer Hill	Worcestershire	hillfort	SO 763 372	M-LIA	*
Siden's Hill, Wollescote	Worcestershire	find site	SO 931 831	MIA	

Appendix 5 – Non-production sites where Essex briquetage has been discovered

Site Name	Site Type	OS Grid Ref	Dating
Asheldham Camp	hillfort	TL 972 012	EIA
Baldock, Hertfordshire	settlement	**TL 252 339**	LIA, RB
Billericay Secondary School	ditches & pits	TQ 675 938	LIA, RB
Burnham-on-Crouch	find site	TQ 9475 9672	LIA
Camulodunum (Sheepen)	*oppidum*	TL 9850 2578	LIA, RB
Camulodunum (Stanway)	mortuary enclosure ditches	TL 955 226	LIA,RB
Corringham	find site	TQ 7105 8332	LBA
Elm Park House, Ardleigh	settlement	TM 055 287	LIA, ERB
Elms Farm, Heybridge	settlement	TL 847 082	LIA,RB
Gosbecks Farm Site B	ditches & pits	TL 968 288	LIA
Great Chesterford	settlement	**TL 503 429**	LIA
Gun Hill, West Tilbury	settlement	TQ 655 778	MIA
Hill Farm Malting Barn, Layer-de-la-Haye	settlement	TL 9792 1953	IA
Kelvedon	settlement	TL 864 190	LIA
Kelvedon Doucecroft KL4	enclosures	TL 862 191	LIA-ERB
Kelvedon Eddy Trench A	ditch & pits	TL 864 186	LIA
Kelvedon Eddy Trench B	ditch	TL 864 187	LIA
Kelvedon Eddy Trench C	ditch	TL 865 188	LIA
Little Dunmow	settlement	TL 662 222	LIA
Little Oakley Roman Villa, Tendring	find site	TM 222 292	E-MIA
Little Waltham	settlement	TL 705 126	MIA
Maldon Landfill Site	settlement	TL 863 057	LIA
Moor Hall Farm, Rainham, London	settlement	**TQ 544 822**	M-LIA
Mucking North Ring	settlement	TQ 6755 8112	LBA, EIA
Mucking South Rings	settlement	TQ 675 804	LBA
Orsett	posthole, ditch	TQ 6535 8135	LIA, RB
Radwinter	settlement	**TL 605 375**	LIA
Skyline 120 Business Park, Great Notley	farmstead	TL 7366 2171	LIA, RB
St Helena School, Sheepen Road, Colchester	pits, ditches & gully	TL 9898 2578	LIA, RB
Stifford Clays	find site	**TQ 615 815**	M-LIA
Woodham Walter	**settlement**	TL 812 081	M-LIA

Appendix 6 – Inland sites where south coast briquetage has been discovered

Site Name	County	Type	Location	Period
Knight's Farm, Burghfield	Berkshire	settlement	SU 678 700	LBA
Maiden Castle	Dorset	hillfort	SY 670 885	IA
Oakley Down, Wimborne St Giles	Dorset	ditched enclosure	SU 016 177	E-MIA; RB
Eastbourne	East Sussex	find site	**TQ 595 005**	IA
Bourton-on-the-Water Primary School	Gloucestershire	find site	SP 1666 2100	MIA
Assizes Court North, Winchester	Hampshire	ditch	**SU 477 294**	IA
Balksbury Camp	Hampshire	hillfort	SU 350 445	LBA; IA
Beggarwood Lane, Basingstoke	Hampshire	enclosure	**SU 616 494**	IA
Bossington	Hampshire	pit	SU 32 30	M-LIA
Brighton Hill South, Basingstoke, Site B/C	Hampshire	enclosure	SU 607 686	MIA-ERB
Bury Hill	Hampshire	hillfort	SU 346 436	M-LIA
Chilbolton Down	Hampshire	enclosure	SU 411 391	E-MIA, MIA
Danebury	Hampshire	hillfort	SU 324 377	IA

Eversley Park Recreation Ground, Kingsworthy	Hampshire	find spot	SU 493 330	IA
Houghton Down, Stockbridge	Hampshire	settlement	SU 341 361	eEIA-MIA,LIA- RB
Micheldever Wood	Hampshire	enclosure	SU 527 370	MIA
Nettlebank Copse, Wherwell	Hampshire	banjo enclosure	SU 341 393	LIA, RB
New Buildings, Longstock	Hampshire	enclosures and pits	**SU 344 374**	IA
Portsdown Hill	Hampshire	find site	**SU 648 068**	MIA
Portsmouth	Hampshire	find site	SU 679 041	IA
Ravager House , Winchester	Hampshire	find site	**SU 47 29**	IA
Rucstalls Hill, Basingstoke	Hampshire	enclosure	SU 651 515	IA
Suddern Farm	Hampshire	ditched enclosure	**SU 282 373**	MIA-RB
Winklebury Camp	Hampshire	hillfort	SU 6135 5290	E-MIA
Winnall Down	Hampshire	enclosed settlement	SU 498 303	MIA; ERB
Woolbury	Hampshire	hillfort	SU 381 353	MIA
Brading Roman Villa	Isle of Wight	find site	**SZ 6000 8625**	LIA
Briddlesford Lodge Farm, Newport	Isle of Wight	domestic activity site	SZ 5317 9009	LIA-RB
Colwell Bay	Isle of Wight	find site	SZ 326 884	LIA
Abingdon Town Centre	Oxfordshire	*oppidum*	SU 4962 9702	LIA-ERB
Castle Hill, Little Wittenham	Oxfordshire	hillfort	**SU 570 925**	MIA
Segsbury Camp	Oxfordshire	hillfort	SU 384 845	IA
The Vineyard, Abingdon	Oxfordshire	settlement	SU 499 972	LIA-ERB
Mill Hill, Shoreham	West Sussex	settlement	**TQ 212 065**	MIA

Appendix 7 – Chapter 5 study area excavated sites without briquetage

Worcestershire briquetage-negative sites

Parish/City	Site Name	Location
Astley & Dunley	Larford Farm, Astley	SO 8075 6955
Droitwich	Netherwich Canal Marina	SO 896 636
Hindlip	Northeast of Hindlip Hall	SO 88140 58750
Hindlip	West Mercia Police HQ, Hindlip Hall	SO 8834 5887
Holt	Top Barn Quarry	SO 822 630
Holt	Northeast of Naunton Farm	SO 82300 62300
Holt	Not named	SO 82450 62260
Holt	Southeast of Naunton Farm	SO 82540 61980
Worcester	Grounds of Old Bishop's Palace	SO 84 54
Worcester	New Magistrate's Court, Castle Street	SO 840 550
Worcester	The Blackfriars, Broad Street Development	**SO 847 549**
Worcester	Broad Street	SO 850 551
Worcester	Lichgate	SO 851 547
Worcester	Lich Street	SO 851 547

Oxfordshire briquetage-negative sites

Parish/City	Site Name	Location
Abingdon	Ashvile Trading Estate	SU 4825 9742
Abingdon	Wyndyke Furlong, Abingdon Business Park	SU 4825 9772
Abingdon	MG Works redevelopment	SU 484 973
Abingdon	Spring Road Cemetery Expansion	SU 4875 9755

Abingdon	West Central Redevelopment Area	SU 4962 9702
Abingdon	Lombard Street/West St Helen Street	SU 497 969
Abingdon	Number 6 Lombard Street	SU 4972 9700
Abingdon	Queen Street	SU 4978 9713
Abingdon	75 Vineyard, former British Gas Works	SU 498 974
Abingdon	Former Station Inn, Station Yard	SU 4981 9727
Abingdon	Abbey Courtyard - west side of Checker Walk	SU 499 971
Abingdon	Thrupp	SU 520 971
Appleford	Appleford Field	SU 520 935
Appleford	Northwest of Appleford Sidings	SU 52184 93519
Appleford	Destroyed site	SU 523 936
Appleford	Amey Gravel Pits	SU 533 933
Aston Upthorpe	Blewburton Hill	SU 547 862
Berinsfield	Housing site east of Allen's Pit	SU 575 961
Berinsfield	Allen's Pit	SU 5750 9628
Berinsfield	Old gravel pit near Mount Farm	SU 575 963
Berinsfield & Dorchester	Wally Corner	SU 581 955
Brightwell-cum-Sotwell	Site 22 Chalgrove to East Ilsley Pipeline	SU 56743 91163
Brightwell-cum-Sotwell	Brightwell Barrow	SU 5761 9190
Brightwell-cum-Sotwell	Mackney Farm water-main ditch	SU 583 908
Cumnor	Farmoor New Reservoir	SP 444 056
Cumnor	Farmoor New Reservoir	SP 444 059
Didcot	Didcot West	SU 5085 9015
Didcot	w of parish church	SU 5194 9050
Didcot	nr Church of All Saints Rectory	SU 5195 9051
Dorchester	Allen's Gravel Pit	SU 55 95
Dorchester	Southwest of Bishop's Court Farm	SU 5716 9440
Dorchester	Allotments	SU 577 940
Dorchester	House at end of Watling Lane	SU 577 941
Dorchester	10 Wittenham Lane	SU 5783 9383
Drayton	Drayton Barrow	SU 4675 9560
Drayton	Manor Farm	SU 47718 94379
Frilford	Cleeve to Hagbourne Hill to Fyfield Thames Water Pipeline Site 54 (extends to SU 439 967)	SU 436 968
Grove	Monks Farm	SU 4060 9070
Kennington	Hinksey Hill	SP 5068 0285
Little Wittenham	Hill Farm	SU 56350 92550
Little Wittenham	Hill Farm	SU 56300 92600
Little Wittenham	Round Hill	SU 5658 9258
Little Wittenham	Sinodun Camp (Wittenham Clumps)	SU 567 924
Little Wittenham	Castle Hill	SU 5695 9262
Long Wittenham	Wigbalds, nr Pearith's Farm	SU 5411 9234
Long Wittenham	Willington Farm	SU 5463 9394
Long Wittenham	College Farm	SU 550 937
Long Wittenham	Northfield Farm	SU 553 953
Marcham-Frilford	Noah's Ark site	SU 4388 9623
Marcham-Frilford	Trendle's Field	SU 440 962
Marcham-Frilford	east of Noah's Ark	SU 4425 9630
Marcham-Frilford	Cleeve to Hagbourne Hill to Fyfield ThamesWater Pipeline Site 50 (to SU 447 962)	SU 445 962
Milton	Cleeve to Hagbourne Hill to Fyfield Thames Water Pipeline Site 37	SU 471 932

Milton	Cleeve to Hagbourne Hill to Fyfield Thames Water Pipeline Site 22 (extends to SU 480 912)	SU 480 909
Milton	Land to the north of Milton Park	SU 4960 9213
North Hinksey/South Hinksey	Hinksey Hill Farm (also SP 496 040 in S.Hinksey)	SP 499 040
Northmoor	Multiple sites, including Pinnock's Farm	SP 417 025
Northmoor	Yew Tree Cottage	SP 4200 0250
Nuneham Courtenay	Abingdon Pipeline Group 152	SP 54020 00120
Oxford	Oxford City Football Ground, Whitehouse Road	SP 5135 0530
Oxford	Magdalen College School	SP 5221 0580
Oxford	Oxford Science Park, Grenoble Road	SP 5390 0210
Oxford	former Morris Motors Sports & Social Club	SP 547 058
Oxford	Blackbird Leys Peripheral Rd & Hsg Area C2	SP 555 020
Oxford	Eastfield House, Brasenose Driftway	SP 55520 04860
Oxford	Cowley - former Rover Car Works	SP 560 040
Radley	A34 bypass n of Sugarworth Farm	SP 5122 0102
Radley	Sugarworth Farm A34 Abingdon Bypass	SP 515 010
Radley	Barton Court Farm	SU 5095 9775
Radley	Thrupp House Farm - Site E 1977	SU 518 971
Radley	Thrupp House Farm	SU 5186 9712
Radley	Tuckwell's Pit	SU 5214 9780
Radley	gravel pit	SU 5215 9888
Radley	Site B 1978	SU 522 970
Radley	destroyed site	SU 5228 9714
Radley	Thrupp	SU 5232 9763
Radley	Thrupp Farm Site C	SU 525 972
Radley	Goose Acre Farm, Tuckwell's Gravel Pit	SU 525 980
Radley	Pumney Farm Area 1	SU 5280 9840
Radley	Abingdon Pipeline Area 2	SU 53084 99836
Stanton Harcourt	Vicarage Field/Beard Mill gravel pits	SP 40 05
Stanton Harcourt	Partridge's Pit	SP 4015 0565
Stanton Harcourt		SP 405 053
Stanton Harcourt	Dix's Pit	SP 405 055
Stanton Harcourt	Hamlin's Site 1	SP 4054 0495
Stanton Harcourt	Hamlin's Site 2	SP 4061 0487
Stanton Harcourt		SP 4065 0483
Stanton Harcourt		SP 4065 0490
Stanton Harcourt		SP 4067 0479
Stanton Harcourt	Blackditch By-pass	SP 408 061
Stanton Harcourt	Devils Quoits	SP 411 047
Stanton Harcourt	Linch Hill barrow complex	SP 4125 0445
Stanton Harcourt	Linch Hill barrow complex	SP 4153 0480
Stanton Harcourt	Linch Hill - Grimes Site 8	SP 4156 0485
Sunningwell	Ridgeway, Lincombe Lane, Boars Hill	SP 4919 0092
Sutton Courtenay	South end of Drayton Cursus	SU 4892 9397
Sutton Courtenay	Warrior burial	SU 50 90
Tubney	Tubney Wood Quarry Extension Area 2	SP 4464 0072
West Hagbourne	Hagbourne Hill Bronze/Iron Age Cemetery	SU 4969 8679
West Hagbourne	Hagbourne Hill Farm	SU 4971 8704

Essex briquetage-negative sites

Parish/City	Site Name	Location
Ardleigh	Wick Farm	TM 032 295
Ardleigh	Wick Farm	TM 033 295
Ardleigh	Elm Park	TM 05 28
Ardleigh	Elm Park, Station Road	TM 054 287
Ardleigh	East of Ardleigh - Ring 5	TM 0548 2850
Ardleigh	East of Ardleigh	TM 055 292
Ardleigh		TM 057 283
Ardleigh	East of Ardleigh	TM 057 287
Ardleigh	Shut Four Acres Field, Vince's Farm	TM 06 28
Colchester	Chitts Hill	TL 957 262
Colchester	Former Brights Public House, Lexden	TL 96 24
Colchester	Shrub End Dyke	TL 96 24
Colchester	Grymes Dyke	TL 963 237
Colchester	Dugard Ave, Oaklands Ave	TL 963 240
Colchester	Adjacent to Kidman's Dyke	TL 964 235
Colchester	Kidman's Dyke	TL 965 231
Colchester	Dugard Avenue	TL 965 236
Colchester	Oaklands Avenue	TL 965 242
Colchester	Park Farm, Berechurch Dyke	TL 9662 2080
Colchester	Gosbecks Farm	TL 966 223
Colchester	126 Lexden Road	TL 9678 2510
Colchester	Gosbecks	TL 969 223
Colchester	Gosbecks Farm Stages I & II	TL 97 22
Colchester	Lexden Tapswoods Dyke	TL 97 25
Colchester	West House Farm	TL 972 260
Colchester	Lexden Dyke Middle Section	TL 973 247
Colchester	Lexden Tumulus	TL 975 247
Colchester	Lexden Belgic Cemetery	TL 975 250
Colchester	Lexden	TL 975 252
Colchester	Gosbecks A	TL 9780 2295
Colchester	Kingswode Hoe School	TL 9835 2528
Colchester	Institute	TL 989 255
Colchester	Former Sports Field, Colchester Institute	TL 99 25
Colchester	Land west of Colchester Gen Hospital	TL 99 26
Colchester	Northern Approaches Road (runs from TL995 267 to TL 990 285)	TL 99 26
Colchester	Crouched Friars, Crouch Street	TL 993 249
Colchester		TL 994 244
Colchester	Friday Wood	TL 995 210
Colchester	Abbey Field	TL 995 241
Colchester	Berechurch Dyke	TL 996 211
Colchester		TM 0005 2355
East Mersea	Bond Green Farm	TM 059 149
Elmstead	Marsh Farm Pit	TM 052 214
Feering	Langley Green	TL 878 218
Fingringhoe		TM 0345 1970
Fingringhoe	Frog Hall Farm	TM 0347 1965
Fingringhoe	Priors Pit, Frog Hall Farm	TM 035 198
Great Horkesley	Pitchbury Ramparts (Pitchbury Rings)	TL 966 290
Kelvedon	Behind Institute Hall, High Street	TL 861 187

Kelvedon	East of Freemasons' Hall	TL 862 185
Kelvedon		TL 862 186
Kelvedon	Lawson Villas, High Street (KL8)	TL 863 189
Kelvedon	Star & Fleece, High Street (KL11)	TL 864 191
Kelvedon	Doucecroft	TL 865 190
Kelvedon	West of Inworth (Kelvedon warrior burial)	TL 875 175
Layer-de-la-Haye	Land adjacent to Water Treatment Works	TL 96269 19787
Stanway	Abbotstone II	TL 94 22
Stanway	Abbotstone Quarry	TL 943 227
Stanway	Church Lane	TL 943 239
Stanway	Stanway Hall Farm	TL 95 22
Stanway	Stanway Gravel Pits (Stanway Hall Farm)	TL 9545 2255
Stanway	Fiveways Fruit Farm, Dyers Road	TL 956 231
Tolleshunt d'Arcy	Hill Farm Gravel Pit	TL 922 117
West Bergholt	St Marys Church	TL 953 280
West Bergholt	Near St Marys Church	TL 955 280
West Bergholt	East of Chitts Hill Road	TL 958 264
West Mersea	42 Fairhaven Avenue	TM 022 124
White Colne	Gravel pits	TL 876 284
Wivenhoe	Keelar's Farm, Wivenhoe Site 1	TM 04 22
Wivenhoe/Elmstead	Keelar's Farm	TM 050 233

Danebury area briquetage-negative sites

Parish/City	Site Name	Location
Amport	Lains Farm - field system	SU 264 442
Amport	Lains Farm - D shaped enclosure	**SU 2690 4453**
Amport	Lains Farm - field system	SU 276 445
Andover	HQ Quartermaster General North Site, RAF Andover	SU 33 45
Andover	Smeaton Road	SU 33 46
Andover	Portway West Industrial Estate	SU 340 465
Andover	Blendon Drive	SU 346 463
Andover	Northwest of Charlton	SU 35 47
Andover	Old Down Farm	SU 356 465
Andover	Enclosed Early Iron Age settlement	SU 36 45
Andover	Vigo Road, St John's Street	SU 36 45
Andover	Area 6, Charlton	SU 36 47
Andover	Chantry Street	SU 363 457
Andover	Vigo Road	SU 366 458
Andover	Knights Enham Hill	SU 367 477
Andover	Viking Way	SU 3675 4765
Andover	East Anton	SU 371 474
Andover	Andover Spine Road Site	SU 3787 4622
Andover	Walworth Barrows	SU 38 46
Andover	Pavey's Grave	SU 387 447
Ashley	Ashley Wood Camp	SU 394 301
Barton Stacey	Southern Feeder Gas Pipeline	SU 425 412
Bossington	Fir Hill, Bossington (Site C)	**SU 323 310**
Bossington	Bossington Farm water pipeline	SU 33 30
Bossington	Water pipeline	SU 335 306
Bossington	DMV site	SU 345 305
Charlton	Portway West Industrial Estate	SU 34 46

Cholderton	Windy Dido, Cholderton	SU 2490 4262
Crawley	Mitigation Area 6, Barton Stacey to Lockerley Pipeline	SU 40 33
Crawley	Plot 5.6 Watching Brief, Barton Stacey to Lockerley Pipeline	SU 41 34
Crawley	Mitigation Area 5, Barton Stacey to Lockerley Pipeline	SU 41 35
Crawley	Plot 5.1 Watching Brief, Barton Stacey to Lockerley Pipeline	SU 42 35
Fyfield	Beehive pits	SU 2920 4845
Kimpton	D-shape enclosure, pits, ditches	SU 27 48
Kimpton	Willis Green Golf Course	SU 27 49
Kimpton	High View Farm	SU 2836 4761
Little Somborne	Somborne Park Farm	**SU 379 330**
Little Somborne	Iron Age enclosed settlement	SU 389 328
Little Somborne	Iron Age pit at Romano-British settlement	SU 392 332
Longstock	Fiveways	SU 337 376
Longstock	Meon Hill	SU 3438 3527
Longparish	Picket Twenty	SU 390 455
Nether Wallop	Ashley's Copse hillfort	SU 2591 3486
Nether Wallop	Linear ditches at Danebury	SU 330 376
Over Wallop	Martin's Clump linear boundary	SU 24 38
Over Wallop	Martin's Clump linear boundary	SU 25 39
Over Wallop	Martin's Clump linear boundary	SU 2500 3879
Quarley	Quarley hillfort	SU 2625 4235
Shipton Bellinger	Snoddington Down	SU 2443 4557
Sparsholt	Near Roman villa	SU 4149 3012
Stockbridge	Inhumation burial of a child	SU 38 35
Thruxton	Four beehive pits	SU 26 45
Thruxton	Roman Villa	SU 29 46
Tidworth	Tidworth lynchets	SU 227 476
Tidworth	East of Andover Plantation	SU 2425 4835
Tidworth	West of Upper Newdown Copse	SU 2608 4800
Upper Clatford	Sparrow Croft	SU 3748 4390
Wherwell	Northeast of Rowbury Farm	SU 35 40
Winterslow	West of Winterslow Church	SU 228 326
Winterslow	East Winterslow	SU 2425 3379

Appendix 8 – Information contained in the sites database

The sites database, provided as a CD with this book, contains two linked datasheets. The first is a gazetteer of sites associated with salt production or ones where briquetage has been found. The majority of the sites are individual locations, although there are a small number of generic sites, such as "Tees Lowlands" where an industry is known to exist based on inland briquetage finds but the production source is as yet unknown. This portion of the data base includes site name, county, brief site description, an OS Grid Reference, dating by period, detail on whether the site was thought to be a production source and whether briquetage was found. The source of briquetage is provided for non-production sites if known; sites with briquetage from two sources appear in the database twice, each listing having a common site number but a unique briquetage find number. Metrics, including the distance from the salt production source, number of briquetage sherds, total weight and mean sherd weight are listed if available. The pottery to briquetage index or Specific Salt Container Index (SSCI) is shown for some sites. There is also a free form comments section, though this is not used on all entries. Due to the nature of the data sources, not all categories of information are available for all sites.

The second portion of the database provides a bibliography for each of the sites. The site name, county and briquetage find number linked with the first database. This sheet provides both a Harvard style citation and the full bibliographic reference for discussions of the site as compiled by the author. The database has been provided to the Bodliean Library as an Access database on a CD with the printed copy of the original DPhil thesis that was the source for this text and as a .csv file with the electronic version for the Oxford Research Archive (ORA).

Please note that the CD referred to above has now been replaced with a download available at www.barpublishing.com/additional-downloads.html

Bibliography

Abbreviations used in bibliography

AHBR – Archaeology and Historic Building Record
ECC – Essex County Council
EHNMR – English Heritage National Monument Record
FAOUN – Food and Agriculture Organisation of the United Nations
HCC – Hampshire County Council
HER – Historic Environment Record
KCC – Kent County Council
OCC – Oxfordshire County Council
UM – University of Michigan

AGRICOLA, G., *De Re Metallica.* Trans. HOOVER, H. C. & HOOVER, L. H. 1950. New York, Dover Publications, Inc.

ALBARELLA, U. 2007. The end of the Sheep Age: people and animals in the Late Iron Age. *In:* HASELGROVE, C. & MOORE, T. (eds.) *The Later Iron Age in Britain and Beyond.* Oxford: Oxbow Books, 389-402.

ALLEN, M. J. & GARDINER, J. 2000. *Our Changing Coast: A survey of the intertidal archaeology of Langstone Harbour, Hampshire,* York, Council for British Archaeology.

ALLEN, T. 2000. The Iron Age Background. *In:* HENIG, M. & BOOTH, P. (eds.) *Roman Oxfordshire.* Thrupp, Stroud, Gloucestershire: Sutton Publishing, 1-33.

ALLEN, T. G. 1990. *An Iron Age and Romano-British Enclosed Settlement at Watkins Farm, Northmoor, Oxon,* Oxford, Oxford University Committee for Archaeology.

AMBROSE, T. M. & WHITE, A. J. 1981. Ingoldmells, a Salt-Making Site. *Lincolnshire History and Archaeology,* 16, 71.

AMMERMAN, A. J. 1979. A study of obsidian exchange networks in Calabria. *World Archaeology,* 11, 95-110.

ANDREWS, A. P. 1980. The Salt Trade of the Maya. *Archaeology,* 33, 24-33.

ANON. 1996. "Sheep Pocket Guide". *AS-989* [Online]. Available: http://www.ext.nodak.edu/extpubs/ansci/sheep/as989-3.htm [Accessed 05/09/2006].

ANON. 2000. "Ideal Small Farm Cows: Dexter Cattle". *Mother Earth News* [Online], December 2000/January 2001. Available: http://www.motherearthnews.com/Sustainable-Farming/2001-12-01/Ideal-Small-Farm-Cows-Dexter-Cattle.aspx [Accessed 13 October 2008].

ANON. 2001. "Salt for Dairy Cattle". Available: http://www.saltinstitute.org/47g.html [Accessed 1 September 2006].

ANON. 2004. *RMIF Glossary.* Online at Winterhill, Milton Keynes: Red Meat Industry Forum. Available at: http://www.redmeatindustryforum.org.uk/supplychain/Glossary.htm#killingoutpercentage [Accessed 13 October 2008].

ANON. 2006. What is Sour Cream and Can You Make It at Home? *Ochef* [Online]. Available: http://www.ochef.com/516.htm [Accessed 17 December 2009].

ANON. 2007. Cold Casting Bronze. *www.scultpture.net/community* [Online]. Available: http://www.sculpture.net/community/showthread.php?t=4416&page=2 [Accessed 17 November 2010].

ANON. 2008. Rubbing salt into the wound. *The Sunday Times*, 23 March,.

ANON. 2010a. Planning Policy Statement 5: Planning for the Historic Environment. Available: http://www.communities.gov.uk/documents/planningandbuilding/pdf/1514132.pdf [Accessed 21 October 2010].

ANON. 2010b. *WA Salt Supply Corporate Website.* Online at Fremantle: WA Salt Supply. Available at: http://wasalt.com.au/index.html [Accessed 28 February 2011].

APPADURAI, A. 1986. Introduction: commodities and the politics of value. *In:* APPADURAI, A. (ed.) *The social life of things.* Cambridge: Cambridge University Press, 3-63.

ARISTOTLE. *The History of Animals, Book VIII, Chapter 10.* Online at: Internet Classics Archive. Available at: http://classics.mit.edu/Aristotle/history_anim.8.viii.html [Accessed 23 January 2009].

ARNOLD, D. E. 1985. *Ceramic theory and cultural process,* Cambridge, Cambridge University Press.

ATKINSON, J. C. 1880. Some Further Notes on the Salting Mounds of Essex. *The Archaeological Journal,* XXXVII, 196-199.

BAILEY, C. J. 1962. An Early Iron Age 'B' Hearth Site Indicating Salt Working on the North Shore of the Fleet at Wyke Regis. *Proceedings of the Dorset Natural History & Archaeological Society,* 84, 132-136.

BAKER, F. T. 1975. Saltmaking Sites on the Lincolnshire Coast before the Romans. *In:* DE BRISAY, K. & EVANS, K. A. (eds.) *Salt-The Study of an Ancient Industry.* Colchester: Colchester Archaeological Group, 31-32.

BARFIELD, L. 2006. Bays Meadow villa, Droitwich: excavations 1967-1977. *In:* HURST, D. (ed.) *Roman Droitwich.* York: Council for British Archaeology, 78-242.

BARFORD, P. M. 1984-85. Early Briquetage from Corringham. *Essex Archaeology and History,* 16 (Third Series), 140-141.

BARFORD, P. M. 1988. Salt production equipment *In:* BOND, D. (ed.) *Excavation at the North Ring, Mucking, Essex.* Chelmsford: Archaeology Section, Essex County Council, 39-41.

BARFORD, P. M. 1995. The Prehistoric Briquetage from Crouch Site 2. *In:* WILKINSON, T. J. & MURPHY, P. L. (eds.) *The Archaeology of the Essex Coast, Volume I: The Hullbridge Survey.* Chelmsford: Essex County Council, 161-164.

BARFORD, P. M. 2000. Marshland-Inland Relationships in Roman Essex: Sheep, Salt-licks and Seasonal Salters - a reply. *Essex Archaeology and History,* 31 (Third Series), 276-279.

BARRETT, J. C. 1989. Food, Gender and Metal: Questions of Social Reproduction. *In:* SØRENSEN, M. L. S. & THOMAS, R. (eds.) *The Bronze Age-Iron Age Transition in Europe.* Oxford: B.A.R., 304-320.

BASSINGDALE, R. 1943. Studies on the Biology of the Bristol Channel XI. *Journal of Ecology* [Online], 31. Available: http://www.jstor.org/stable/2256783 [Accessed 22 June 2011].

BELL, M. 1977. Excavations at Bishopstone. *Sussex Archeological Collections,* 115, 1-299.

BERRY, E. K. 1975. Medieval Droitwich and the Salt Trade. *In:* DE BRISAY, K. W. & EVANS, K. A. (eds.) *Salt the Study of an Ancient Industry.* Colchester: Colchester Archaeological Group, 76-80.

BERSU, G. 1940. Excavations at Little Woodbury, Wiltshire. *Proceedings of the Prehistoric Society,* 6, 30-111.

BRADFORD, J. S. P. 1942. Notes on Unpublished Early Iron Age Pottery from the Oxford District. *Oxoniensia,* VII, 53-60.

BRADLEY, R. 1971. Trade competition and artefact distribution. *World Archaeology,* 2, 347-52.

BRADLEY, R. 1975. Salt and Settlement in the Hampshire Sussex Borderland. *In:* DE BRISAY, K. W. & EVANS, K. A. (eds.) *Salt the Study of an Ancient Industry.* Colchester: Colchester Archaeological Society, 20-25.

BRADLEY, R. 1978. *The Prehistoric Settlement of Britain,* London, Routledge & Kegan Paul.

BRADLEY, R. 1992. Roman Salt Production in Chichester Harbour: Rescue Excavations at Chidham, West Sussex. *Britannia,* 23, 27-44.

BRADLEY, R., LOBB, S., RICHARDS, J. & ROBINSON, M. 1980. Two Late Bronze Age settlements on the Kennet gravels: excavations at Aldermarston Wharf and Knight's Farm, Burghfield, Berkshire. *Proceedings of the Prehistoric Society,* 46, 217-295.

BRIDBURY, A. R. 1955. *England and the Salt Trade in the Later Middle Ages,* Oxford, Oxford University Press.

BRITNELL, W. 1989. The Collfryn hillslope enclosure, Llansantffraid Deuddwr, Powys: excavations 1980—82. *Proceedings of the Prehistoric Society,* 55, 89-134 and microfiche 29-31.

BROOKS, H. 1 January 2010. *RE: Your Article: " Effect of Diet on the Behaviour and Welfare of Pigs."* (Personal communication to KINORY, J.)

BROOKS, P. H. 2009. Effect of Diet on the Behaviour and Welfare of Pigs. Available: http://gov.mb.ca/agriculture/livestock/pork/pdf/bab19s03.pdf [Accessed 23 December 2009].

BROSSLER, A., GOCHER, M., LAWS, G. & ROBERTS, M. 2002. Shorncote Quarry: excavations of a Late Prehistoric Landscape in the Upper Thames Valley, 1997 and 1998. *Transactions of the Bristol and Gloucestershire Archaeological Society,* 120, 37-87.

BROTHWELL, D. & BROTHWELL, P. 1969. *Food in Antiquity,* London, Thames and Hudson.

BROWN, D. L. 1991a. *Evaluation at 97, Friar Street, Droitwich.* Online at http://worcestershire.whub.org.uk/DocRepository/wr1196.pdf. [Accessed 1 May 2009].

BROWN, L. 1991b. The Iron Age Pottery. *In:* CUNLIFFE, B. W. & POOLE, C. (eds.) *Danebury: an Iron Age Hillfort in Hampshire, Volume 5,The excavations 1979-1988: the finds.* London: Council for British Archaeology, 277-319.

BROWN, L. 1995. Pottery Production. *In:* CUNLIFFE, B. W. (ed.) *Danebury, an Iron Age Hillfort in Hampshire, Volume 6, A hillfort community in perspective.* York: Council for British Archaeology, 53-65.

BROWN, N. R. 1999. *The Archaeology of Ardleigh, Essex: Excavations 1955-1980,* Chelmsford, Essex County Council.

BULLEID, A. 1914. Romano-British Potteries in Mid-Somerset. *Proceedings of the Society of Antiquaries of London,* XXVI, Second Series, 137-145.

BURCHELL, J. P. T. 1925. The Shell Mound Industry of Denmark as Represented at Lower Halstow, Kent. *Proceedings of the Prehistoric Society of East Anglia,* V, 73-78.

CALKIN, J. B. 1948. The Isle of Purbeck in the Iron Age. *Proceedings of the Dorset Natural History & Archaeological Society,* 70, 29-59.

CARDON, D. 2007. *Natural Dyes Sources, Tradition, Technology and Science,* London, Archetype Publications Ltd.

CARROLL, R. 1986. "Beginning Cheese Making". *Mother Earth News* [Online], March/April 1986. Available: http://www.motherearthnews.com/print-article.aspx?id=70290 [Accessed 18 October 2007].

CARTER, C. O. 1975. Man's Need of Salt. *In:* DE BRISAY, K. W. & EVANS, K. A. (eds.) *Salt the Study of an Ancient Industry.* Colchester: Colchester Archaeological Society, 13.

CARVER, E. 2001. *The Visibility of Imported Wine and its Associated Accoutrements in Later Iron Age Britain,* Oxford, BAR Publishing.

CASE, H. 1963. Notes on the Finds and on Ring-Ditches in the Oxford Region. *Oxoniensia,* XXVIII, 19-52.

CASE, H. & STURDY, D. 1961/2 Archaeological Notes - Stanton Harcourt. *Oxoniensia* XXVI/XXVII, 338.

CASE, H. J., BAYNE, N., STEELE, S., AVERY, G. & SUTERMEISTER, H. 1964/1965. Excavations at City Farm, Hanborough, Oxon. *Oxoniensia,* XXIX/XXX, 1-98.

CASTELL, M. 5 April 2009. *RE: Milking Sheep.* (Personal communication to KINORY, J.)

CATO, M. P., *De Agricultura.* Trans. HOOPER, W. D. & ASH, H. B. 1934, Loeb Classical Library. London, William Heinemann Ltd.

CHAMPION, T. 1984. Chilbolton Down: the pottery. *In:* CATHERALL, P. D., BARNETT, M. & MCCLEAN, H. (eds.) *The Southern Feeder: The Archaeology of a Gas Pipeline.* London: British Gas, 114-115.

CHAPMAN, J. 2000. *Fragmentation in Archaeology,* London, Routledge.

CHAPMAN, P. 2004. Iron Age settlement and Romano-British enclosures at Coventry Road, Hinckley, Leicestershire. *Transactions of the Leicestershire Archaeological and Historical Society,* 78, 35-82.

CHOWNE, P., CLEAL, R. M. J., FITZPATRICK, A. P. & ANDREWS, P. 2001. *Excavations at Billingborough, Lincolnshire, 1975-8: A Bronze—Iron Age Settlement and Salt-working Site,* Salisbury, The Trust for Wessex Archaeology Ltd.

CHRISTY, M. 1906. A History of Salt-Making in Essex. *The Essex Naturalist,* XIV, 193-204.

CLAY, P. 1992. An Iron Age Farmstead at Grove Farm, Enderby, Leicestershire. *The Leicestershire Archaeological and Historical Society Transactions,* LXVI, 1-82.

CLEAL, R. M. J. & BACON, J. K. F. 2001. Briquetage. *In:* CHOWNE, P., CLEAL, R. M. J., FITZPATRICK, A. P. & ANDREWS, P. (eds.) *Excavations at Billingborough, Lincolnshire, 1975—8: a Bronze—Iron Age Settlement and Salt-working Site.* Salisbury: The Trust for Wessex Archaeology Ltd, 56-65.

CLEAL, R. M. J. & CHOWNE, P. 2001. Phase 2: Late Bronze Age—Early Iron Age. *In:* CHOWNE, P., CLEAL, R. M. J. & FITZPATRICK, A. P. (eds.) *Excavations at Billingborough, Lincolnshire, 1975—8: a Bronze—Iron Age Settlement and Salt-Working Site.* Salisbury: The Trust for Wessex Archaeology Ltd., 14-16.

COLE, W. 1889. Notes on Results of Examination of a Red-hill near Burnham, Essex. *Essex Naturalist,* III, 163-164.

COLE, W. 1904. Visit to Colchester and the Wigborough and Virley Districts. *Essex Naturalist,* XIII, 243-245.

COLE, W. 1906. Exploration of Some "Red-Hills" in Essex, with remarks upon the objects found. *The Essex Naturalist,* XIV, 170-83.

COLEMAN, L. 2001. Land to the North of Milton Park. *South Midlands Archaeology,* 31, 41.

COLEMAN, L., WATTS, M. & MCSLOY, E. 2003. *Wormington to Tirely Gas Pipeline Gloucestershire and Worcestershire: Post Excavation Assessment and Updated Project Design.* Online at http://worcestershire.whub.org.uk/DocRepository/wr865.pdf. [Accessed 18 May 2009].

COLES, J. M. & HARDING, A. F. 1979. *The Bronze Age in Europe,* London, Methuen & Co Ltd.

COLLINS, A. E. P. 1947. Excavations at Blewburton Hill 1947. Online at http://ads.ahds.ac.uk/catalogue/adsdata/arch-787-1/dissemination/pdf/BAJ050_PDFs/BAJ050_A02_collins.pdf. [Accessed 12 August 2010].

COLLIS, J. 1994. The Iron Age. *In:* VYNER, B. (ed.) *Building on the Past.* London: The Royal Archaeological Institute, 123-148.

COLUMELLA, L. J. M., *De Re Rustica Book VII.* Trans. FORSTER, E. S. & HEFFNER, E. H. 1954, Loeb Classical Library. London, William Heinemann Ltd.

COLUMELLA, L. J. M., *De Re Rustica Book VI.* Trans. FORSTER, E. S. & HEFFNER, E. H. 1954, Loeb Classical Library. London, William Heinemann Ltd.

COLUMELLA, L. J. M., *De Re Rustica Book VIII.* Trans. FORSTER, E. S. & HEFFNER, E. H. 1954, Loeb Classical Library. London, William Heinemann Ltd.

COOPER, L. 1994. Kirby Muxloe, A46 Leicester Western Bypass (SK 530 050). *Transactions of the Leicestershire Archaeological and Historical Society,* LXVIII, 162-65.

COPLEY, M. S., BERSTAN, R., DUDD, S. N., STRAKER, V., PAYNE, S. & EVERSHED, R. P. 2005. Dairying in antiquity. I. Evidence from absorbed lipid residues dating from the British Iron Age. *Journal of Archaeological Science,* 32, 485-503.

COWELL, R. W. 2000. Brook House Farm, Halewood. *In:* COWELL, R. W. & PHILPOTT, R. A. (eds.) *Prehistoric, Romano-British and Medieval Settlement in Lowland North West England: Archaeological excavations along the A5300 road corridor in Merseyside.* Liverpool: National Museums and Galleries on Merseyside, 27-66.

COX, P. W. & HEARNE, C. M. 1991. *Redeemed from the Heath: The Archaeology of the Wytch Farm Oilfield (1987 - 90),* Dorchester, Dorset Natural History and Archaeological Society.

CRAWFORD, R. J. M. 1990. The technology of traditional milk products in developing countries. Available: http://www.fao.org/docrep/003/t0251e/T0251E04.htm#ch2 [Accessed 13 October 2008].

CRONIN, T., DOWNEY, L., SYNNOTT, C., MCSWEENEY, P., KELLY, E. P., CAHILL, M., ROSS, R. P. & STANTON, C. 2007. Composition of ancient Irish bog butter. *International Dairy Journal,* 17, 1011-1020.

CROSBY, A. 2001. Briquetage Containers from Ingoldmells Beach. *In:* LANE, T. & MORRIS, E. L. (eds.) *A Millennium of Saltmaking; Prehistoric and Romano-British Salt Production in the Fenland.* Sleaford: Heritage Trust of Lincolnshire, 410-424.

CRUMMY, N. 2007. The Salt Briquetage. *In:* CRUMMY, P., BENEFIELD, S., CRUMMY, N., RIGBY, V. & SHIMMIN, D. (eds.) *Stanway: An Élite Burial Site at Camulodunum.* London: Society for the Promotion of Roman Studies, 375-377.

CRUMMY, P. 1978. Colchester, Gryme's Dyke. *Essex Archaeology and History,* 10, 242.

CRUMMY, P., BENFIELD, S., CRUMMY, N., RIGBY, V. & SHIMMIN, D. 2007. *Stanway: An Élite Burial Site at Camulodunum,* London, Society for the Promotion of Roman Studies.

CUNLIFFE, B. 1978. *Iron Age Communities in Britain,* London, Routledge & Kegan Paul.

CUNLIFFE, B. 1982. Britain, the Veneti and beyond. *Oxford Journal of Archaeology,* 1, 39-68.

CUNLIFFE, B. 1984a. *Danebury, An Iron Age Hillfort in Hampshire, Volume 1 the excavations, 1969-1978: the site,* London, Council for British Archaeology.

CUNLIFFE, B. 1984b. *Danebury, An Iron Age Hillfort in Hampshire, Volume 2 The finds 1969-1978,* London, Council for British Archaeology.

CUNLIFFE, B. 1984c. Iron Age Wessex: Continuity and Change. *In:* CUNLIFFE, B. W. & MILES, D. (eds.) *Aspects of the Iron Age in Central Southern Britain.* Oxford: Oxford University Committee for Archaeology, 12-45.

CUNLIFFE, B. 1987. *Hengistbury Head, Dorset, Volume 1,* Oxford, Oxford University Committee for Archaeology.

CUNLIFFE, B. 1991. *Iron Age Communities in Britain,* London, Routledge.

CUNLIFFE, B. 1995. *Danebury, an Iron Age Hillfort in Hampshire, Volume 6, A hillfort community in perspective,* York, Council for British Archaeology.

CUNLIFFE, B. 2000. *The Danebury Environs Programme, Volume 1: Introduction,* Oxford, Institute of Archaeology.

CUNLIFFE, B. 2004. *Facing the Ocean,* Oxford, Oxford University Press.

CUNLIFFE, B. 2005. *Iron Age Communities in Britain,* London, Routlege.

CUNLIFFE, B. & POOLE, C. 1991a. *Danebury: an Iron Age Hillfort in Hampshire, Volume 4, The excavations, 1979-1988: the site,* London, the Council for British Archaeology.

CUNLIFFE, B. & POOLE, C. 1991b. *Danebury: an Iron Age Hillfort in Hampshire, Volume 5, The excavations 1979-1988: the finds,* London, Council for British Archaeology.

CUNLIFFE, B. & POOLE, C. 2000a. *The Danebury Environs Programme Volume 2 - Part 1: Woolbury and Stockbridge Down, Stockbridge, Hants, 1989,* Oxford, Institute of Archaeology.

CUNLIFFE, B. & POOLE, C. 2000b. *The Danebury Environs Programme Volume 2 - Part 2: Bury Hill, Upper Clatford, Hants, 1990,* Oxford, Institute of Archaeology.

CUNLIFFE, B. & POOLE, C. 2000c. *The Danebury Environs Programme Volume 2 - Part 3: Suddern Farm, Middle Wallop, Hants, 1991 and 1996,* Oxford, Institute of Archaeology.

CUNLIFFE, B. & POOLE, C. 2000d. *The Danebury Environs Programme Volume 2 - Part 4: New Buildings, Longstock, Hants, 1992 and Fiveways, Longstock, Hant, 1996,* Oxford, Institute of Archaeology.

CUNLIFFE, B. & POOLE, C. 2000e. *The Danebury Environs Programme Volume 2 - Part 5: Nettlebank Copse, Wherwell, Hants, 1993,* Oxford, English Heritage and Oxford University Committee for Archaeology.

CUNLIFFE, B. & POOLE, C. 2000f. *The Danebury Environs Programme Volume 2 - Part 6: Houghton Down, Stockbridge, Hants, 1994,* Oxford, Institute of Archaeology.

CUNLIFFE, B. & POOLE, C. 2000g. *The Danebury Environs Programme Volume 2 - Part 7: Windy Dido, Cholderton, Hants, 1995,* Oxford, English Heritage and Oxford University Committee for Archaeology.

DARVILL, T. 1987. *Prehistoric Britain,* London, B. T. Batsford Ltd.

DAVIES, H. F. 1936. The Shale Industries at Kimmeridge, Dorset. *The Archaeological Journal,* XCIII, 200-219.

DE BRISAY, K. 1972. Preliminary Report on the Exploration of the Red Hill at Osea Road, Maldon, Essex. *Colchester Archæological Group Annual Bulletin,* 15, 24-43.

DE BRISAY, K. 1975. The Red Hills of Essex. *In:* DE BRISAY, K. W. & EVANS, K. A. (eds.) *Salt the Study of an Ancient Industry.* Colchester: Colchester Archaeological Group, 5-11.

DE BRISAY, K. 1978. The Excavation of a Red Hill at Peldon, Essex, with Notes on Some Other Sites. *The Antiquaries Journal,* LVIII, 31-60.

DE BRISAY, K. W. & EVANS, K. A. (eds.) 1975. *Salt, the Study of an Ancient Industry,* Colchester: Colchester Archaeological Group.

DE JERSEY, P. 2001. *Celtic Coinage in Britain,* Princes Risborough, Shire Publications Ltd.

DINN, J. & EVANS, J. 1990. Aston Mill Farm, Kemerton: Excavations of a Ring-ditch, Middle Iron Age Enclosures, and a *Grubenhaus. Transactions of the Worcestershire Archaeological Society,* Third Series 12, 5-66.

DIODORUS, *The Library of History of Diordorus Siculus.* Trans. OLDFATHER, C. H. 1935, Loeb Classical Library. Cambridge, MA, Harvard University Press.

DRENNAN, R. D. 1984. Long-Distance Transport Costs in Pre-Hispanic Mesoamerica. *American Anthropologist,* 86, 105-112.

DRURY, P. J. & RODWELL, W. J. 1973. Excavations at Gun Hill, West Tilbury. *Essex Archaeology and History,* 5 (Third Series), 48-113.

DUDD, S. N., EVERSHED, R. P. & GIBSON, A. M. 1999. Evidence for Varying Patterns of Exploitation of Animal Products in Different Prehistoric Pottery Traditions Based on Lipids Preserved in Surface and Absorbed Residues. *Journal of Archaeological Science,* 26, 1473-1482.

ECC 2003. Essex HER 11563 - Hill Farm Malting Barn, Layer-de-la-Haye Online at http://unlockingessex.essexcc.gov.uk/custom_pages/monument_detail.asp?content_page_id=68&previous_page_id=48&monument_id=33567. [Accessed 28 April 2010].

EDDY, M. R. 1982. *Kelvedon: the Origins and Development of a Roman Small Town,* Chelmsford, Essex County Council.

EDWARDS, H. J. 1961. Reflections and Ruminations Regarding Red Hills. *Colchester Archaeological Group Quarterly Bulletin,* IV, 81-84.

EHNMR 2007a. PastScape 1395435 - Berechurch Dyke. Online at http://pastscape.english-heritage.org.uk/hob.aspx?hob_id=1395435&sort=2&type=&typeselect=c&rational=a&class1=None&period=None&county=None&district=None&parish=None&place=&recordsperpage=10&source=text&rtype=monument&rnumber=1395435. [Accessed 3 Sept 2010].

EHNMR 2007b. PastScape 1397160 - Lexden Middle Dyke. Online at http://pastscape.english-heritage.org.uk/SearchResults.aspx?rational=a&type=&class1=None&period=None&county=None&district=None&parish=None&place=&recordsperpage=10&source=text&sort=2&rtype=monument&rnumber=1397160&typeselect=c. [Accessed 3 Sept 2010].

EHNMR 2009a. Excavation Index for England 658173 - Ex Royal Eastern Cos. Hospital, Lexden Straight Road. Online at http://ads.ahds.ac.uk/catalogue/search/fr.cfm?RCN=EHNMR-658173 [Accessed 3 Sept 2010].

EHNMR 2009b. Excavation Index for England 658619 - Tapswoods, Lexden. Online at http://ads.ahds.ac.uk/catalogue/search/fr.cfm?rcn=EHNMR-658619. [Accessed 3 Sept 2010].

ELSDON, S. M. 1991. The Iron Age and Anglo-Saxon Pottery. *The Leicestershire Archaeological and Historical Society Transactions,* LXV 1991, 10-11.

ELSDON, S. M. 1992. The Iron Age Pottery. *The Leicestershire Archaeological and Historical Society Transactions,* LXVI 1992, 38-52.

ELSDON, S. M. 1994. The Iron Age Pottery. *The Leicestershire Archaeological and Historical Society Transactions,* LXVIII, 35-45.

ERITH, F. 1975. Ardleigh Ring Five. *Colchester Archaeological Group Annual Bulletin,* 18, 2-4.

ERITH, F. H. 1960. Winter Ploughing 1959 - 60 at Ardleigh. *Colchester Archaeological Group Quarterly Bulletin,* 3, 2-7.

ERLANDSON, K. 2003. *Home Smoking and Curing,* London, Ebury Press.

EVERETT, R. N. & EELES, B. M. G. 1999. Investigations at Thrupp House Farm, Radley, near Abingdon. *Oxoniensia,* 64, 117-152.

FAIRBURN, N., BONNER, D., CARRUTHERS, W., GALE, G. R., MATTHEWS, K. J., MORRIS, E. & WARD, M. 2003. Brook House Farm, Bruen Stapleford. *Journal of the Chester Archaeological Society,* 77, 9-57.

FAOUN. 1990. Manual on simple methods of meat preservation, "Chapter 2. Simple techniques for production of dried meat". FAO Animal Production and Health Paper 79. Available: http://www.fao.org/docrep/003/x6932e/X6932E02.htm#ch2 [Accessed 21 October 2008].

FARRANDS, R. H. 1959. Essex Red Hills in the Hamford Water Area. *Colchester Archaeological Group Quarterly Bulletin,* II, 25--27.

FARRAR, R. A. H. 1962a. Miscellaneous Discoveries and Accessions. *Proceedings of the Dorset Natural History & Archaeological Society,* 84, 111-116.

FARRAR, R. A. H. 1962b. A Note on the Prehistoric and Roman Salt Industry in Relation to the Wyke Regis Site, Dorset. *Proceedings of the Dorset Natural History & Archaeological Society,* 84, 137-144.

FARRAR, R. A. H. 1975. Prehistoric and Roman Saltworks in Dorset. *In:* DE BRISAY, K. W. & EVANS, K. A. (eds.) *Salt the Study of an Ancient Industry.* Colchester: Colchester Archaeological Society, 14-20.

FAULKNER, N. 2008. From Bronze Age briquetage to Saxon spearheads. *Current Archaeology,* XIX, no. 6, 40-44.

FAWN, A. J., EVANS, K. A., MCMASTER, I. & DAVIES, G. M. R. 1990. *The Red Hills of Essex,* Colchester, Colchester Archaeological Group.

FELL, C. I. 1961. Shenberrow Hill Camp, Stanton, Gloucestershire. *Transactions of the Bristol and Gloucestershire Archaeological Society,* LXXX, 16-41.

FEREDAY, G. 2003. *Natural Dyes,* London, The British Museum Press.

FIELDING, A. & FIELDING, A. 2006. *The Salt Industry,* Princes Risborough, Shire Books.

FITTS, R. L., HASELGROVE, C. C., LOWTHER, P. C. & WILLIS, S. H. 1999. Melsonby Revisited: Survey and Excavation 1992-95 at the Site of Discovery of the "Stanwick", North Yorkshire, Hoard of 1843. *Durham Archaeological Journal,* 14-15, 1-52.

FLETCHER, J. M. J. 1926. Third Summer Meeting at Poole and Poole Harbour. *Proceedings of the Dorset Natural History and Antiquarian Field Club,* XLVII, xxxvii-xli.

FOSTER, J. 1990. Briquetage objects (Figs 116-120). *In:* BELL, M. (ed.) *Brean Down excavations 1983-1987.* London: English Heritage, 165-173.

FOX, C. F. 1935. Salt Works at Hook, Warsash, Hants. *Papers and Proceedings of the Hampshire Field Club and Archaeological Society,* XIII, 105-109.

FOXWELL, M. 1988. *A Self-Sufficient Larder,* London, Macdonald & Co (Publishers) Ltd.

FRANKS, A. W. 1850. Exhibition of a Guernsey hand-brick. *The Archaeological Journal,* VII, 70.

FREND, W. H. C. 1949. Some Further Iron Age and Roman Sites in the Isle of Purbeck. *Proceedings of the Dorset Natural History and Archaeological Society,* 71, 51-53.

FULFORD, M. 1987. *Calleva Atrebatum*: An Interim Report on the Excavation of the Oppidum, 1980-86. *Proceedings of the Prehistoric Society,* 53, 271-278.

GALE, R. 2001. The Derivation and Uses of Some Plant-Based Fuels. *In:* LANE, T. & MORRIS, E. L. (eds.) *A Millennium of Saltmaking: Prehistoric and Romano-British Salt Production in the Fenland.* Sleaford: Heritage Trust of Lincolnshire, 456-459.

GARDINER, J. 2000. Briquetage. *In:* ALLEN, M. J. & GARDINER, J. (eds.) *Our Changing Coast: A survey of the intertidal archaeology of Langstone Harbour, Hampshire.* York: Council for British Archaeology, 160.

GARDNER, W. & SAVORY, H. N. 1964. *Dinorben A Hillfort Occupied in Early Iron Age and Roman Times,* Cardiff, The National Museum of Wales.

GELL, A. 1992. The Technology of Enchantment and the Enchantment of Technology. *In:* COOTE, J. & SHELTON, A. (eds.) *Anthropology, Art and Aesthetics.* Oxford: Clarendon Press, 40-63.

GELL, A. 1998. *Art and Agency,* Oxford, Clarendon Press.

GELLING, P. S. & STANFORD, S. C. 1965. Dark Age Pottery or Iron Age Ovens? *Transactions and Proceedings of the Birmingham Archaeological Society,* 82, 77-91.

GENT, H. 1983. Centralized Storage in Later Prehistoric Britain. *Proceedings of the Prehistoric Society,* 49, 243-267.

GERMANY, M. 2004. A Middle Iron Age Red Hill at Tollesbury Creek, Tollesbury. *Essex Archaeology and History,* 35 (Third Series), 192-196.

GIBSON, A. & WOODS, A. 1997. *Prehistoric Pottery for the Archaeologist,* London, Leicester University Press.

GILL, D. R. undated. "Limiting Feed Intake with Salt". *Beef Cattle Handbook* [Online]. Available: http://www.iowabeefcenter.org/pdfs/bch/05350.pdf [Accessed 1 September 2006].

GILMAN, P. J. 1992. Excavations in Essex 1991. *Essex Archaeology and History,* 23, 98-113.

GODELIER, M. 1977. 'Salt money' and the circulation of commodities among the Baruya of New Guinea. *In:* GODELIER, M. (ed.) *Perspectives in Marxist Anthropology.* Cambridge: Cambridge University Press, 127-151.

GOULETQUER, P. & DAIRE, M. Y. 1994. Le sel de la Préhistoire et de la Protohistoire. *In:* DAIRE, M.-Y. (ed.) *Le Sel Gaulois: Bouilleurs de sel et ateliers de briquetage armoricains à l'Age du Fer.* Saint-Malo: Editeur Centre Régional d'Archéologie d'Alet, 5-13.

GOULETQUER, P. L. 1974. The Development of Salt Making in Prehistoric Europe. *Essex Journal,* Nine, 2-14.

GOULETQUER, P. L. 1975. Niger, Country of Salt. *In:* DE BRISAY, K. W. & EVANS, K. A. (eds.) *Salt the Study of an Ancient Industry.* Colchester: Colchester Archaeological Group, 47-51.

GOURMETSLEUTH.COM. 2007. *How to Make Cheese.* Online at: GourmetSleuth.com. Available at: http://www.gourmetsleuth.com/cheeserecipes.htm [Accessed 6 March 2008].

GREEN, M. J. 2005. *Exploring the World of the Druids,* London, Thames and Hudson.

GREGORY, C. A. 1982. *Gifts and Commodities,* London, Academic Press.

GRIFFIN, N. 2006. *Kingsnorth Power Station, Hoo, Isle of Grain.* Online at: Archaeology South-East. Available at: http://www.archaeologyse.co.uk/04-Projects/Kent/Kingsnorth/index.htm [Accessed 26 January 2009].

GRIFFIN, S., GRIFFIN, L. & JACKSON, R. 2005. *Salvage Recording and Evaluation at Throckmorton Airfield, Throckmorton, Worcestershire.* Online at http://worcestershire.whub.org.uk/DocRepository/wr10190.pdf. [Accessed 29 April 2009].

GROENMAN-VAN WAATERINGE, W., KILIAN, M. & VAN LONDEN, H. 1999. The curing of hides and skins in European prehistory. *Antiquity,* 73, No. 282, 884-90.

GURNEY, D. 1980. Evidence of Bronze Age salt-production at Northey, Peterborough. *Northamptonshire Archaeology,* 15, 1-11.

GURNEY, D. 1986. A Salt-Production Site at Denver; Excavations by Charles Green, 1960. *In:* GURNEY, D. (ed.) *Settlement, Religion and Industry on the Fen-edge; Three Romano-British Sites in Norfolk.* Gressenhall: The Norfolk Archaeological Unit, 93-145.

HALL, D. 1987. *The Fenland Project, Number 2: Fenland Landscapes and Settlement between Peterborough and March,* Cambridge, Cambridgeshire Archaeological Committee.

HALL, D. & COLES, J. 1994. *Fenland Survey,* London, English Heritage.

HALSTEAD, P., HODDER, I. & JONES, G. 1978. Behavioural Archaeology and Refuse Patterns: A Case Study. *Norwegian Archaeological Review,* 11, 118-131.

HAMILTON, J. 2000. Animal bones. *In:* CUNLIFFE, B. & POOLE, C. (eds.) *The Danebury Environs Programme, Volume 2 - Part 2: Bury Hill, Upper Clatford, Hants, 1990.* Oxford: English Heritage and Oxford University Committee for Archaeology, 67-73.

HAMLIN, A. 1963. Excavations of Ring-Ditches and Other Sites at Stanton Harcourt. *Oxoniensia,* XXVIII, 1-18.

HANFMANN, G. M. A. 1972. *Letters from Sardis,* Cambridge, MA, Harvard University Press.

HARDING, A. 1998. Resources and their distribution in the European Bronze Age. *In:* HÄNSEL, B. (ed.) *Mensh und Umwelt in der Bronzezeit Europas.* Keil: Oetker-Voges- Verlag, 149-155.

HARDING, D. W. 1972. *The Iron Age in the Upper Thames Basin,* Oxford, Clarendon Press.

HARDING, D. W. 2004. *The Iron Age in Northern Britain,* London, Routledge.

HARDY, A. & BARCLAY, A. 2007. Uffington Village Project Excavation 2002. *South Midlands Archaeology,* 37, 53-60.

HASELGROVE, C. 1982. Wealth, prestige and power: the dynamics of late iron age political centralisation in south-east England. *In:* RENFREW, C. & SHENNAN, S. (eds.) *Ranking, Resource and*

Exchange. Cambridge: Press Syndicate of the University of Cambridge, 79-88.

HASELGROVE, C. 1999. The Iron Age. *In:* HUNTER, J. & RALSTON, I. (eds.) *The Archaeology of Britain.* London: Routledge, 113-134.

HASELGROVE, C. 2006. Early potin coinage of Britain: an update. *In:* DE JERSEY, P. (ed.) *Celtic Coinage: New Discoveries, New Discussion.* Oxford: BAR Publishing, 17-27.

HASELGROVE, C. & MOORE, T. 2007. New narratives of the Later Iron Age. *In:* HASELGROVE, C. & MOORE, T. (eds.) *The Later Iron Age in Britain and Beyond.* Oxford: Oxbow Books, 1-15.

HASELGROVE, C. & POPE, R. 2007. Characterising the Earlier Iron Age. *In:* HASELGROVE, C. & POPE, R. (eds.) *The Earlier Iron Age in Britain and the Near Continent.* Oxford: Oxbow Books, 1-23.

HAWKES, C. 1932. Early Settlement at Runcton Holme, Norfolk. Part II. The Second Occupation: A Peasant Settlement of the Iceni. *Proceedings of the Prehistoric Society of East Anglia,* VII, 231-262.

HAWKES, C. 1959. The A B C of the British Iron Age. *Antiquity,* XXXIII, 170-182.

HAWKES, C. F. C. & CRUMMY, P. 1995. *Camulodunum 2,* Colchester, Colchester Archaeological Trust Ltd

HAWKES, C. F. C. & HULL, M. R. 1947. *Camulodunum,* Oxford, The Society of Antiquaries.

HAWKES, J. 1987. The Briquetage. *In:* SUNTER, N. & WOODWARD, P. J. (eds.) *Romano-British Industries in Purbeck.* Dorchester: Dorset Natural History and Archaeological Society, 158-9.

HAYES, P. P. & LANE, T. W. 1992. *The Fenland Project Number 5: Lincolnshire Survey, The South-West Fens,* Sleaford, Heritage Trust of Lincolnshire.

HAYS-GILPIN, K. & WHITLEY, D. S. (eds.) 1998. *Reader in Gender Archaeology,* London: Routledge.

HCC 2006a. Hampshire AHBR 21992 - *"Fareham".* Online at http://historicenvironment.hants.gov.uk/ahbresults.aspx. [Accessed 29 May 2009].

HCC 2006b. Hampshire AHBR 23490 - *"Langstone Harbour, Havant".* Online at http://historicenvironment.hants.gov.uk/ahbresults.aspx. [Accessed 29 May 2009].

HCC 2006c. Hampshire AHBR 55723 - "Pit SE of Fir Hill". Online at http://historicenvironment.hants.gov.uk/ahbdetails.aspx. [Accessed 21 April 2010].

HEALY, J. F. 1999. *Pliny the Elder on Science and Technology,* Oxford, Oxford University Press.

HENDERSON, J. 1991. Industrial Specialization in Late Iron Age Britain and Europe. *The Archaeological Journal,* 148, 104-148.

HENDERSON, J. C. 2007. The Atlantic West in the Early Iron Age. *In:* HASELGROVE, C. & POPE, R. (eds.) *The Earlier Iron Age in Britain and the near Continent.* Oxford: Oxbow Books, 306-327.

HILL, J. D. 1993. Can we recognise a different European past? A contrastive archaeology of later prehistoric settlements in southern England. *Journal of European Archaeology,* 1, 57-75.

HILL, J. D. 1995. *Ritual and Rubbish in the Iron Age of Wessex,* Oxford, BAR Publishing.

HILL, J. D. 2002. Pottery and the Expression of Society, Economy and Culture. *In:* WOODWARD, A. & HILL, J. D. (eds.) *Prehistoric Britain: The Ceramic Basis.* Oxford: Oxbow Books, 75-84.

HINCHLIFFE, J. & THOMAS, R. 1980. Archaeological Investigations at Appleford. *Oxoniensia,* XLV, 9-111.

HINDMARCH, E. 2002. Pumney Farm. *South Midlands Archaeology,* 32, 67.

HINGLEY, R. 1997. Iron, ironworking and regeneration: a study of the symbolic meaning of metalworking in Iron Age Britain. *In:* GWILT, A. & HASELGROVE, C. (eds.) *Reconstructing Iron Age Societies.* Oxford: Ox bow, 9-18.

HODDER, I. 1977. The distribution of material culture items in the Baringo district, western Kenya. *Man,* 12, 239-269.

HODDER, I. 1982. *Symbols in Action,* Cambridge, Cambridge University Press.

HODDER, I. & HUDSON, S. 2003. *Reading the Past,* Cambridge, Cambridge University Press.

HODDER, I. & ORTON, C. 1976. *Spatial analysis in archaeology,* Cambridge, Cambridge University Press.

HOOKE, D. 1985. *The Anglo-Saxon landscape: the kingdom of the Hwicce,* Manchester, Manchester University Press.

HOPKINSON, B. 1975. Archaeological Evidence of Saltmoulding at important European Saltsites and its Relationship to the Distribution of Urnfields. *The Journal of Indo-European Studies,* 3, 1-52.

HOPKINSON, B. 1979. The Ancient Salt Industry and the lost Upwich Pit at Droitwich. *Council for British Archaeology Group 8,* 22, 83-87.

HUGHES, J. & HUNT, A. 1992. Friar Street (HWCM 605), the excavation. *In:* WOODIWISS, S. (ed.) *Iron Age and Roman salt production and the medieval town of Droitwich.* London: Council for British Archaeology, 115-131.

HUGHES, M. 1973. Excavations at Brownwich Farm, Titchfield 1971. *Rescue Archaeology in Hampshire,* 5-28.

HUNT, A. M., SHOTLIFF, A. & WOODHOUSE, J. 1986. A Bronze Age Barrow Cememtery and Iron Age Enclosure at Holt. *Transactions of the Worcestershire Archaeological Society,* Third Series, 10, 7-46.

HURST, D. 1992a. Pottery. *In:* WOODIWISS, S. (ed.) *Iron Age and Roman salt production and the medieval town of Droitwich.* London: Council for British Archaeology, 132-154.

HURST, D. 2001. Dobunnic Tribal Centres, Commodities and Trade: the South Worcestershire Hoard, Salt and Pottery. *West Midlands Archaeology,* 44, 84-93.

HURST, J. D. 1992b. *Savouring the Past: The Droitwich Salt Industry,* Worcester, Hereford and Worcester County Council.

HURST, J. D. (ed.) 1997. *A multi-period salt production site at Droitwich,* York: Council for British Archaeology.
HURST, J. D. 2006. Overall Conclusions. *In:* HURST, D. (ed.) *Roman Droitwich.* York: Council for British Archaeology, 243-245.
HURST, J. D. & PEARSON, E. A. 1996. An Iron Age site at Madeley Heath, Belbroughton, and other sites on the Fairfield to Frankley Green gas pipeline. *Transactions of the Worcestershire Archaeological Society,* Third Series 15, 127-134.
HURST, J. D., ROBERTS, R. & WOODIWISS, S. 1988. A Possible Second Roman Fort at Droitwich. *Transactions of the Worcestershire Archaeological Society,* Third Series 11, 75-80.
ISHIKAWA, K. 1983. *Guide to Quality Control,* Tokyo, Asian Productivity Organisation.
JACKSON, R., HURST, D., JONES, L. & PEARSON, E. 2000. Evaluation of the Proposed Environmental Recycling Area at Wellington Quarry, Wellington, Herefordshire. Online at http://worcestershire.whub.org.uk/DocRepository/wr5307.pdf. [Accessed 27 April 2009].
JAMES, S. & RIGBY, V. 1997. *Britain and the Celtic Iron Age,* London, Trustees of the British Museum.
JEFFERIES, R. S. & BARFORD, P. M. 1990. The Pottery of the Red Hills. *In:* FAWN, A. J., EVANS, K. A., MCMASTER, I. & DAVIES, G. M. R. (eds.) *The Red Hills of Essex.* Colchester: Colchester Archaeological Group, 35-36.
JENKINS, J. H. B. 1908. The Chemical Examination of some Substances from the Red Hills of Essex. *Proceedings of the Society of Antiquaries,* XXII, 182-186.
JENKINS, J. H. B. 1910. Remarks on Dr. Flinders Petrie's Theory. *Proceedings of the Society of Antiquaries of London,* XXIII, 90-96.
JESSUP, R. F. 1930. *The Archaeology of Kent,* London, Methuen & Co.
JODŁOWSKI, A. 1977. Techniques of Salt Production in Europe in Prehistoric and Early Medieval Times. *Recherches Archeologiques Krakow de 1976*, 72-74.
JONES, E. 1923. The Symbolic Significance of Salt in Folklore and Superstition. *In:* JONES, E. (ed.) *Essays in applied psycho-analysis.* London: The International Psycho-Analytical Press, 112-203.
JONES, L. & EVANS, C. J. 2006. The Excavation of an Iron Age and Romano-British Site at Stoke Lane, Wychbold. *Transactions of the Worcestershire Archaeological Society,* Third Series 20, 11-35.
JONES, M. U. 1977. Prehistoric salt equipment from a pit at Mucking, Essex. *The Antiquaries Journal,* LVII, 317-319.
JONES, M. U. 1978. The Mucking Excavations 1977. *Panorama, The Journal of the Thurrock Local History Society,* 21, 48-58.
JONES, M. U. 1980. Mucking Post-Excavation. *Panorama, The Journal of the Thurrock Local History Society,* 23, 47-57.
JONES, S. 1997. *The Archaeology of Ethnicity,* London, Routledge.
KAMASH, Z. 12 January 2011. *RE: Strabo.* (Personal communication to KINORY, J.)
KCC 2009. Kent HER Report TQ 86 NE 3 - "Possible Neolithic Settlement at Lower Halstow". Online at http://extranet7.kent.gov.uk/ExploringKentsPast/SingleResult.aspx?uid=MKE2990. [Accessed 10 November 2009].
KIRKHAM, B. 1975. Salt Making Sites found in North East Lincolnshire since 1960. *In:* DE BRISAY, K. & EVANS, K. A. (eds.) *Salt the Study of an Ancient Industry.* Colchester: Colchester Archaeological Group, 41-42.
KIRKHAM, B. 1981. The Excavation of a Prehistoric Saltern at Hogsthorpe, Lincolnshire. *Lincolnshire History and Archaeology,* 16, 5-10.
KIRKHAM, B. 2001. Iron Age and Roman Saltmaking on the Lindsey Coast and Marshland. *In:* LANE, T. & MORRIS, E. L. (eds.) *A Millennium of Saltmaking: Prehistoric and Romano-British Salt Production in the Fenland.* Sleaford: Heritage Trust of Lincolnshire, 405-410.
KURLANSKY, M. 2002. *Salt: A World History,* London, Jonathan Cape.
KUT, G. 1996. Turkish Culinary Culture. *In:* PEKIN, E. & SÜMER, A. (eds.) *Timeless Tastes Turkish Culinary Culture.* Istanbul: Divan, 38-71.
LAMBRICK, G. & ALLEN, T. 2004. *Gravelly Guy, Stanton Harcourt, Oxfordshire,* Oxford, Oxford Archaeology.
LAMBRICK, G. & ROBINSON, M. 1979. *Iron Age and Roman riverside settlements at Farmoor, Oxfordshire,* Oxford, the Oxfordshire Archaeological Unit and the Council for British Archaeology.
LAMBRICK, G. & ROBINSON, M. 2009. *The Thames through Time Late Prehistory: 1500 BC - AD 50,* Oxford, Oxford Archaeology.
LANE, T. 1994. Hoe Hills, Dowsby. *Fenland Research,* 9, 16-19.
LANE, T. 2001. Invisible and Irretrievable? Consideration of Some Non-ceramic Aspects of Fenland Salt Production. *In:* LANE, T. & MORRIS, E. L. (eds.) *A Millennium of Saltmaking: Prehistoric and Romano-British Salt Production in the Fenland.* Sleaford, Lincs: Heritage Trust of Lincolnshire, 459-465.
LANE, T. & MORRIS, E. L. (eds.) 2001. *A Millennium of Saltmaking: Prehistoric and Romano-British Salt Production in the Fenland,* Exeter: Heritage Trust of Lincolnshire.
LASZLO, P. 2001. *Salt,* New York, Columbia University Press.
LAVER, P. G. 1927. The Excavation of a Tumulus at Lexden, Colchester. *Archaeologia,* Second Series, XXVI, 241-254
LEEDS, E. T. 1931. Chastleton Camp, Oxfordshire, a hill-fort of the Early Iron Age. *The Antiquaries Journal,* XI, 382-398.
LEGGE, A. J. 1981. The Agricultural Economy. *In:* MERCER, R. J. (ed.) *Grimes Graves, Norfolk, Excavations 1971-72: Volume I.* London: HMSO, 79-103.

LEGGE, T. 2005. Milk use in prehistory: the osteological evidence. *In:* MULVILLE, J. & OUTRAM, A. K. (eds.) *The Zooarchaeology of Fats, Oils, Milk and Dairying.* Oxford: Oxbow Books, 8-13.

LOCK, G. 1984. The computing of the briquetage. *In:* CUNLIFFE, B. W. (ed.) *Danebury, An Iron Age Hillfort in Hampshire, Volume 2: the finds 1969-1978.* London: Council for British Archaeology, Microfiche 11:C4.

LUKIS, F. C. 1850. Communication regarding 'hand-bricks' from Channel Islands. *The Archaeological Journal,* VII, 175-7.

MACKINNON, J. J. & KEPECS, S. M. 1989. Prehispanic Saltmaking in Belize: New Evidence. *American Antiquity,* 54, 522-533.

MACPHERSON-GRANT, N. 2002. The Pottery. *Surrey Archaeological Collections,* 89, 75-86.

MAJOR, H. J. 1999. Fired Clay. *In:* BROWN, N. R. (ed.) *The Archaeology of Ardleigh, Essex: Excavations 1955-1980.* Chelmsford: Essex County Council, 157-58.

MALTBY, J. M. 1985. Patterns in Faunal Assemblage Variability. *In:* BARKER, G. & GAMBLE, C. (eds.) *Beyond Domestication in Prehistoric Europe.* London: Academic Press, 33-74.

MALTBY, M. 1981. Iron Age, Romano-British and Anglo-Saxon Animal Husbandry - a Review of the Faunal Evidence. *In:* JONES, M. & DIMBLEBY, G. (eds.) *The Environment of Man: the Iron Age to the Anglo-Saxon Period.* Oxford: B.A.R., 155-203.

MALTBY, M. 2006. Salt and Animal Products: linking Production and Use in Iron Age Britain. *In:* MALTBY, M. (ed.) *Integrating Zooarchaeology.* Oxford: Oxbow Books, 117-122.

MARSDEN, P. 2004. The Iron Age and Roman Pottery. *Transactions of the Leicestershire Archaeological and Historical Society,* 78, 24-25.

MARSDEN, P. & MORRIS, E. L. 2004. The Iron Age and Roman Pottery. *Transactions of the Leicestershire Archaeological and Historical Society,* 78, 12-13.

MASEFIELD, R., BRANCH, N., COULDREY, P., GOODBURN, D. & TYERS, I. 2003. A Later Bronze Age Well Complex at Swalecliffe, Kent. *Antiquaries Journal,* 83, 47-121.

MATTHEWS, K. J. 1999. The Iron Age of North-west England and the Irish Sea Trade. *In:* BEVAN, B. (ed.) *Northern Exposure: interpretative devolution and the Iron Ages in Britain.* Leicester: School of Archaeological Studies University of Leicester, 173-195.

MATTHEWS, K. J. 2002. The Iron Age of North-West England. *Journal of the Chester Archaeological Society,* 76, 1-51.

MCAVOY, F. 2006. Dodderhill, Droitwich: excavations 1977—85. *In:* HURST, D. (ed.) *Roman Droitwich: Dodderhill fort, Bays Meadow villa, and roadside settlement.* York: Council for British Archaeology, 3-20.

MCGRAIL, S. 1990. Early boats of the Humber basin. *In:* ELLIS, S. & CROWTHER, D. R. (eds.) *Humber Perspectives.* Hull: Hull University Press, 109-130.

MILES, A. 1975. Salt-Panning in Romano-British Kent. *In:* DE BRISAY, K. W. & EVANS, K. A. (eds.) *Salt the Study of an Ancient Industry.* Colchester: Colchester Archaeological Group, 26-31.

MILLER, D. 1987. *Material Culture and Mass Consumption,* Oxford, Basil Blackwell.

MILLER, D., GRIFFIN, L. & PEARSON, E. 2003. *Archaeological Investigations at Stonebridge Cross, Westwood, Worcestershire.* Online at http://worcestershire.whub.org.uk/DocRepository/wr2398.pdf. [Accessed 26 May 2009].

MILLETT, M. & MCGRAIL, S. 1987. The Archaeology of the Hasholme Logboat. *Archaeological Journal,* 144, 69-155.

MISCHLER, G. & SOSORBARAM, C.-E. 2005-2006. *Byaslag.* Online at: G. Mischler & C-E Sosorbaram. Available at: http://www.mongolfByaslag [Accessed 18 March 2008].

MOORE, J. 2003. Abingdon, Former Station Inn, Station Yard. *South Midlands Archaeology,* 33, 59-60.

MOORE, T. 2006. *Iron Age Societies in the Severn-Cotswolds,* Oxford, BAR Publishing.

MOORE, T. 2007. Life on the edge? Exchange, community, and identity in the Later Iron Age of the Severn-Cotswolds. *In:* HASELGROVE, C. & MOORE, T. (eds.) *The Later Iron Age in Britain and Beyond.* Oxford: Oxbow Books, 41-61.

MORRIS, E. L. 1979. *A comparative study of the Iron Age salt trade and ceramic networks of North Wales, the West Midlands and Hampshire.* Masters Masters, University of Southampton.

MORRIS, E. L. 1983. *Salt and Ceramic Exchange in Western Britain during the First Millennium B.C.* PhD, University of Southampton.

MORRIS, E. L. 1985. Prehistoric Salt Distributions: Two Case Studies from Western Britain. *The Bulletin of the Board of Celtic Studies,* XXXII, 336-379.

MORRIS, E. L. 1991. Report on the Prehistoric Ceramics found at Wall Camp, Kynnersley. *Transactions of the Shropshire Archaeological and Historical Society,* LXVII, 106-107.

MORRIS, E. L. 1994a. The Organisation of Salt Production and Distribution in Iron Age Wessex. *In:* FITZPATRICK, A. P. & MORRIS, E. L. (eds.) *The Iron Age in Wessex: Recent Work.* Salisbury: the Trust for Wessex Archaeology Ltd, 14-16.

MORRIS, E. L. 1994b. Production and Distribution of Pottery and Salt in Iron Age Britain: a Review. *Proceedings of the Prehistoric Society,* 60, 371-393.

MORRIS, E. L. 1999. Other ceramic materials. *Derbyshire Archaeological Journal,* 119, 183-185.

MORRIS, E. L. 2001a. Briquetage. *In:* LANE, T. & MORRIS, E. L. (eds.) *A Millennium of Saltmaking: Prehistoric and Romano-British Salt Production in the Fenlands.* Sleaford: Heritage Trust of Lincolnshire, 33-63.

MORRIS, E. L. 2001b. Briquetage and Salt Production and Distribution Systems: A Comparative Study. *In:*

LANE, T. & MORRIS, E. L. (eds.) *A Millennium of Saltmaking: Prehistoric and Romano-British Salt Production in the Fenland.* Sleaford: Heritage Trust of Lincolnshire, 389-404.

MORRIS, E. L. 2003a. The VCP. *Journal of the Chester Archaeological Society,* 77, 31-33.

MORRIS, E. L. 2003b. *Wormington to Tirley Gas Pipeline Gloucestershire/Worcestershire: Post-Excavation Assessment and Updated Project Design, Appendix 4: Briquetage*. Online at http://worcestershire.whub.org.uk/DocRepository/wr865.pdf. [Accessed 3 July 2009].

MORRIS, E. L. 2004. Briquetage. *In:* LAMBRICK, G. & ALLEN, T. (eds.) *Gravelly Guy, Stanton Harcourt, Oxfordshire.* Oxford: Oxford Archaeology, 357-360.

MORRIS, E. L. 2007a. *Briquetage and Cheshire Salt Containers.* Online at: Mellor Archaeological Trust. Available at: http://www.mellorarchaeology.org.uk [Accessed 29 Sept 2007].

MORRIS, E. L. 2007b. Making magic: later prehistoric and early Roman salt production in the Lincolnshire fenland. *In:* HASELGROVE, C. & MOORE, T. (eds.) *The Later Iron Age in Britain and Beyond.* Oxford: Oxbow Books, 430-443.

MORRIS, E. L. 2009. The Briquetage. *In:* DANIEL, P. (ed.) *Archaeological Excavations at Pode Hole Quarry.* Oxford: BAR Publishing, 74-82, 175-177.

MORRIS, E. L. 2010. Briquetage. *Internet Archaeology* [Online], 28. Available: http://intarch.ac.uk/journal/issue28/3/4.3.html [Accessed 20 September 2010].

MORRIS, E. L. & CROSBY, A. 2007. Pottery and Briquetage. *Transactions of the Bristol and Gloucestershire Archaeological Society,* 125, 51-72.

MORRIS, E. L., MARSDEN, P. & WILLIAMS, D. F. 2005. Pottery and briquetage. *In:* THOMAS, N. (ed.) *Conderton Camp, Worcestershire: a small middle Iron Age hillfort on Bredon Hill.* York: Council for British Archaeology, 117-147 and 279-315.

MOUZE, C. 2008. How to make ghee. *Online Yoga* [Online]. Available: http://www.yoga-magazine.net/ghee.htm [Accessed 17 December 2009].

MUDD, A. 1993. Radley, Goose Acre Farm. *South Midlands Archaeology,* 23, 76-77.

MULHALL, I. 2010. The Peat Men from Clonycavan and Oldcroghan. *British Archaeology,* January February 2010, 34-41.

MULTHAUF, R. P. 1978. *Neptune's Gift: A History of Common Salt,* London, Johns Hopkins University Press.

MULVILLE, J. 2008. Foodways and Social Ecologies from the Middle Bronze Age to the Late Iron Age. *In:* POLLARD, J. (ed.) *Prehistoric Britain.* Oxford: Blackwell Publishing, 225-247.

MURDOCK, G. P. & PROVOST, C. 1973. Factors in the Division of Labor by Sex: A Cross-Cultural Analysis. *Ethnology,* 12, 203-225.

MUSSON, C. R., BRITNELL, W. J. & SMITH, A. G. 1991. *The Breiddin Hillfort,* London, the Council for British Archaeology.

MYRES, J. N. L. 1937. A Prehistoric and Roman Site on Mount Farm, Dorchester. *Oxoniensia,* II, 12-40.

NAPTHAN, M., PEARSON, E. & RATKAI, S. 1997. Evaluation of the Proposed Hampton Lovett Industrial Estate. Online at http://worcestershire.whub.org.uk/DocRepository/wr8375.pdf. [Accessed 7 May 2009].

NEEDHAM, S. 2007. 800 BC, The Great Divide. *In:* HASELGROVE, C. & POPE, R. (eds.) *The Earlier Iron Age in Britain and the near Continent.* Oxford: Oxbow Books, 39-63.

NELSON, S. M. 1997. *Gender in Archaeology,* Walnut Creek, AltaMira Press.

NENQUIN, J. 1961. *Salt, A Study in Economic Prehistory,* Brugge, De Tempel.

NICHOLS, P. 2001. Bourton-on-the Water Primary School. *Transactions of the Bristol and Gloucestershire Archaeological Society,* 119, 187.

NICHOLSON, W. A. 1850. Report on finds made near Ingoldmells. *The Archaeological Journal,* VII, 69-70.

NORTHOVER, P. 18 January 2010. *RE: Re: Can you provide advice to a DPhil student?* (Personal communication to KINORY, J.)

O'NEIL, B. H. S. J. 1942. Excavations at Ffridd Faldwyn Camp, Montgomery, 1937-39. *Archaeologia Cambrensis,* XCVII, 1-57.

OCC 2009. Site 22 on Chalgrove to East Ilsley pipeline - PRN 26000. Online at http://www.oxfordshire.gov.uk/wps/portal/publicsite/!ut/p/c5/hc7LCoMwEAXQLypzTUiiy9KH0UWk9qXZSCgiAR9dlEL_vrqTUtuZ5bnDHbI0bu-evnEPP_SupYKsrGTIc5FEHJmJFBLog2HgAU5s9HLmYDICuxiTK7ULjlr8ub5OfbNEiHhKyM32bFLEmfzpEOrDv_RPjoVZg4weuppKsmrxz5RR2daNu73o3hXwyWr_BgTCEuQ!/dl3/d3/L0lJSklna21DU1EhIS9JRGpBQU15QUJFUkNKRXFnLzRGR2dzbzBWdnphOUlBOW9JQSEhLzdfNjgzUjVJOTMwT045NzBJMEhRTjIwMzEwSjIvMTpNS3c1MTkyMDAyMy9zYS5zcGZfQWN0aW9uTGlzdGVuZXI!/?PC_7_683R5I930ON970I0HQN20310J2012425_spf_strutsAction=!2frecordSearch.do!3fmethod%3drecordSearch!26offset%3d0. [Accessed 10 August 2010].

OSGOOD, R. 1995. Study 6. Activity patterns. *In:* CUNLIFFE, B. W. (ed.) *Danebury, an Iron Age Hillfort in Hampshire, Volume 6, A hillfort community in perspective.* York: Council for British Archaeology, Microfiche 33:A3-B11.

OVID, *Fasti.* Trans. FRAZER, S. J. G. 1976, Loeb Classics Library. London, William Heinemann Ltd.

PALMER-BROWN, C. 1993. Bronze Age salt production at Tetney. *Current Archaeology,* XII, No. 4, 143-145.

PARRINGTON, M. 1978. *The excavation of an Iron Age settlement, Bronze Age ring-ditches and Roman features at Ashville Trading Estate, Abingdon (Oxfordshire) 1974-76*, the Oxfordshire Archaeological Unit and the Council for British Archaeology.

PATTEN, R. 2009. *Excavations at Eye Quarry: The Southern Extension,* Cambridge, Cambridge Archaeological Unit.

PEACOCK, D. P. S. 1968. A Petrological Study of Certain Iron Age Pottery from Western England. *Proceedings of the Prehistoric Society,* XXXIV, 414-427.

PEACOCK, D. P. S. 1981. Archaeology, ethnology and ceramic production. *In:* HOWARD, H. & MORRIS, E. L. (eds.) *Production and Distribution: a ceramic viewpoint.* Oxford: BAR, 187-194.

PETERSON, R., POLLARD, J. & MORRIS, E. 2006. Prehistoric pottery. *In:* POLLARD, J., HOWELL, R., CHADWICK, A. & LEAVER, A. (eds.) *Lodge Hill Camp, Caerleon and the Hillforts of Gwent.* Oxford: BAR Publishing, 33-37.

PETRIE, W. F. 1910. Suggested Origin of the Red Hills. *Proceedings of the Society of Antiquaries of London,* 1909-1911, 88-90.

PHILLIPS, C. W. 1934. The present state of archaeology in Lincolnshire. *The Archaeological Journal,* XCI, 97-187.

PHILPOTT, R. A. 2007. The Iron Age. *In:* GRIFFITHS, D., PHILPOTT, R. A. & EGAN, G. (eds.) *Meols The Archaeology of the North Wirral Coast.* Oxford: Institute of Archaeology, University of Oxford, 379-387.

PIMENTEL, M. 1998. *Living High Off Half the Hog.* Online at Medford, WI: Countryside and Small Stock Journal. Available at: http://www.countrysidemag.com/issues/88/88-5/Marcia_Pimentel.html [Accessed 23 February 2009].

PIRES-FERREIRA, J. W. & FLANNERY, K. V. 1976. Ethnographic Models for Formative Exchange. *In:* FLANNERY, K. V. (ed.) *The Early Mesoamerican Village.* London: Academic Press, Inc. (London) Ltd., 286-292.

PLINY, *The Elder Pliny's Chapters on Chemical Subjects Part I.* Trans. BAILEY, K. C. 1929. London, Edward Arnold & Co.

PLINY, *Natural History Volume IX.* Trans. RACKHAM, H. 1968, Loeb Classical Library. London, William Heinemann Ltd.

POLLARD, J. 2002. The Nature of Archaeological Deposits and Finds Assemblages. *In:* WOODWARD, A. & HILL, J. D. (eds.) *Prehistoric Britain: The Ceramic Basis.* Oxford Oxbow Books, 22-33.

POLLARD, J., HOWELL, R., CHADWICK, A. & MCFADYEN, L. 2006. Discussion: Lodge Hill Camp and the hillforts of Gwent. *In:* POLLARD, J., HOWELL, R., CHADWICK, A. & LEAVER, A. (eds.) *Lodge Hill Camp, Caerleon and the Hillforts of Gwent.* Oxford: BAR Publishing, 47-61.

POLLARD, J. & MORRIS, E. 2006. Droitwich briquetage. *In:* POLLARD, J., HOWELL, R., CHADWICK, A. & LEAVER, A. (eds.) *Lodge Hill Camp, Caerleon and the Hillforts of Gwent.* Oxford: BAR Publishing, 37-38.

POOLE, C. 1984a. 4.2.4 The structural use of daub, clay and timber. *In:* CUNLIFFE, B. (ed.) *Danebury, an Iron Age Hillfort in Hampshire, Volume 1.* London: the Council for British Archaeology, 110-123 and fiche 3:D1-G11.

POOLE, C. 1984b. Briquetage containers. *In:* CUNLIFFE, B. (ed.) *Danebury, An Iron Age Hillfort in Hampshire, Volume 2 1969-1978: the finds.* London: the Council for British Archaeology, 426-430.

POOLE, C. 1987. Salt Working. *In:* CUNLIFFE, B. W. (ed.) *Hengistbury Head, Dorset, Volume 1.* Oxford: Oxford Committee for Archaeology, 178-180.

POOLE, C. 1991. Briquetage containers *In:* CUNLIFFE, B. W. & POOLE, C. (eds.) *Danebury: an Iron Age Hillfort in Hampshire, Volume 5 the excavations 1979-1988: the finds.* London: the Council for British Archaeology, 404-407 and microfiche 30:A3-4.

POOLE, C. 2000a. The briquetage. *In:* CUNLIFFE, B. & POOLE, C. (eds.) *The Danebury Environs Programme Volume 2 - Part 3: Suddern Farm, Middle Wallop, Hants, 1991 and 1996.* Oxford: Institute of Archaeology, 142.

POOLE, C. 2000b. Briquetage. *In:* CUNLIFFE, B. & POOLE, C. (eds.) *The Danebury Environs Programme Volume 2 - Part 6: Houghton Down, Stockbridge, Hants, 1994* Oxford: English Heritage and Oxford University Committee for Archaeology, 119-120 and Microfiches 11, 12 14:A12.

POOLE, C. 2000c. Briquetage. *In:* CUNLIFFE, B. & POOLE, C. (eds.) *The Danebury Environs Programme Volume 2 - Part 4: New Buildings, Longstock, Hants, 1992 and Fiveways, Longstock, Hant, 1996.* Oxford: Institute of Archaeology, 74.

POOLE, C. 2000d. Briquetage. *In:* CUNLIFFE, B. & POOLE, C. (eds.) *The Danebury Environs Programme Volume 2 - Part 5: Nettlebank Copse, Wherwell, Hants, 1993.* Oxford: English Heritage and Oxford University Committee for Archaeology, Microfiche 10:D5.

POOLE, C. 2000e. Briquetage. *In:* CUNLIFFE, B. & POOLE, C. (eds.) *The Danebury Environs Programme Volume 2 - Part 1: Woolbury and Stockbridge Down, Stockbridge, Hants, 1989.* Oxford: Institute of Archaeology, 60 and Fiche 1:F6.

POOLE, C. 2000f. Briquetage. *In:* CUNLIFFE, B. & POOLE, C. (eds.) *The Danebury Environs Programme, Volume 2 - Part 2: Bury Hill, Upper Clatford, Hants, 1990.* Oxford: Institute of Archaeology, 62, fiche 2:B7 to C8.

POOLE, C. 2005. *The Daub and Fired Clay and Briquetage.* Online at. Available at: http://www.arch.ox.ac.uk/_data/assets/pdf_file/1527/daubreport.pdf [Accessed 25 December 2008].

POPE, R. 2007. Ritual and the roundhouse: a critique of recent ideas on the use of domestic space in later British prehistory. *In:* HASELGROVE, C. & POPE, R. (eds.) *The Earlier Iron Age in Britain and the near Continent.* Oxford: Oxbow Books, 204-228.

PROCTOR, J. 2002. Late Bronze Age/Early Iron Age placed deposits from Westcroft Road, Carshalton: their meaning and interpretation. *Surrey Archaeological Collections,* 89, 65-103.

PRYOR, F. M. M. 1980. *Excavation at Fengate, Peterborough, England: The Third Report,* Toronto,

Northamptonshire Archaeological Society and Royal Ontario Museum.
PRYOR, S. & BROWN, L. 1984. Description and illustration of selected pits. *In:* CUNLIFFE, B. (ed.) *Danebury, an Iron Age Hillfort in Hampshire, Volume 1 The excavations, 1969-1978: the site.* London: Council for British Archaeology, fiche 4:B1-5:F8.
QIU, Z. 1993. Molten Salt and Slag in Ancient China. *Transactions of Nonferrous Metals Society of China,* 3, 91-94.
QUINNELL, H. 2006. Prehistoric and Roman Pottery from Lundy. *Council for British Archaeology South-West,* 18, 15-17.
RAINBIRD-CLARKE, R. 1939. The Iron Age in Norfolk and Suffolk. *The Archaeological Journal,* XCVI, 1-113.
RAINER, C. 2010a. In Sahara, Salt-Hauling Camel Trains Struggle On. *National Geographic News* [Online]. Available: http://news.nationalgeographic.com/news/2003/05/0528_030528_saltcaravan.html [Accessed 20 June 2011].
RAINER, C. 2010b. Sahara Salt Trade Camel Caravans. *National Geographic News* [Online]. Available: http://news.nationalgeographic.com/news/2003/05/photogalleries/salt/photo6.html [Accessed 20 June 2011].
READER, F. W. 1910a. Notes on the Briquetage found in 1908-9. *Proceedings of the Society of Antiquaries of London,* XXIII, 86-88.
READER, F. W. 1910b. Report of the Red Hills Exploration Committee, 1908-9. *Proceedings of the Society of Antiquaries of London,* XXIII, 69-83.
READER, F. W. & WILMER, H. 1908. Report on the Red Hills Exploration Committee 1906-7. *Proceedings of the Society of Antiquaries,* XXII, 164-214.
REED, R. 1972. *Ancient Skins Parchments and Leathers,* London, Seminar Press Ltd.
REES, H. 1986. Ceramic Salt Working Debris from Droitwich. *Transactions of the Worcestershire Archaeological Society,,* Third Series 10, 47-54.
REES, H. 1995. Briquetage. *In:* WAINWRIGHT, G. J. & DAVIES, S. M. (eds.) *Balksbury Camp Hampshire.* London: English Heritage, 58-59.
RENFREW, C. 1969. Trade and Culture Process in European Prehistory. *Current Anthropology,* 2-3, 151-169.
RENFREW, C. 1977. Alternative Models for Exchange and Spatial Distribution. *In:* EARLE, T. K. & ERICSON, J. E. (eds.) *Exchange Systems in Prehistory.* London: Academic Press Inc. (London) Ltd., 71-90.
RENFREW, C., DIXON, J. E. & CANN, J. R. 1968. Further Analysis of Near Eastern Obsidians. *Proceedings of the Prehistoric Society,* New Series, 34, 319-331.
REYNARDSON, R. J. B. 1850. "Catalogue of Antiquities" in Memoirs Illustrative of the History and Antiquities of the County and City of Lincoln Communicated at the Annual Meeting of the Archaeological Institute of Great Britain and Ireland, held at Lincoln, July, 1848. *Proceedings of the Archaeological Institute,* 1848, xxvii-lviii.
REYNOLDS, P. 1995. The Food of the Prehistoric Celts. *In:* WILKINS, J., HARVEY, D. & DOBSON, M. (eds.) *Food in Antiquity.* Exeter: University of Exeter Press, 303-315.
RICE, P. M. 1987. *Pottery Analysis A Sourcebook,* Chicago, University of Chicago Press.
RICHARDSON, K. M. & YOUNG, A. 1951. An Iron Age Site on the Chilterns. *The Antiquaries Journal,* XXI, 132-148.
RICHARDSON, K. W. 1972. Report on the Suggested Laboratory Examination of the Osea Road Red Hill Material. *Colchester Archæological Group Annual Bulletin,* 15, 37-40.
RIEHM, K. 1961. Prehistoric Salt-Boiling. *Antiquity,* XXXV, 181-191.
RINGGAARD, M. & SCHARFF, A. B. 2010. The Impact of Dyes and Natural Pigmentatioin of Wool on the Preservation of Archaeological Textiles. *In:* STRAND, E. A., GLEBA, M., MANNERING, U., MUNKHOLT, C. & RINGGAARD, M. (eds.) *North European Symposium for Archaeological Textiles X.* Oxford: Oxbow Books, 221-224.
RIPPON, S. 2000. The Romano-British Exploitation of Coastal Wetlands: Survey and Excavation on the North Somerset Levels, 1993-7. *Britannia,* 31, 69-200.
RIPPON, S. 2006. *Landscape, Community and Colonisation,* York, Council for British Archaeology.
RIXSON, D. 2000. *The History of Meat Trading,* Thrumpton, Nottingham University Press.
ROBERTS, M. R. 1995. Abingdon, Abingdon Business Park, Wyndyke Furlong. *South Midlands Archaeology,* 25, 48.
RODWELL, K. A. 1988. *The prehistoric and Roman settlement at Kelvedon, Essex,* London, Chelmsford Archaeological Trust and Council for British Archaeology.
RODWELL, W. J. 1974. The Orsett 'Cock' Cropmark Site. *Transactions of the Essex Archaeological Society,* 6 (Third Series), 13-39.
RODWELL, W. J. 1976. Coinage, Oppida and the rise of Belgic Power in South-Eastern Britain. *In:* CUNLIFFE, B. & ROWLEY, T. (eds.) *Oppida: the Beginnings of Urbanisation in Barbarian Europe.* Oxford: British Archaeological Reports, 181-367.
RODWELL, W. J. 1979. Iron Age and Roman Salt-Winning on the Essex Coast. *In:* BURNHAM, B. C. & JOHNSON, H. B. (eds.) *Invasion and Response.* Oxford: B.A.R., 133-175.
RUDKIN, L. 2007. *Natual Dyes,* London, A & C Black.
RYDER, M. L. 1983a. Milk Products. *In:* JONES, M. (ed.) *Integrating the Subsistence Economy.* Oxford: B.A.R., 239-250.
RYDER, M. L. 1983b. *Sheep & Man,* London, Duckworth.
RYKA, W. & WERNER, Z. 1985. Ten Centuries of Salt Mining in Poland. *Episodes: Journal of International Geoscience* [Online], 8, No. 3. Available:

http://www.episodes.org/backissues/83/ARTICLES--169.pdf [Accessed 22 September 2010].
SALWAY, P. 2000. Preface. *In:* HENIG, M. & BOOTH, P. (eds.) *Roman Oxfordshire.* Thrupp, Stroud, Glos.: Sutton Publishing, ix-x.
SAVORY, H. N. 1971. *Excavations at Dinorben, 1965-9,* Cardiff, The National Museum of Wales.
SAWLE, J. 1978. Excavation of the Old Bowling Green, Ricketts Lane, Droitwich Interim Report. *West Midlands Archaeological News Sheet,* 21, 77-78.
SAWLE, J. 1984. Ceramic salt-making debris from Droitwich. *Bulletin of the Experimental Firing Group,* 2, 1983-4, 5-12.
SCHIFFER, M. B. 1987. *Formation Processes of the Archaeological Record,* Albequerque, University of New Mexico Press.
SCULLARD, H. H. 1999. *Roman Britain Outpost of the Empire,* London, Thames and Hudson.
SEALEY, P. R. 1995. New light on the salt industry and Red Hills of prehistoric and Roman Essex. *Essex Archaeology and History,* 26 (Third Series), 65-81.
SEALEY, P. R. 2007. *A Late Iron Age Warrior Burial from Kelvedon, Essex,* Colchester, Colchester Museum.
SELKIRK, A. 1981. Claydon Pike. *Current Archaeology,* VII, No. 4, 110-111.
SERJEANTSON, D. 2007. Intensification of animal husbandry in the Late Bronze Age? The contribution of sheep and pigs. *In:* HASELGROVE, C. & POPE, R. (eds.) *The Earlier Iron Age in Britain and the near Continent.* Oxford: Oxbow Books, 80-93.
SHARPLES, N. 2010. *Social Relations in Later Prehistory,* Oxford, Oxford University Press.
SHERLOCK, R. L. 1921. *Rock-Salt and Brine,* London, HMSO.
SHERRATT, A. & SHERRATT, S. 1991. From Luxuries to Commodities: The Nature of Mediterranean Bronze Age Trading Systems. *In:* GALE, N. H. (ed.) *Bronze Age Trade in the Mediterranean.* Jonsered: Paul Åströms Förlag, 351-386.
SHIMWELL, D. W. & OGLE, M. I. 2000. Macro-botanic Evidence. *In:* COWELL, R. W. & PHILPOTT, R. A. (eds.) *Prehistoric, Romano-British and Medieval Settlement in Lowland North West England.* Liverpool: National Museums and Galleries on Merseyside, 50-57.
SHORE, T. W. 1892. Hampshire Mudlands and Other Alluviums. *Papers and Proceedings of the Hampshire Field Club,* II, 181-200.
SIMMEL, G. & FRISBY, D. 1990. *The Philosophy of Money,* London, Routledge.
SIMMONS, M. P. & BREM, G. F. 1979. The Analysis and Distribution of Volcanic Ash-Tempered Pottery in the Lowland Maya Area. *American Antiquity,* 44, 79-91.
SMITH, H. P. 1930. The Occupation of the Hamworthy Peninsula in the late Celtic and Romano-British Periods *Proceedings of the Dorset Natural History and Archaeological Society,* LII, 96-130.
SMITH, H. P. 1948. *The History of the Borough and County of the Town of Poole,* Poole, J. Looker, Ltd.
SMITH, R. A. 1918. The Essex Red Hills as Salt-works. *Proceedings of the Society of Antiquaries of London,* Second Series, Vol. XXX, 36-54.
SOUTHAMPTON_CITY_COUNCIL 2006. Southampton HER MSH 1544 - "East Park Terrace/St Andrew's Road," Online at http://www.heritagegateway.org.uk/Gateway/Results_Single.aspx?uid=MSH1544&resourceID=1016. [Accessed 24 June 2009].
STANFORD, S. C. 1967. Croft Ambrey hill-fort - some interim conclusions. *Transactions of the Woolhope Naturalists' Field Club,* XXXIX, 31-39.
STANFORD, S. C. 1970. Credenhill Camp, Herefordshire: An Iron Age Hill-Fort Capital. *The Archaeological Journal,* CXXVII, 82-129.
STANFORD, S. C. 1972. Welsh Border Hill Forts. *In:* THOMAS, C. (ed.) *The Problems of the Iron Age in the Irish Sea Province.* London: The Council for British Archaeology, 25-36.
STANFORD, S. C. 1974. *Croft Ambrey,* Hereford, S. C. Stanford.
STEAD, I. M. & RIGBY, V. 1986. *Baldock: the Excavation of a Roman and Pre-Roman Settlement, 1968—72,* London, The Society for the Promotion of Roman Studies.
STEVENS, C. 2005. Bronze Age and Iron Age settlement at Meriden Quarry, Solihull. *Transactions of the Birmingham and Warwickshire Archaeological Society,* 109, 1-25.
STOPES, H. 1879. The Salting Mounds of Essex. *The Archaeological Journal,* XXXVI, 369-372.
STOPES, H. 1887. The Salting Mounds of Essex. *The Essex Naturalist,* I, 96-105.
STRABO, *The Geography of Strabo, Book II.* Trans. JONES, H. L. 1923. London, William Heinemann Ltd.
STRABO, *The Geography of Strabo, Books III-V.* Trans. JONES, H. L. 1969, Loeb Classical Library. London, William Heinemann Ltd.
SWINNERTON, H. H. 1932. The Prehistoric Pottery Sites of the Lincolnshire Coast. *The Antiquaries Journal,* XII, 239-253.
SWORN, S. & GRIFFIN, L. 2004. *Archaeological Watching Brief at 1A High Street, Droitwich, Worcestershire.* Online at http://worcestershire.whub.org.uk/DocRepository/wr9773.pdf. [Accessed 28 May 2009].
SYKES, P. 2009. Nutrient requirements for horses. Available: http://www.ranvet.com.au/feeding_by_dr_percy_sykes.htm [Accessed 23 December 2009].
TACITUS, P. C. *The Annals.* Online at New York: The Modern Library. Available at: http://mcadams.posc.mu.edu/txt/ah/tacitus/TacitusAnnals14.html [Accessed 1 February 2010].
TANNAHILL, R. 2006. Former Morris Motors Sports and Social Club Phase II. *South Midlands Archaeology,* 36, 50.
THOMAS, N. 2005. *Conderton Camp, Worcestershire: a small middle Iron Age hillfort on Bredon Hill,* York, Council for British Archaeology.

THOMPSON, I. & BARFORD, P. M. 1986. Late Iron Age Pottery and Briquetage from Elm Park House, Ardleigh, 1981. *Essex Archaeology and History,* 17 (Third Series), 166-170.

THORPE, R., SHARMAN, J. & CLAY, P. 1994. An Iron Age and Romano-British Enclosure System at Normanton le Heath. *The Leicestershire Archaeological and Historical Society Transactions,* LXVIII 1994, 1-63.

TODD, M. 2001. Iron Age Material from Charterhouse on Mendip. *Somerset Archaeology and Natural History,* 145, 53-55.

TOMALIN, D. 1989. An Early Roman Cliff-Top Salt-Working Site at Redcliff Battery, Sandown, Isle of Wight. *Proceedings of the Isle of Wight Natural History and Archaeological Society,* 9, 91-120.

TROTT, K. & TOMALIN, D. J. 2003. The maritime role of the island of Vectis in the British Pre-Roman Iron Age. *The International Journal of Nautical Archaeology,* 32, 158-181.

TYLECOTE, R. F. 1962. *Metallurgy in Archaeology,* London, Edward Arnold (Publishers) Ltd.

TYLECOTE, R. F. 1986. *The Prehistory of Metallurgy in the British Isles,* London, The Institute of Metals.

TYLECOTE, R. F. 1992. *A History of Metallurgy,* London, The Institute of Materials.

UM. 2004. *Taste Buds.* Online at Ann Arbor: University of Michigan. Available at: http://www.reachoutmichigan.org/funexperiments/agesubject/lessons/tastebud.html [Accessed 17 April 2008].

VAN ARSDELL, R. D. 1989. *Celtic Coinage of Britain,* London, Spink and Son Ltd.

VANDEN BERGHE, I., GLEBA, M. & MANNERING, U. 2010. Dyes: to be or not to be. An Investigation of Early Iron Age Dyes in Danish Peat Bog Textiles. *In:* STRAND, E. A., GLEBA, M., MANNERING, U., MUNKHOLT, C. & RINGGAARD, M. (eds.) *North European Symposium for Archaeological Textiles X.* Oxford: Oxbow Books, 247-251.

VARLEY, W. J. 1950. Excavations of the Castle Ditch, Eddisbury, 1935-38. *Transactions of the Historic Society of Lancashire and Cheshire,* 102, 1-68.

VARLEY, W. J. 1976. A Summary of the Excavations at Castle Hill, Almondbury 1939-1972. *In:* HARDING, D. W. (ed.) *Hillforts Later Prehistoric Earthworks in Britain and Ireland.* London: Academic Press, 119-131.

VARRO, M. T., *De Re Rustica, Book II.* Trans. HOOPER, W. D. & ASH, H. B. 1967, Loeb Classical Library. London, William Heinemann Ltd.

VEITCH, A. & EADES, S. C. 2006. Equine Nutrition for Health and Happiness. Available: http://www.evrp.lsu.edu/healthtips/EquineNutrition.htm [Accessed 23 December 2009].

VERGIL, *Georgics, Book III.* Trans. FAIRCLOUGH, H. R. & GOOLD, G. P. 2001. London, Harvard University Press.

WAINWRIGHT, G. J. 1979. *Gussage All Saints,* London, HMSO.

WAINWRIGHT, G. J. & DAVIES, S. M. 1995. *Balksbury Camp Hampshire,* London, English Heritage.

WAINWRIGHT, J. 2010. Archaeological Investigations at West Mercia Police HQ, Hindlip, Worcestershire. Worcester: Historic Environment and Archaeology Service, Worcestershire County Council.

WALKER, I. 1958. Excavations on a Romano-British Site at Astley, 1956-58. *Transactions of the Worcestershire Archaeological Society,* New Series, XXXV, 29-57.

WARREN, S. H. 1932. Prehistoric Timber Structures associated with a Briquetage Site in Lincolnshire. *The Antiquaries Journal,* XII, 254-256.

WEAVER, P. 2008. *To Eat or Not to Eat,That is the Question for Soay Sheep Owners.* Online at: Priscilla Weaver. Available at: http://priscilla.saltmarshranch.com/category/butchering/ [Accessed 23 February 2009].

WEEDMAN, K. 2006. Gender and Ethnoarchaeology. *In:* NELSON, S. M. (ed.) *Handbook of Gender in Archaeology.* Lanham, Maryland: Alta Mira Press, 247-294.

WELLS, P. S. 1980. Iron Age Central Europe. *Archaeology,* 33, Number 5, 6-11.

WELLS, P. S. 1981. *The Emergence of an Iron Age Economy,* Cambridge, MA, Harvard University Press.

WHEELER, R. E. M. & WHEELER, T. V. 1936. *Verulamium,* London, The Society of Antiquaries.

WHIMSTER, R. 1977. Iron Age burial in southern Britain. *Proceedings of the Prehistoric Society,* 43, 317-327.

WILD, J. P. 2003. *Textiles in Archaeology,* Princes Risborough, Shire Publications Ltd.

WILKINS, J. M. & HILL, S. 2006. *Food in the Ancient World,* Oxford, Blackwell Publishing.

WILKINSON, T. J. & MURPHY, F. 1986. Archaeological Survey of an Intertidal Zone: the Submerged Landscape of the Essex Coast, England. *Journal of Field Archaeology,* 13, 177-194.

WILKINSON, T. J. & MURPHY, P. L. 1995. *The Archaeology of the Essex Coast, Volume I: The Hullbridge Survey,* Chelmsford, Essex County Council Archaeology Section.

WILLIAMS-FREEMAN, J. P. 1933. The Coast of Hampshire. *Papers and Proceedings of the Hampshire Field Club,* XII, 215-220.

WILLIS, S. 2007. Sea, coast, estuary, land, and culture in Iron Age Britain. *In:* HASELGROVE, C. & MOORE, T. (eds.) *The Later Iron Age in Britain and Beyond.* Oxford: Oxbow Books, 107-129.

WILLIS, S. H. 1994. The ceramic assemblage. *Durham Archaeological Journal,* 10, 27-31.

WILLIS, S. H. 1995. The Briquetage. *Durham Archaeological Journal,* 11, 15-16.

WILMER, H. 1914. Comments on the Red Hills of Essex. *Proceedings of the Society of Antiquaries of London,* Second Series, Vol. XXVI, 143.

WILSON, B. 1993. Reports on the Bones and Oyster Shell. *In:* ALLEN, T. G. & ROBINSON, M. A. (eds.) *The Prehistoric Landscape and Iron Age Enclosed*

Settlement at Mingies Ditch, Hardwick-with-Yelford, Oxon. Oxford: Oxford University Committee for Archaeology, 123-139.

WILSON, R. & ALLISON, E. 1990. The Animal and Fish Bones. *In:* ALLEN, T. G. (ed.) *An Iron Age and Romano-British Enclosed Settlement at Watkins Farm, Northmoor, Oxon.* Oxford: Oxford University Committee for Archaeology, 57-61.

WOOD, J. 2007. *Prehistoric Cooking,* Stroud, Tempus.

WOODIWISS, S. (ed.) 1992. *Iron Age and Roman Salt Production and the Medieval Town of Droitwich,* London: Council for British Archaeology.

WOODWARD, A. 2002. Sherds in Space: Pottery and the Analysis of Site Organisation. *In:* WOODWARD, A. & HILL, J. D. (eds.) *Prehistoric Britain The Ceramic Basis.* Oxford: Oxbow Books, 62-74.

WOODWARD, A. & HUGHES, G. 2007. Deposits and doorways: patterns within the Iron Age settlement at Crick Covert Farm, Northamptonshire. *In:* HASELGROVE, C. & POPE, R. (eds.) *The Earlier Iron Age in Britain and the Near Continent.* Oxford: Oxbow Books, 185-203.

www.ingramcontent.com/pod-product-compliance
Lightning Source LLC
LaVergne TN
LVHW072256300926
849160LV00018B/1679

* 9 7 8 1 4 0 7 3 0 9 7 2 9 *